Breeding Field Crops II
Advances

by

V.L. Chopra

Chancellor, Central University of Kerala
(Formerly) Member Planning Commission, Govt. of India
Secretary (DARE) Govt. of India and
Director General, Indian Council of Agriculture Research
Member, Science Council Consultative Group of International Agric. Research (CGIAR)

2019

Studium Press (India) Pvt. Ltd.

Breeding Field Crops II
Advances

©*First Indian Edition* 2019
First Published 2014 (Studium Press LLC, USA)

ISBN (Vol. 1) : 978-93-85046-56-8
ISBN (Vol. 2): 978-93-85046-57-5
ISBN (Set): 978-93-85046-55-1

Published by:

Studium Press (India) Pvt. Ltd.
4735 / 22, 2nd Floor, Prakash Deep Building
(Near Delhi Medical Association)
Ansari Road, Darya Ganj, New Delhi-110 002
Tel.: + 91-11-43240200-15 (15 lines); Fax: 91-11-43240215
E-mail: pubdir@studiumpress.in

Printed in India

List of Contributors

Aladakatti, Y.R.: *Cotton*, University of Agricultural Sciences, Dharwad, India, *E-mail*: yraladakatti@rediffmail.com

Bandyopadhyay, A.: *Groundnut*, National Fund for Basic Strategic and Frontier Research in Agriculture, Krishi Anusandhan Bhawan, New Delhi, India, *E-mail*: abandyo@gmail.com

Bhat, S.R.: *Plant Breeding*, Challenges, expectations and opportunities in the 21st Century, National Research Centre on Plant Biotechnology, New Delhi, India, *E-mail*: srbhat22@gmail.com

Dhillon, B.S.: *Maize*, Punjab Agricultural University, Ludhiana, India, *E-mail*: dhillonbaldevsingh@gmail.com

Govila, O.P.: *Pearl millet*, Formerly, Division of Genetics, Indian Agricultural Research Institute, New Delhi, India, *E-mail*: govilaop@gmail.com

Kadirvel, P.: *Sunflower*, Directorate of Oilseeds Research, Hyderabad, India, *E-mail*: kadirvel.palchamy@gmail.com

Kapur, Raman: *Sugarcane*, Division of Crop Improvement, Indian Institute of Sugarcane Research, Lucknow, India, *E-mail*: kapurraman 2@gmail.com

Khadi, B.M.: *Cotton*, University of Agricultural Sciences, Dharwad, India, *E-mail*: bmkhadi@rediffmail.com

Lal, S.K.: *Soybean*, Division of Genetics, Indian Agricultural Research Institute, New Delhi, India, *E-mail*: sklal68@gmail.com

Manivel, P.: *Groundnut*, Directorate of Medicinal & Aromatic Plants Research, Anand, India, *E-mail*: manivelp@yahoo.com

Nadarajan, N.: *Pigeonpea*, Indian Institute of Pulses Research, Kanpur, India, *E-mail*: iiprr@rediffmail.com

Patil, J.V.: *Sorghum*, Directorate of Sorghum Research, Hyderabad, India, *E-mail*: jvp@sorghum.res.in

Patil, Rajesh: *Cotton*, University of Agricultural Sciences, Dharwad, India, *E-mail*: rajeshpatil68@rediffmail.com

Prabhu, K.V.: *Oilseed Brassicas*, Division of Genetics, Indian Agricultural Research Institute, New Delhi, India, *E-mail*: kvinodprabhu@ rediffmail.com; kv4515 gpb@gmail.com

Prasanna, B.M.: *Maize*, CIMMYT (International Maize and Wheat Improvement Center), Nairobi, Kenya, *E-mail*: b.m.prasanna@cgiar.org bmprasanna@gmail.com

Rakshit, Sujay: *Sorghum*, Directorate of Sorghum Research, Hyderabad, India, *E-mail*: rakshit@sorghum.res.in

Sapra, R.L.: *Soybean*, Division of Genetics, Indian Agricultural Research Institute, New Delhi, India, *E-mail*: saprarl@gmail.com

Sendhil, R.: *Wheat*, Directorate of Wheat Research, Karnal, India, *E-mail*: r.sendhil@gmail.com

Siddiq, E.A.: *Rice*, Directorate of Rice Research, Hyderabad, India, *E-mail*: siddiq@rediffmail.com

Singh, A.K.: *Mungbean*, Department of Genetics and Plant Breeding, Institute of Agricultural Sciences, Banaras Hindu University, Varanasi, India.

Singh, M.N.: *Mungbean*, Department of Genetics and Plant Breeding, Institute of Agricultural Sciences, Banaras Hindu University, Varanasi, India, *E-mail*: mnsbhu@yahoo.co.in

Singh, Naveen: *Oilseed Brassicas*, Division of Genetics, Indian Agricultural Research Institute, New Delhi, India, *E-mail*: n.singhk@rediffmail.com

Singh, R.M.: *Chickpea*, Department of Genetics and Plant Breeding, Institute of Agricultural Sciences, Banaras Hindu University, Varanasi, India, *E-mail*: singhrishimuni@yahoo.co.uk

Sujatha, M.: *Sunflower*, Directorate of Oilseeds Research, Hyderabad, India, *E-mail*: sujatha@dor-icar.org.in

Swapna, M.: *Sugarcane*, Division of Crop Improvement, Indian Institute of Sugarcane Research, Lucknow, India, *E-mail*: sarikkath@yahoo.com

Tiwari, R.: *Wheat*, Directorate of Wheat Research, Karnal, India, *E-mail*: tiwari1964@yahoo.com

Vausdev, Sujata: *Oilseed Brassicas*, Division of Genetics, Indian Agricultural Research Institute, New Delhi, India, *E-mail*: sujatavasudev@iari.res.in

Venkatesh, K.: *Wheat*, Directorate of Wheat Research, Karnal, India, *E-mail*: kv4515gpb@gmail.com; r.sendhil@gmail.com

Viraktamath, B.C.: *Rice*, Directorate of Rice Research, Hyderabad, India, *E-mail*: viraktmath123@rediffmail.com

Yadava, D.K.: *Oilseed Brassicas*, Division of Genetics, Indian Agricultural Research Institute, New Delhi, India, *E-mail*: dkygenet@gmail.com

About the Editor

Educated at the University of Delhi, the Indian Agricultural Research Institute (IARI) and the University of Edinburgh, Chopra has been associated with IARI in various capacities that include active research in basic and applied genetics, Headship of the Genetics Division, National Professorship of the ICAR and founder of the Biotechnology Centre. He has published extensively in scientific journals of high repute and impact. His scientific standing has been recognized by national and international science academies and he is elected Fellow of the Indian National Science Academy, the Indian Academy of Science, the National Academy of Sciences India and the National Academy of Agricultural Sciences. He has also been elected to Fellowship of the European Academy of Science, Arts & Humanities, Paris and The World Academy of Science (TWAS), Trieste. In the arena of international agricultural research, Chopra was member of the Research Council of the CGIAR and on the Boards of Trustees of IRRI, ICRISAT and CIMMYT.

Chopra has held key positions in the Govt. of India including that of Secretary (DARE) & Director General, ICAR and Member, Planning Commission. Currently, he is Chancellor of the Central University of Kerala and Chairs the Science Advisory Committee to the Chief Minister of Jammu & Kashmir.

Chopra is a much decorated scientist and science administrator; his major Awards include Padma Bhushan by the Govt. of India, Aryabhata Medal by Indian National Science Academy, Dr. B.P. Pal Memorial Award by National Academy of Agricultural Sciences, Outstanding Achievement in Research in Science & Technology Award by FICCI, Om Prakash Bhasin Award and the Borlaug Award. He has been President of the International Genetics Federation and the National Academy of Agricultural Sciences.

Preface

For an agrarian society like India whose more than sixty percent population derives livelihood from pursuit of agriculture, the health and stability of the crop sector is of critical importance. This is particularly so in the context of progressively shrinking and degrading natural resources of biodiversity, water and land and the uncertainties created by changes imposed by global and local climate changes. Crop varieties for the coming days will need to be innately more productive and resilient to stresses of both biotic and abiotic kind. The changing patterns of national income distribution, the altered scenario of food preferences of the consuming public and demands of international competitiveness of globalised agriculture, of which we have become a part, will likewise require that crop varieties of the future possess many different combinations of the required new attributes. Fortunately, many technological advances for gaining deeper understanding and management of genetic apparatus of plants have now become available and these innovations are being captured for crafting 'new generation' crop varieties. Developments in this area have been particularly rapid in the last decade. This book has attempted to record this progress and evaluated its implications. Each crop has been handled by an acclaimed specialist to present an authentic critique.

I am very grateful to all the authors for their contributions and the patience with which they have favoured me on this project. My special gratitude is due to Dr. S.R. Bhat but for whose efforts and help this project would never have been completed. I am also grateful to Shri Mukhtiar Singh for providing a smooth conduit to the publishers.

V.L. Chopra

Table of Contents

1

Plant Breeding: Challenges, Expectations and Opportunities in the 21st Century

S.R. Bhat

National Research Centre on Plant Biotechnology, New Delhi, India

The Twenty First Century is said to be the century of biology just as the Twentieth Century is regarded as the century of physical sciences. This century, however, poses the greatest challenge to humankind, particularly with respect to our ability to sustain on the limited and diminishing resources of this planet. The global climate change and population scenarios of the next 30-40 years do not portend a happy situation with respect to agriculture and food production. Our ability to breed crop varieties that can withstand the vagaries of weather and still be productive and nutritious will be critical to achieving food and nutritional security in the coming decades. It is just a fortunate coincidence that plant breeders now have the best tools and techniques ever available to face such a challenge. In the following pages we examine the problems facing agriculture and crop production, the role of plant breeding in finding solutions to the problems, and review the recent technological prowess that has become available to plant breeders to tailor varieties to suit various demands and needs.

The history of plant breeding could be traced back to the dawn of agriculture when humans started domesticating plants. Despite spectacular success in developing new and novel cultivars of diverse crops, plant breeding remained an art or hobby until the end of the 19th century. Following the rediscovery of Mendel's laws of heredity

V.L. Chopra (Ed.): Breeding Field Crops II: Advances, Studium Press (LLC), USA.

in 1900 and subsequent elaboration of principles of genetics, plant breeding gained sound scientific footing and became an academic discipline.

1. PLANT BREEDING: AN INTEGRATED APPROACH TO PLANT IMPROVEMENT

Plant breeding can be regarded as applied genetics, and draws its primary strength from developments in Genetics. For instance, concepts like polygenic inheritance, epistasis, heterosis, genotype x environment interaction, and principles of quantitative and population genetics have helped refining various plant breeding methods and strategies. Further, for the improvement of traits ranging from morphological (*e.g.,* height, seed size), physiological (*e.g.,* tolerance to cold, drought, flooding), biochemical (*e.g.,* nutritional content of seeds/fruits), to more complex biotic (*e.g.,* antagonistic or symbiotic) and abiotic interactions, plant breeders often seek help from allied subjects such as chemistry, plant physiology, biochemistry, entomology, microbiology etc. All these disciplines have witnessed explosive growth in the past century and are consequently, greatly impacting plant breeding. The benefits of scientific plant breeding are best illustrated by the dramatic gains in productivity of crops and the spectrum of traits that have been modified in different crops to suit specific needs. For instance, the global cereal production increased from 877 million tons in 1961 to 2351 million tons in 2007 with a rate of growth of 32 million tons/year (Tester and Langridge, 2010). The cultivated area grew by 12% or 67 million ha during this period (FAO Statistical Yearbook 2012, http://www.fao.org/docrep/015/i2490e/i2490e00.htm). Improved technologies contributed the most to higher yields. Thanks to the success of plant breeding, crop production could keep pace with the food requirements of the growing population through the past century. Nearly one half of the productivity increase realized from the application of Green Revolution technologies is attributed to improvements emanating from breeding. Likewise, breeding for disease and pest resistance have helped avert epiphytotics, and provided secure harvests in major staples such as rice, wheat and maize. The fact that epiphytotics of the scale of the Irish potato famine (1845), the Bengal rice famine (1940s) and the Southern corn blight (1970s) are not repeated bears testimony to the success of resistance breeding and plant protection techniques. It is ironical that success of plant breeding and associated agritechnologies has given room to complacency that has led to declined investments in agricultural research (Alston *et al.*, 2009).

1.1 Food Demand of the 21st Century

The world population today stands close to 7 billion and is increasing @1.4% (98 million) per year. By 2025 AD world population is expected to touch 8 billion. To meet the food needs of nearly 9 billion people in 2050, food production needs to be increased by 70% (Tester and Langridge, 2010). This translates into annual increase of 44 million tons to be sustained for the next 40 years. India with 1.2 billion people is the second most populous country today and is projected to occupy the top slot by 2030 with 1.53 billion people and surpass China (projected population of 1.46 b by 2030). India also has the world's largest cattle and other farm animal population. Hence, production of both food and fodder needs to be doubled in the next two decades. In other words, production needs to be increased from the current rate of <2% to 3.5%. The depleting fossil fuels has spurred interest in alternative renewable sources of energy and agriculture-based biodiesel production is being explored as an option in several countries. There is growing concern that shift to biodiesel crops by the developed countries could severely compromise global food availability. This adds additional burden on crop production to meet this new demand.

2. CHALLENGES OF FOOD PRODUCTION: RESOURCE LIMITATIONS, CLIMATE CHANGE, SUSTAINABILITY

Considering the fact that most arable land in the world is already under plough, increasing crop productivity is the only option for securing higher production. Dramatic productivity gains obtained during the Green Revolution era were due to both improved varieties and high dependence on inputs (fertilizer and irrigation). However, the input-intensive technologies are not entirely relevant today. All over the world, the quality and quantity of water available for irrigation is declining (Hanjra and Qureshi, 2010). Similarly, the high cost of chemical nutrients and adverse consequences of high use of fertilizers on soil and environment make this approach unsustainable. Further, climate change with greatly fluctuating weather conditions including extremes of temperature, flood, drought etc., paint a grim picture of agriculture in the coming decades. Given these constraints, how can we approach meeting the challenge of increasing crop productivity with lesser inputs and ensure sustained food supply in the future? India is the first country in the world to recognize 'Right to Food' and to pass a legislation for its implementation. This puts additional responsibility of ensuring adequate food production to fulfill this obligation. While an integrated approach involving investments in research and training, technology infusion and policy interventions would be necessary to drive higher agricultural growth, steady supply of improved varieties would be the key driver of crop productivity.

3. EXPECTATIONS FROM PLANT BREEDING: FOOD AND NUTRITIONAL SECURITY, POVERTY ALLEVIATION, GENERATING EMPLOYMENT OPPORTUNITIES, PROMOTING PROFITABILITY

Worldwide, those undernourished have increased from 894 m in 1990-92 to 925 m in 2010 (http://www.worldhunger.org). In India, nearly a third of the population depends on agriculture for employment and livelihood, and this situation is expected to continue in the near future. Further, a large section of urban and rural population is undernourished and suffers from nutrition-related ailments; blindness due to vitamin A deficiency, anemia from iron- and zinc-deficiency being the major ones. Hence, improving agriculture with better crops has great implications for the rural economy and health of poor people. Further, farming needs to be made more remunerative and less risky. Also, to check movement of people from villages to urban areas for seeking employment, agricultural growth should include larger opportunities for rural employment. Plant breeding can play an important role in meeting these national goals. For instance, placing emphasis on crop breeding for both quantity and quality can contribute to attaining calorie and nutritional security. The work on quality protein maize (QPM) shows that both nutritional and yield improvements can go hand in hand (Gupta *et al.*, 2009; Badu-Apraku and Fontem Lum, 2010; Gutiérrez-Rojas *et al.*, 2010). Similarly, value addition, alternative or efficient use of crop residues could make crops more remunerative. For example, a large quantity of protein-rich mustard seed cake is currently exported. By breeding low/zero glucosinolate varieties, this byproduct could be turned into high-value feed for the poultry industry thereby increase profitability of the crop to the farmer. Efforts in this direction are being made through molecular marker assisted breeding (Bisht *et al.*, 2009). Similarly, India is the largest producer and exporter of castor. However, castor seed cake is at present of little food value due to the presence of highly toxic ricin protein. If ricin-free varieties could be bred, the value of castor crop would be greatly enhanced. The significance of introduction of modern technologies in agriculture is best illustrated by the *Bt*-cotton story in India. The commercial cultivation of transgenic *Bt*-cotton since 2002 has not only led to significant increase in productivity (from 308 kg/ha to > 526 kg/ha) and production (Choudhary and Gaur, 2010) but also reduced the use of pesticides (Kouser and Qaim, 2011; Krishna and Qaim, 2012). Prior to introduction of transgenics, nearly half of the quantum of pesticides used in agriculture in India was applied on cotton. Following widespread adoption of *Bt*-cotton, use of pesticides on cotton came down by nearly 50% and consequently, the incidence of pesticide poisoning was also reduced drastically (Kouser and Qaim, 2011; Krishna and Qaim, 2012). Thus, besides direct economic benefit in the form of yield, introduction of *Bt*-cotton has also contributed to the health of humans and environment.

4. OPPORTUNITIES: OMICS TECHNOLOGIES (GENOMICS, PHENOMICS, TRANSCRIPTOMICS, METABOLOMICS ETC.), MOLECULAR MARKERS, TRANSGENICS

Modern societies rely on technological solutions to the problems they face and hence advanced countries invest heavily in science and technology. In agriculture, plant breeding solutions are popular over other options because seed-embedded technologies have a significantly higher frequency of success and adoption especially under diverse field conditions. Given the challenges stated above, what are the current opportunities and strengths of plant breeding? How can they be used to find practicable solutions to problems at hand?

A plant breeder attempts to bring together desirable allelic combinations from two or more parents to create a new variety with desirable characteristics. Often, this is a multi-step process involving several generations of crossing between various parents and selection of the desirable ones from among the progenies. The two key components of plant breeding that determine success are i) availability of genetic variability for the trait to be manipulated and ii) the reliability of phenotypic selection of individuals for the correct genetic makeup responsible for the given phenotype. As parents often differ for a number of loci, the frequency of progenies carrying desired combination of alleles from the two parents is low. Therefore, plant breeders use ingenious mating strategies to maximize the chance of obtaining progenies with desirable allelic combinations. Similarly, multilocation and multi-year testing are undertaken to minimize the environmental effect interfering with phenotypic selection. However, as almost all agronomic traits including yield are controlled by polygenes, it is impossible to infer the genotype, and phenotype-based selections, especially in the early segregating populations are unreliable. Therefore, the progress is slow and unpredictable. Two major advances during the last quarter of the previous century have provided undreamt of opportunities to the plant breeders. These are i) genetic transformation of plants and ii) genomics and other omics technologies.

Transgenics

Genetic diversity provides the building blocks for plant breeding. In general, plant breeders depend on the natural genetic variation available in the primary or secondary genepool for improving the desired traits. However, for many important traits such as pest and pathogen resistance non-availability of adequate genetic variation is the limiting factor. Although mutation breeding can generate novel variation, it is of limited significance as novel traits/phenotypes not found in the species are hardly ever obtained. Plant genetic transformation techniques developed in the 1980s have opened unique opportunities to engineer any trait. Gene cloning techniques now make it possible to obtain genes and regulatory sequences from diverse organisms (viruses, bacteria, animals) and DNA fragments can be routinely custom synthesized. Further, such cloned genes can be mobilized and expressed in plants to engineer specific traits. A variety of systems for DNA delivery into plant cells utilizing gene/particle gun, *Agrobacterium tumefaciens*, nanoparticles, laser etc., have been devised and are being routinely used for plant transformation (Bhat, 2000). The plant tissue culture and transformation techniques are now well standardized for many crop plants and such facilities can be established with modest investment. Following the first commercial approval of Flavr Savr tomato in 1994, a number of transgenics in crops such as cotton, maize, soybean, canola etc. carrying genes derived from bacteria, viruses and conferring herbicide tolerance, pest or disease resistance have been commercialized. In fact, transgenic crops are making big impact on the agricultural economy and trade of some countries. As stated earlier, adoption of pest-resistant *Bt*-cotton hybrids in India since 2002, has increased the average productivity and total cotton production and transformed India from an importer to an exporter of cotton (Choudhary and Gaur, 2010). Despite opposition to transgenic crops is some countries, mainly in Europe, area under transgenic crops has grown at 8%. In 2011, 29 countries planted transgenic crops on 160 million ha (James, 2011).

The ability to mobilize genes across species barrier at will through transformation has led to the development of crops with novel traits not feasible through conventional plant breeding. For example, in the past, herbicide tolerant crops were available only where natural variation existed for the concerned chemical agent. Such natural variation, however, could not be exploited to breed other crops for the same feature. Discovery and cloning of bacterial genes conferring herbicide tolerance and genetic transformation technology have made genetic engineering feasible of any crop for the desired herbicide tolerance. For example, gene encoding EPSP synthase (5-enolpyruvylshikimate-3-phosphate synthase) isolated from *A. tumefaciens* strain CP4 has been mobilized into several crop plants including corn, cotton, soybean, canola to create glyphosate tolerant varieties (Green, 2009). Similarly, the *bar* gene from *Streptomyces hygroscopicus* capable of detoxifying the herbicide phosphinothricin or Basta has been widely deployed to produce commercial transgenic crops (Tsaftaris, 1996). In 2011, herbicide resistance trait accounted for more than 80% of the total area under transgenic crops (James, 2011).

Transgenics also hold promise for cracking some of the intractable problems of crop improvement. The pest problem (particularly boll worm) is a major constraint in cotton cultivation. Absence of resistance source for boll worm precluded breeding cotton for this trait. The development and deployment of transgenic cotton hybrids carrying the *cry* gene sourced from *Bacillus thuringiensis* has provided a unique solution. Similarly, breeding for abiotic stress tolerance (heat, cold, drought, flood, salinity etc.) is imperative in view of the anticipated water shortages and unpredictable weather conditions likely to result from global warming. However, traditional breeding approaches can not tackle this problem effectively as available genetic variability for these traits in most crops is inadequate and phenotyping for such traits is difficult (Fleury *et al.*, 2010). Molecular biology studies have uncovered a number of genes governing different pathways involved in stress tolerance. In particular, these studies have shown that fine regulation of key genes involved in stress signaling will be essential to achieve the desired change without adverse effects. Recently, maize transgenic lines carrying cold shock protein gene *CspA* from *Escherichia coli* or *CspB* from *Bacillus subtilis* have been field tested and found to confer water stress tolerance (Castiglioni *et al.*, 2008).

Yield improvement achieved so far has mainly come from better partitioning of photosynthates towards economic products rather than real increase of biomass. Transgenic approaches are being explored for improving photosynthetic efficiency and biomass enhancement (Demura and Ye, 2010). Transgenic approach is not considered appropriate for enhancing crop yield because yield is a complex polygenic trait. Studies have, however, shown promise of yield enhancement though introduction of single genes (reviewed in Camp 2005; Bhat, 2010). A novel approach to increasing productivity has been demonstrated in sugarcane. Sugarcane accumulates sucrose in vacuoles. An internal mechanism prevents sucrose accumulation beyond 500 μ mol/g FW. To overcome this limitation, Wu and Birch (2007) expressed a bacterial gene coding for sucrose isomerase (SI) and capable of converting sucrose to its isomer isomaltulose. This led to accumulation of both isomaltulose and sucrose in the vacuoles and thereby doubled the total sugar content of the canes. This study also highlights that by overcoming negative feedbacks in source-sink relations, productivity of crops can be enhanced. In cotton, ovule-specific expression of auxin biosynthesis gene *iiaM* was found to increase lint yield by 15% (Zhang *et al.*, 2011). Similarly, manipulation of dwarf4 gene involved in brassinosteroid synthesis pathway was shown to increased yield in rice (Sakamoto *et al.*, 2006).

Nearly a third of produced vegetables and fruits are lost during transit due to poor storability. This puts farmers at disadvantage as they cannot hold back their produce for obtaining better market price and become victims of distress selling. Transgenic approach could provide solution to this problem. In many fruits and vegetables, the plant hormone ethylene regulates ripening. By regulating internal production of this hormone in fruits, ripening process can be slowed down thereby extending shelf-life of vegetables and fruits (Mayer *et al.*, 2008; Matas *et al.*, 2009). Widespread adoption of this technology has the potential to alleviate the economic conditions of resource-poor farmers and also provide consumers better quality products.

Improving nutritional quality of crops is receiving increasing attention. Our understanding of the biochemical pathways of various nutraceuticals has advanced greatly in recent years. Further, key genes governing the pathways have been cloned. This has paved the way for transgenic manipulation of nutritional quality of crops. The 'Golden Rice' that accumulates high levels of provitamin A (β-carotene) in seeds was made possible by assembling genes and regulatory sequences from bacteria and maize (Al-Babili and Beyer, 2005). This could not have been possible through conventional breeding as variation for this trait is not available in gene pool nor could it have been obtained through mutagenesis. Likewise, biofortification for important minerals such as iron and zinc (Mayer *et al.,* 2008), carotenoids (Farre´ *et al.*, 2011) and ω-3 fatty acids (Napier and Graham, 2010) through genetic engineering have been demonstrated. Availability of such products will have a major impact on combating malnutrition in developing countries (Zhao and

Shewry, 2011). Plants have also been engineered to produce vaccines, antibodies and other therapeutics for humans and farm animals. Thus transgenic crops have the potential to make agriculture both remunerative and competitive.

Transgenics are also highly relevant for developing genetic stocks that aid plant breeding. For instance, pollination control systems are the basis of hybrid technology and are not available in several crops. Mariani *et al.* (1990, 1992) devised a generic pollination control system utilizing bacterial gene barnase for inducing male sterility through tissue specific cell ablation. Barstar was deployed for inhibiting barnase gene product for restoring male fertility. This system has been commercialized in *Brassica napus*. Recently, genetic engineering of centromere histone protein (CENH3) has been shown to be useful to produce haploids (Ravi and Chan, 2010). In this scheme, transgenically expressed chimeric CENH3 protein replaces the native CENH3 proteins at the centromeres. Such plants when pollinated by wild type plants yield paternal haploids due to selective elimination of chromosomes bearing chimeric CENH3 proteins. This novel haploid-inducing technology has the potential to accelerate breeding of varieties and hybrids. Genetic engineering is also permitting manipulation of meiosis. For example, by ingenious manipulation of genes governing meiotic recombination and chromosome separation, d'Erfurth *et al.* (2009) succeeded in converting meiotic pathway into mitosis to develope plants with fully functional unreduced gametes. By combining the techniques of haploid induction and unreduced gamete formation, Marimuthu *et al.* (2011) have provided the first proof-of-concept of engineering apomixis in plants. Similarly, critical genes governing apomixis are expected to be cloned in the near future . Such developments are most likely to bring about major changes in plant breeding and permit rapid development of new varieties and hybrids.

Genomics

The other major breakthrough to impact plant breeding is the explosion of 'omics' science. As stated earlier, polygenes where each one makes a small contribution to the total variability for a trait could not be individually identified and assigned to specific loci through conventional genetic approaches. Hence, such polygenes remained anonymous conceptual entities. Consequently, most yield traits were handled using quantitative genetic models. This generally proved unsatisfactory when dealing with individuals, while individuals are the units on which plant breeders make decisions while making selections. Adoption of molecular biology and r-DNA techniques by plant scientists and breeders during the 1980s can be regarded as a turning point in the history of plant breeding. Molecular markers (RFLP, AFLP, SSR, RAPD and several others) for the first time provided a powerful handle to deal with polygenes. These DNA-based markers are highly heritable and are not influenced by the environment. Therefore, these markers could be treated just like any other morphological trait for genetic analysis. Further, unlike morphological traits, there was no limit for the number of molecular markers that could be developed for each crop species. This permitted construction of dense molecular marker linkage maps of crops covering all the linkage groups. Combining segregation data of molecular markers with the segregation data of traits, molecular markers linked to the trait could be identified. Thus, what were hither to anonymous polygenes governing a quantitative trait could be assigned with specific DNA markers, and were given the name Quantitative Trait Loci (QTLs). Molecular markers soon found numerous other applications spanning the entire gamut of plant breeding such as assessment of genetic variation, fingerprinting, interrelatedness, hybrid/variety purity test etc. Using closely linked molecular markers flanking a gene of interest, map based cloning of genes has now become feasible, particularly in crops with fully sequenced genomes.

The ability to precisely genotype individuals using molecular markers has opened doors for precision breeding. Indeed varieties have started flowing from approaches based on molecular marker assisted selection (MAS) (see chapter on Rice in this book). Molecular markers have proved particularly useful in improving traits that need special conditions to phenotype (*e.g.,* disease resistance, abiotic stress tolerance) or those that could be assessed only after the harvest (*e.g.,* lint characters in cotton). For example, using markers tightly linked to the genes controlling disease resistance, one can make selections in the absence of disease and still be fairly certain to incorporate the desired genes. Incorporation of additional resistance (R) genes when a genotype already carries one R gene poses problem as progenies carrying one or more R genes cannot be distinguished based on phenotype. Using molecular markers each linked to different R genes, it is now possible to pyramid genes for securing durable resistance. For instance, in rice, MAS for bacterial blight has been successfully employed to breed improved versions of proven varieties. Similarly, wheat lines carrying multiple genes for rust resistance have been bred through MAS. In Indian mustard, breeding for low glucosinolate content was difficult through conventional methods because glucosinolate content is controlled by 6-7 unlinked genes. However, availability of molecular markers has accelerated breeding of double low (*i.e.,* low glucosinolate and low erucic acid) varieties (Bisht *et al.*, 2009). Additional examples of MAS in different crops are provided in different chapters of this book.

Plant breeding has always been a long-term project and it generally takes 10-12 years to breed a new variety. However, to keep pace with the emerging demands, this period needs to be shortened. Molecular markers have provided an ideal solution for accelerating plant breeding. For instance, using a set of polymorphic markers evenly distributed across the entire genome, it is now possible to identify genotypes with desired combinations of genes in the early generations (F_2 or BC_1). Such individuals could then be either intermated or selfed to derive homozygous lines in 1-3 generations. This approach is currently being applied in crops such as maize, rice, sorghum, pigeonpea, chickpea, cotton etc. where extensive genomic information has now become available.

The multinational human genome sequencing project of the 1990 was the harbinger of the era of genomics. It spurred innovations in instrumentation, computation, automation and brought about a paradigm shift in methods of sequencing. As a result, genome sequencing has now become common-place, and the cost and time have been reducing drastically (Metzker, 2010). This has directly facilitated plant genome sequencing projects. Since the first complete genome sequence of the model plant *Arabidopsis thaliana* in 2000, whole genome sequences have become available for more than 50 crop plants and the list is expanding on a weekly basis. The availability of complete genome sequence of several crop plants has dramatically enlarged our understanding of the genes and their regulatory sequences. Comparison of whole genome sequences of different species has revealed that linear order of genes are conserved (called synteny) during evolution. For example, chromosomes of various members of the grass family could be represented in the form of crop circles based on synteny (Devos, 2005). Thus, availability of whole genome sequence of one species now permits deeper enquiries in related species lacking such information. For example, genome complexity and size of bread wheat (17.0 Gbp) and barley (5.1 Gbp) have discouraged whole genome sequencing of these otherwise most important crops. By using comparative genomic analysis based on synteny information, array hybridization and advanced bioinformatics tools, a comprehensive genomic picture has been constructed (Mayer *et al.*, 2011). Similarly, sequence information of specific genes makes it possible to quickly identify and clone new alleles or orthologous genes from other genotypes or species, respectively. This is called allele mining and is now extensively employed to search new R genes against various pathogens/pests (Edwards and Batley, 2010).

Knowledge of gene sequence has given new impetus to mutation breeding also. The random and undirected nature of induced mutations makes mutation breeding a slow and less rewarding exercise. The new reverse genetics technique with the acronym TILLING (Targeting Induced Local Lesions IN Genome) devised by McCallum *et al.* (2000) utilizes gene sequence information for high throughput screening of mutants to identify individuals carrying mutation in the genes of interest. Here, DNA from individuals of early generation of mutagenised population is extracted separately. Systematic pools of DNA are prepared and analyzed for change in DNA sequence of specific genes under test using high throughput methods. If a pool indicates mutation, the individuals of the pool are analyzed to pick the one(s) carrying mutation. In this way, a number of mutations in the gene of interest can be quickly identified and then studied in detail (Haughn and Gilchrist, 2006). TILLING is particularly relevant in perennial crops with long generation time. Similarly, large scale transcriptome queries based on microarray, SAGE (Serial Analysis of Gene Expression) or sequencing have now become common. Also, metabolomic and proteomic tool kits are providing additional power and precision to plant breeding (Mochida and Shinozaki, 2010). Complete genome sequences of crops is providing a rich source of polymorphic markers including Single Nucleotide Polymorphism (SNPs) markers for plant breeding (Feuillet *et al.*, 2011). The automation of genotyping with such precise markers has indeed opened doors for designer crop breeding (Schneeberge and Weigel, 2011).

Accurate phenotyping for various traits has always been a challenge particularly for traits that show strong genotype x environment interaction. Therefore, with the now available ease of genotyping offered by the new genomics technologies, phenotyping remains a bottleneck to efficient plant breeding. New phenomics facilities are being devised to meet this challenge (http://www.plantphenomics.com/partners). Here, using modern, state-of-the-art technologies of imaging, cultivation under controlled conditions, bioinformatics etc., plants are monitored continuously to obtain a precise picture of phenotype as it unfolds and to link genotype with phenotype. This information is then used to screen and select genotypes that match with a desired phenotype (Furbank and Tester, 2011).

Plant breeding has benefitted greatly from statistics. In fact, Mendel's success in arriving at the laws of inheritance is attributed to his mathematical approach to the interpretation of data. Extensive and deep genotyping facilities now made possible by new technologies have given new impetus to quantitative and population genetics. New models of Genome Wide Association (GWA) analysis of DNA markers and traits, using natural genetic variation, permits identification of genes/markers linked to complex traits (Mackay *et al.*, 2009; Ingvarsson and Street, 2011). This is especially useful in perennial crops where conventional genetic analyses and breeding are hampered by long generation time and other constraints. Likewise, there is resurgence of quantitative genetic analysis tools to handle large volume of complex data to help interpret results (Mochida and Shinozaki, 2010).

5. IMPLICATIONS: POLICY SUPPORT, ADMINISTRATIVE REFORMS, TRAINING NEEDS, RESOURCES, NEWER WAYS FOR FAST TRACKING RESEARCH AND IMPLEMENTATION

In recent years, there is widely voiced lament that dedicated, well-trained and proficient plant breeders have become scarce all over the world as the discipline is not attracting the best talents (Ransom *et al.*, 2006; Stuber and Hancock, 2008; Bhat and Bandyopadhyay, 2011). Considering that the challenges faced by agriculture today are daunting and urgent, it is imperative that the best minds are pressed into service to tackle the complex problems. Therefore, steps to make plant breeding an attractive learning experience as well as a lucrative career option for the youngsters are needed to be taken. Some measures to correct the situation have been suggested (Bhat and Bandyopadhyay, 2011). Also, steps have been initiated to attract bright people to take up the challenge of plant breeding. For instance, CIMMYT has instituted special fellowships (Borlaug Fellowship) to attract young breeders to work on wheat and corn.

While many new technologies are unfolding novel opportunities and providing new impetus to plant breeding, their adoption in routine plant breeding would need major overhaul of the current practices and funding mechanism for plant breeding programmes. The new genotyping and phenotyping technologies are resource intensive and need skilled personnel. Therefore, training of next generation of plant breeders needs to be more broad based. Even the current plant breeders need training to upgrade their skills in molecular breeding for a harmonious interaction with molecular biologists. In fact, this problem of lack of mutual understanding and trust between plant breeders and molecular biologists appears to be widespread in both developed and developing world.

Funding of plant breeding currently has largely been limited to meet the expenses for the conduct of field trials and at best, to support laboratory facilities for quality testing. The approach now demands a thorough revision. Heavy infusion of funds to develop laboratory facilities for large scale genotyping would be necessary. Considering that genotyping services are becoming popular, outsourcing or centralized genotyping facilities might be considered. Precision phenotyping facilities and platforms are imperative for breeding varieties to cope with problems associated with climate change (Furbank and Tester, 2011). Also, the composition of plant breeding team needs to be broadened to include personnel with molecular biology and computational skills to gather and interpret large volume of data that is becoming common in modern biology. The past few years have seen a rapid surge of new technologies and equipments. In fact, the equipments and technologies are becoming obsolete faster than ever before. Hence, administrative reforms are imperative regarding use of funds for purchase of equipments and creation of facilities. Instead of the current practice of planning and budgeting five years in advance for specific equipments, funds should be allocated for specific purpose leaving flexibility to decide the most appropriate ways to use the funds at the implementation stages.

In India private seed industries (both indigenous and multinational) are major players especially in crops where hybrids are popular. For instance in crops such as sunflower, cotton, maize and vegetables like tomato, cauliflower, pepper etc., private sector bred hybrids dominate the market. Until recently, public private partnership in plant breeding in India was almost non-existent. While public sector has strengths in basic skills, germplasm resources, field and lab facilities etc., private sector is better placed at converting ideas and preliminary observations/ findings into successful commercial products. This is particularly true with regard to transgenics. Scientists in universities and public sector laboratories are adept at construction of suitable gene cassettes and generation of transgenics. However, event selection, biosafety testing, and seeking regulatory clearance for commercialization of transgenics is quite beyond the capability of most plant breeders/molecular biologists who operate in public sector academic environment. Private sector could play a decisive role in this sphere. Thus public-private partnership in plant breeding could be complementary and fruitful. Indeed some successful partnerships are already operating. For instance, quick and large scale dissemination of new, fine-grain hybrid rice cultivars developed by the Indian Agricultural Research Institute has been realized through strategic partnership with the private sector firm Mahyco. Likewise, transgenic cotton events developed by IIT, Kharagpur were commercialized through licensing to JK Agriseeds. The *Bt*-cotton story in India has an important message for public private participation in plant breeding. Introduction of *Bt*-cotton has led to wider adoption of hybrids and the propriety nature of the technology has meant that public sector hybrids lacking the *Bt*-technology were pushed out of the market. A similar scenario is seen in the US with soybean. The advent of propriety herbicide tolerant Roundup Ready soybeans has led to disappearance of public sector soybean varieties from the market. Adoption of new technologies of propriety nature should be considered even in the public sector to prevent such situations. New models for fostering healthy public-private partnerships need to be evolved and would need change of mindset and policy initiatives and practices.

6. PATENTS, BREEDERS RIGHTS, FARMERS RIGHTS AND OWNERSHIP ISSUES, BIOSAFETY ISSUES

International and national policy changes with regard to sharing and ownership of germplasm and varieties has been impacting plant breeding practices. Sharing of germplasm now often comes with restriction on their commercial use. Plant breeders, therefore, have to know not only about the patent protection of technologies that they need to use but also should know how to protect their products and varieties. Again, this is an area where ignorance abounds and requires intensive efforts on training and facilitation. Similarly, biosafety and ethical issues are becoming a major problem for commercializing transgenics. Many plant breeders are not conversant with these issues and are unable to take informed scientific decisions on such matters, thus contributing to public confusion. Thus, next generation plant breeders will need to be trained to not only deal with scientific aspects of breeding better varieties but also in developing capablity of handling ticklish issues surrounding commercialization.

In summary, the Twenty First century poses great challenge to plant breeders to develop varieties for increasing productivity of crops to ensure sustainable production of food for nearly 9 billion population. Fortunately, recent developments in genetics and genomics promise unprecedented opportunities to precisely tailor plants to face the problems posed by climate change and progressively diminishing natural resources. However, policy interventions and investment to support human resource and infrastructure development would be essential to realize the promise.

REFERENCES

Alston, J.M., Beddow, J.M. and Pardey, P.G. (2009). Agricultural research, productivity, and food prices in the long run. *Science*, 325: 1209–1210.

Al-Babili, S. and Beyer, P. (2005). Golden rice – five years on the road – five years to go? *Trends in Plant Science*, 10: 565–573.

Badu-Apraku, B. and Fontem Lum, A. (2010). The pattern of grain yield response of normal and quality protein maize cultivars in stress and nonstress environments. *Agron. J.*, 102: 381–394.

Bhat, S.R. (2000). Plant transformation for directed genetic change. *In*: Chopra, V.L. (*Ed.*) Plant Breeding: Theory and Practice, 2nd Ed. Oxford and IBH Publishing Co. Pvt. Ltd. New Delhi, Calcutta, pp. 439–459.

Bhat, S.R. (2010). Transgenics for increasing productivity of crops. *J. Plant Biochemistry and Biotechnology*, 19: 1–7.

Bhat, S.R. and Bandyopadhyay, A. (2011). Reinvigorating plant breeding to meet the challenges of food production in the 21st century. *Ind. J. Genet.,* 71: 95–101.

Bisht, N.C., Gupta, V., Ramchiary, N., Sodhi, Y.S., Mukhopadhyay, A., Arumugam, N., Pental, D. and Pradhan, A.K. (2009). Fine mapping of loci involved with glucosinolate biosynthesis in oilseed mustard (*Brassica juncea*) using genomic information from allied species. *Theor. Appl. Genet.,* 118: 413–421.

Camp W.V. (2005). Yield enhancement genes: Seeds for growth. *Curr. Opin. Biotechnol.,* 16: 147–153.

Castiglioni, P., Warner, D., Bensen, R.J., Anstrom, D.C., Harrison, J., Stoecker, M., Abad, M., Kumar, G., Salvador, S., D'Ordine, R., Navarro, S., Back, S., Fernandes, M., Targolli, J., Dasgupta, S., Bonin, C., Luethy, M.H. and Heard, J.E. (2008). Bacterial RNA chaperones confer abiotic stress tolerance in plants and improved grain yield in maize under water-limited conditions. *Plant Physiol.,* 147: 446–455.

Choudhary, B. and Gaur, K. (2010). *Bt* Cotton in India: A country profile. ISAAA Series of Biotech Crop Profiles. ISAAA: Ithaca, NY.

Demura, T. and Ye, Z.H. (2010). Regulation of plant biomass production. *Curr. Opin. Plant Biol.,* 13: 299–304.

d'Erfurth I., Jolivet S., Froger N., Catrice O., Novatchkova M. and Mercier, R.I. (2009). Turning meiosis into mitosis. *PLoS Biol.,* 7: 1–10.

Devos, K.M. (2005). Updating the 'crop circle'. *Curr. Opin. Plant Biol.,* 8: 155–162.

Edwards, D. and Batley, J. (2010). Plant genome sequencing: Applications for crop improvement. *Plant Biotech. J.,* 8: 2–9.

Farre´, G., Bai, C., Twyman, R.M., Capell, T., Christou, P. and Zhu, C. (2011). Nutritious crops producing multiple carotenoids – a metabolic balancing act. *Trends Plant Sci.,* 16: 532–540.

Jan Leemans

Feuillet, C., Leach, J.E., Rogers, J., Schnable, P.S. and Eversole, K. (2011). Crop genome sequencing: Lessons and rationales. *Trends Plant Sci.,* 16: 77–88.

Fleury, D., Jefferies, S., Kuchel, H. and Langridge, P. (2010). Genetic and genomic tools to improve drought tolerance in wheat. *J. Exp. Bot.,* 61: 3211–22.

Furbank, R.T. and Tester, M. (2011). Phenomics – technologies to relieve the phenotyping bottleneck. *Trends Plant Sci.,* 16: 635–644.

Green, J.M. (2009). Evolution of glyphosate-resistant crop technology. *Weed Science*, 57: 108–117.

Gupta, H.S., Agrawal, P.K., Mahajan, V., Bisht, G.S., Kumar, A., Verma, P., Srivastava, A., Saha, S., Babu, R., Pant, M.C. and Mani, V.P. (2009). Quality protein maize for nutritional security: Rapid development of short duration hybrids through molecular marker assisted breeding. *Curr. Sci.,* 96: 230–237.

Gutiérrez-Rojas A., Betrán, J., Paul Scott, M., Atta, H. and Menz, M. (2010). Quantitative trait loci for endosperm modification and amino acid contents in quality protein maize. *Crop Science*, 50: 870–879.

Hanjra M A. and Qureshi, M.E. (2010.) Global water crisis and future food security in an era of climate change. *Food Policy*, 35: 365–377.

Haughn, G.W. and Gilchrist, E.J. (2006). Tilling in the botanical garden: A reverse genetic technique feasible for all plant species. *In*: Floriculture, Ornamental and Plant Biotechnology, Volume I, Global Science Books, UK. pp. 476–482.

Ingvarsson, P.K. and Street, N.R. (2011). Association genetics of complex traits in plants. *New Phytologist*, 189: 909–922.

James, C. (2011). Global status of commercialized biotech/GM crops: 2011. *ISAAA Brief* No. 43. ISAAA: Ithaca, NY.

Kouser, S. and Qaim, M. (2011). Impact of *Bt* cotton on pesticide poisoning in smallholder agriculture: A panel data analysis. *Ecological Economics*, 70: 2105–2113.

Krishna, V.V. and Qaim, M. (2012). *Bt* cotton and sustainability of pesticide reductions in India. *Agricultural Systems*, 107: 47–55.

Mackay, T.F.C., Stone, E.A. and Ayroles, J.F. (2009). The genetics of quantitative traits: Challenges and prospects. *Nature Review Genetics*, 10: 565–577.

Mariani, C., Gossele, V., De Beuckeleer, M., De Block, M., Goldberg, R.B., De Greef, W. and Leemans, J. (1992). A chimaeric ribonuclease-inhibitor gene restores fertility to male sterile plants. *Willy De Greef and Nature*, 357: 384–387.

Mariani, C., De Beuckeleer, M.,Truettner, J., Leemans, J. and Goldberg, R.G. (1990). Induction of male sterility in plants by a chimaeric ribonuclease gene. *Nature*, 347: 737–741.

Marimuthu, M.P.A., Jolivet, S., Ravi, M., Pereira, L., Davda, J.N., Cromer, L., Wang, L., Nogué, F., Chan, S.W.L., Siddiqi, I. and Mercier, R. (2011). Synthetic clonal reproduction through seeds. *Science*, 331: 876.

Matas, A.J., Gapper, N.E., Chung, Mi.Y., Giovannoni, J.J. and Rose, J.K.C. (2009). Biology and genetic engineering of fruit maturation for enhanced quality and shelf-life. *Curr. Opin. Biotech.,* 20: 197–203.

Mayer, J.E., Pfeiffer, W.H. and Beyer, P. (2008). Biofortified crops to alleviate micronutrient malnutrition. *Curr. Opin. Plant Biol.,* 11: 166–170.

Mayer K.F.X., Martis, M. and Hedley, P. (2011). Unlocking the barley genome by chromosomal and comparative genomics. *Plant Cell*, 23: 1249–1263.

McCallum, C.M., Comai, L., Greene, E.A. and Henikoff, S. (2000). Targeting induced local lesions in genomes (TILLING) for plant functional genomics. *Plant Physiol.,* 123: 439–442.

Metzker, M.L. (2010). Sequencing technologies —the next generation. *Nature Reviews Genetics*, 11: 31–46.

Mochida, K. and Shinozaki, K. (2010). Genomics and bioinformatics resources for crop improvement. *Plant Cell Physiol.,* 51: 497–523.

Napier, J.A. and Graham, I.A. (2010). Tailoring plant lipid composition: Designer oilseeds come of age. *Curr. Opin. Plant Biol.,* 13: 330–337.

Ransom, C., Patricia, C., Ando, K. and Olmstead, J. (2006). Report of Breakout Group I: What kind of training do plant breeders need and how can we most effectively provide that training? *HortScience*, 41: 53–54.

Ravi, M. and Chan, S.W.L. (2010). Haploid plants induced by centromere-mediated genome elimination. *Nature*, 464: 615–619.

Sakamoto, T., Morinaka, Y., Ohnishi, T., Sunohara, H., Fujioka, S., Ueguchi-Tanaka, M., Mizutani, M., Sakata, K., Takatsuto, S., Yoshida, S., Tanaka, H., Kitano, H. and Matsuoka, M. (2006). Erect leaves caused by brassinosteroid deficiency increase biomass production and grain yield in rice. *Nature Biotechnology*, 24: 105–109.

Schneeberge, K. and Weigel, D. (2011). Fast-forward genetics enabled by new sequencing technologies. *Trends in Plant Science*, 16: 282–288.

Stuber, C.W. and Hancock, J. (2008). Sustaining plant breeding: National workshop. *Crop Science*, 48: 25–29.

Tester, M. and Langridge, P. (2010). Breeding technologies to increase crop production in a changing world. *Science*, 327: 818–22.

Tsaftaris, A. (1996). The development of herbicide-tolerant crops. *Field Crops Research*, 45: 115–123.

Wu, L. and Birch, R.G. (2007). Doubled sugar content in sugarcane plants modified to produce a sucrose isomer. *Plant Biotechnology J.*, 5: 109–117.

Zhang, M., Zheng, X., Song, S., Zeng, Q., Hou, L., Li, D., Zhao, J., Wei, Y., Li, X., Luo, M., Xiao, Y., Luo, X., Zhang, J., Xiang, C. and Pei, Y. (2011). Spatiotemporal manipulation of auxin biosynthesis in cotton ovule epidermal cells enhances fiber yield and quality. *Nature Biotechnology*, 29: 453–458.

Zhao, F.J. and Shewry, P.R. (2011). Recent developments in modifying crops and agronomic practice to improve human health. *Food Policy*, 36: S94–S101.

2

Rice

E.A. Siddiq and B.C. Viraktamath

Directorate of Rice Research, Hyderabad, India

Rice breeding research was intensified all over the country for overcoming problems that have remained intractable and for tackling new and emerging bottlenecks. The All India Coordinated Rice Improvement Project (AICRIP) located at the Directorate of Rice Research (DRR) has played a pivotal role in planning, generating material and multilocation testing at 47 publicly funded and 80 Voluntary Cooperative Centres. Besides upgrading genetic yield potential, attention has been directed to breeding for resistance/ tolerance to biotic and abiotic stresses and improvement of grain cooking and nutritive qualities.

The status of progress made employing conventional and innovative breeding approaches is summarized in this chapter.

VARIETAL BREEDING

Since the inception of the AICRIP, 946 cultivars including 46 hybrids have been evolved for different rice ecologies. Among these, 107 varieties and 24 hybrids have been released through Central Sub Committee on Crop Standards, Notification and Release of Varieties (CSCS & NRV) while the State Variety Release Committees released 793 varieties and 22 hybrids. The different ecologies for which these releases are recommended and their special attributes such as

V.L. Chopra (Ed.): Breeding Field Crops II: Advances, Studium Press (LLC), USA.

Table 1: Improved varieties and hybrids popular in the irrigated and rainfed ecologies in India

Ecosystem	***Variety name***
Irrigated	
Irrigated Early (IRE)	*Varieties*: IR 36, Jyothi, MTU 1010 (Cottondora Sannalu), Tella Hamsa, Richa *Hybrids*: RH 204, DRRH 2, PA 6129, GK 5003, Sahyadri 4
Irrigated Mid Early (IRME)	*Varieties*: IR64, Pant Dhan 19, Lalat, Gontra Bidhan-1, Varadhan, Indira Rajeshwari *Hybrid*: DRH-775
Irrigated Medium (IRM)	*Varieties*: Jaya, Sarjoo 52, NDR 359, MTU 1001 (Vijetha), PR 114, Pant Dhan 10, Pant Dhan 12, Mahamaya, Poornima, Narendra dhan 8002, Akshaydhan, Pushyami, CO(R)50, DRR Dhan 38, Durgeshwari, Kaushalya *Hybrids*: DRRH-3, Ganga, HRI 120, KRH 2, Suruchi, JKRH 401, HRI 157, PAC 835, PAC 837, US 312
Irrigated Saline & Alkaline soils (IRSA)	*Varieties*: CSR 23, Sumati, Jarava, Naina, Bhuthnath, Amal Mana, DRR Dhan 39, Narendra Usar Dhan 2008
Irrigated Hills (HRIR)	*Varieties*: Vivekdhan 82, VL Dhan 86
Scented (SCR)	*Varieties*: Pusa Basmati 1, Pusa Sugandh 5, Sugandhamati, Improved Pusa Basmati 1, Pusa Basmati 6, Vasumati, CSR 30 *Hybrids*: Pusa RH 10
Boro	*Varieties*: Krishna Hamsa, CR Boro Dhan 2, CR Dhan 601, *Hybrid*: Rajalaxmi
Rainfed	
Rainfed Upland (RUP)	*Varieties*: Anjali, Narendra Dhan 97, Govind, Satabdi, Pant Dhan 16, Abhishek, Shusk Samrat, Virender, CR Dhan 40, RC Maniphou 11, Sahbhagi Dhan
Rainfed Shallow Lowland (RSL)	*Varieties*: Samba Mahsuri, Swarna, Improved Samba Mahsuri, *Hybrid*: CRHR 32 (CR Dhan 701),
Semi Deep Water (SDW)	*Variety*: CR Dhan 501
Upland Hills (HRUR)	*Variety*: Sukardhan 1

resistance to biotic and abiotic stresses and quality traits are given in Tables 1, 2 and 3.

High Yielding Basmati Rices

Among the high priced basmati rices, the semidwarf Pusa Basmati 1 combines high yield and good grain. It has provided a strong agronomic base for developing still higher yielding and early maturing varieties/ hybrids with improved milling characteristics and resistance to diseases.

These improved basmaties include high yielding Pusa Sugandh 5, Pusa Sugand 6, Pusa Basmati 6, Improved Pusa Basmati 1 (MAS derived product with BLB resistance), Pusa RH 10 (hybrid). Among these quality

Table 2: Rice varieties (2001-10) resistant/tolerant to pests and diseases

Blast
Ganga**,** Krishna Hamsa, Pant Dhan 16, Pant Dhan 19, Pusa Sugandh 2, Pusa Sugandh 3, Vivekdhan 82, Basmati CSR 30 (Yamini), Apurva, Shanthi, GR 7, Rashmi,Dhanrasi, ***RH 204, Dandi, GR 104, SYE 2001, Jagabandhu, ADT (R) 46, Sukardhan 1, Hazaridhan, PKV Ganesh, Vivek Dhan 154, Jaldi Dhan 13, Richa, Sugandhamati, ***Suruchi,*** Bhogavati, **DRRH 2,** Jarava, Naina, Narendra dhan 8002, HKR 47, HPR 1068, HPR 2143, Shalimar Rice 1, Karjat 5, Karjat 6, **Sahyadri 2, Sahyadri 3,** Sidhanta, Uphar, VL Dhan 85, VL Dhan 207, VL Dhan 208, Abhishek, Shusk Samrat, Virender, VL Dhan 86, Pardhiva, Sree Kurma, J aldubi, Shalimar Rice 2, Maniphou 10, VL Dhan 65, VL Dhan 209, ***PA 6129,*** Chandrama, Varun Dhan, Madhuri, Akshaydhan, CR Boro Dhan 2, ***GK 5003,*** Sampada, CR Boro Dhan 2, **DRH-775,** ***NK 5251, PAC 835, PAC 837,*** Ratnagiri 24, Mandakini, Mrunalini, Tejaswini, Narendra Dhan, CR Dhan 601, **CRHR 32,** IGKVR-1, RC Maniphou 11, ***US 312,*** Sonal, Sugandha Samba, Pratheeksha, Bhalum 4, Megha SA 1, Megha SA 2, Luna Sampad, Luna Suvarna, Phalguni, Reeta (CR Dhan 401), NDR 2065, Heera, Cross 116, Asha
Bacterial leaf blight
Phule Radha, GR 7, Dandi, SYE 2001, PKV Ganesh, Richa, **Sahyadri 2,** CR Boro Dhan 2, Bapatla Sannalu, Bhudeb, Pusa Basmati 1121, PR 118, PMK (R) 3, Kali Khasa, **Narendra Usar Sankar, Dhan 3,** Indra, Improved Pusa Basmati 1, Improved Samba Mahsuri, PR 120, Punjab Mehak 1, Malviya Dhan 36, Nellore Mahsuri
Rice tungro virus
Pant Sankar Dhan 3, Dhanrasi, Sidhanta, Uphar, Chandrama, ***PAC 835,*** Mrunalini, CO(R)50, DRR Dhan 39, **Sahyadri 2,** ***PA 6444,*** Bhuthnath, ***JKRH 401,*** **CORH 3,** CO(R)48, Pusa Basmati 6, **Sahyadri 4,** Manaswini, CO(R)49, **DRRH 3,** Karjat 5, Akshaydhan, ***GK5003,*** **DRH-775,** ***PAC837,*** Mandakini, Phalguni, **Pusa RH10, HRI 157,** Jagabandhu, Tejaswini, Annalakshmi, Varadhan,Nua Kalajeera, Nua Chinikamini, **DRRH 2,** Naina,CR Dhan 601, Warangal Samba, Narendra Dhan
Sheath blight
Uphar, Chandrama, Mrunalini, Bhuthnath, ***JKRH401,*** CO(R)48, Manaswini, ***GK5003,*** Mandakini, Phalguni, Tejaswini, Rajendra, Sweta, Jogesh, Karimnagar Samba, Rajendra Mahsuri 1, Ketekijoha, Richa, Bhudeb, **Narendra , Usar Sankar, Dhan 3,** Swetha, Pratikshya, Amal Mana, Thanu, CR Dhan 70, Anashwara (PTB 58), Akshaya, Maniphou 10, Pratheeksha, Luna Sampad, Luna Suvarna, Gouri, Pant Dhan 18, Narendra Dhan 2026, ADT (R) 47, Rajendra, Bhagwati, Narendra Dhan , Manair Sona, Warangal Samba, Shusk Samrat, CR Boro Dhan 2
Stem borer
Manaswini, Pratikshya, AmalMana, Shusk Samrat, Sidhanta, PMK (R) 3, SKL 8, Birsamati, Rashmi, Virender, VL Dhan 86, Birsadhan 108
Brown plant hopper
Chandrama, Mrunalini, Mandakini, Phalguni, Jogesh, Rajendra Mahsuri 1, **Narendra, Usar Sankar Dhan 3,** CR Dhan 70, Luna Sampad, Luna Suvarna, **Pant Sankar Dhan 3,** ***PA 6444,*** **Ajay, Rajalaxmi, Sahyadri 3,** ADT (R) 45, Rajendra Suwasini, Rajendra Kasturi, Pratikshya, Sidhanta, Pratheeksha, **CORH 3, Pusa RH 10,** CSR

Table 2: (*Contd....*)

Table 2: *(Contd....)*

23, Karjat 7, Malaviya Sugandh 4-3, IGKVR-2, Jagtial Mahsuri, Godavari, Tholakari, Chandrahasini, Nandyal Sannalu, Gontra Bidhan-1, Bhudeb, Manaswini, Virender, VL Dhan 86, SYE 2001, PKV Ganesh, Indra, Sonal, Pushyami, Amara
White backed plant hopper
Chandrama, Phalguni, **Pant Sankar Dhan 3, Ajay, *Ganga,*** Pratikshya , Sidhanta, **CORH 3,** IGKVR-2, Chandrahasini, Sonal, ***JKRH 401,*** Swetha, ADT (R) 47, PNR 546, GR 7, Dandi, **HKRH 1, *GK 5003,*** Akshaydhan, **DRRH 2,** DRR Dhan 38, Vasumati, Sumati, Chinsurah Rice, Shanthi, ***Suruchi,*** Sampada, CR Dhan 70, ***PA 6444,*** Manaswini, Virender, SYE 2001, Pushyami, PMK (R) 3, Richa, Varadhan, Narendra dhan 8002, Reeta (CR Dhan 401), NDR 2065, GR 12
Gall midge
Ramappa, Chandrama, **Ajay, *Ganga, JKRH401,*** PNR546, Vasumati, CRDhan70, ***PA 4444,*** Mrunalini, Mandakini, **Rajalaxmi,** Pratheeksha, Jagtial Mahsuri, Tejaswini, Ketekijoha, ADT (R) 48, Pariphou, Nua Kalajeera, PKV Makarand, Sumati, Varalu,CR Dhan 10, CR Sugandh Dhan 3, Phalguni, Pratikshya, Sidhanta, Chandrahasini, Sonal, Manaswini, Virender, SYE2001, NDR 2065, Jogesh, ADT (R) 45, Bhudeb, PKV Ganesh, Indra, Uphar, Dhanrasi, Nua Chinikamini, Phou-oibi, Pant Dhan 16, Shusk Samrat, SKL 8, Rashmi, Narendra Dhan, Madhuri, Indira Sugandhit, Dhan 1, **Indira Sona,** Samleshwari, Karma Mahsuri, CR Dhan 40, Abhishek, Sree Kurma, Jaldubi, ***PA 6129,*** IGKVR-1,Anjali, Jagtial Sannalu, Narendra Lalmati, Vallabh Basmati – 22, GMR 17, Manair Sona, Jagtial Samba, Srisatya, Taramati

*Hybrids are shown in bold

rices are also the tall statured and relatively low yielding varieties with special traits such as Pusa Basmati 1121 (with highest elongation on cooking) from IARI; Vasumati and Sugandhamati from DRR and CSR 30 (adapted to sodic soils) from CSSRI, Karnal (Table 4).

Hybrid Breeding

Development, evaluation and commercial planting of hybrids

Evaluation of rice hybrids for yield and adaptability is done through the three tier multilocation testing system. Under the Hybrid Rice Network System, experimental hybrids developed by network Centres, and private seed companies are grouped into Early (<120 days), Mid-Early (121-130 days) and Medium (131-140 days) along with two special groups- *viz.*, (medium slender) grain type and tolerant to abiotic stress Fifty trials are evaluated in Initial Hybrid Rice Trial (IHRT) at 25 to 30 locations Each test entry is assigned an IET (Initial Evaluation Testing) number to trace its identity. Promising entries are promoted successively to two Advanced Yield Trails *viz.*, AVTF (more than 5% yield advantage over the best hybrid check) and AVTI (10% yield

Table 3: Multiple pest and disease resistant varieties (2001-10)

Variety name	*BL*	*BLB*	*RTV*	*ShBl*	*BS*	*GLH*	*SB*	*BPH*	*WBPH*	*GM*	*LF*
Ganga	R					MR	MR	MR	MR	MR	MR
PA 6444	MR		MR				MR	MR	R	MR	
GR 7	R	R					MR		MR		MR
Dhanrasi	R	MR	MR				MR			R	
Rajendra mahsuri 1		MR		MR	MR		MR	MR			MR
Dandi	R	R					MR		MR		MR
Swetha	MR			MR	MR		MR		MR		
SYE 2001	R	R						R	R	R	
Bhudeb		R		MR			MR	R		R	
CSR 23	MR	MR			MR			MR			MR
Sumati	MR				MR	MR			MR		
PNR 546		MR			MR		MR		MR	MR	
Richa	R	R		MR	MR				R		
Rajendra sweta	MR	MR		MR	MR		MR				MR
Pant sankar dhan 3	**MR**	**MR**	**MR**		**MR**		**MR**	**MR**	**MR**		
Ajay	MR	MR					MR	MR	MR	MR	
Jogesh	MR	MR		MR	R		MR	MR		R	
Pratikshya	MR			MR	R		R	MR	MR	R	R
Rajalaxmi	MR	MR					MR	MR		MR	
Sidhanta	R	MR	MR		R		R	MR	MR	R	R
Uphar	R	MR	MR	MR			MR			R	MR
Narendra usar sankar dhan 3		**R**		**MR**		**MR**	**MR**	**MR**			
Bhuthnath	MR		MR	MR	MR		MR				
JKRH 401	MR		MR	MR	MR		MR		MR	MR	MR
Shusk samrat	R			R	R		R			R	
Virender	R				R		R	R	R	R	

Table 3: (*Contd....*)

Table 3: *(Contd....)*

Variety name	***BL***	***BLB***	***RTV***	***ShBl***	***BS***	***GLH***	***SB***	***BPH***	***WBPH***	***GM***	***LF***
VL dhan 86	R						R	R			R
HKRH 1	MR				MR		MR		MR		MR
CORH 3	**MR**		**MR**			**MR**		**MR**	**MR**		
PA 6129	*R*				*MR*	*MR*				R	
Chandrama	R	MR	MR	MR			MR	MR	MR	MR	
Amal mana	MR			MR	MR		R				R
GK 5003	R		MR	MR	MR				MR		
CR Dhan 70	MR			MR			MR	MR	R	MR	
Manaswini	MR		MR	MR	R		R	R	R	R	R
Nua kalajeera	MR		R		MR					MR	
Mandakini	R		MR	MR			MR	MR		MR	R
Mrunalini	R	MR	MR	MR			MR	MR		MR	MR
Tejaswini	R	MR	R	MR			MR			MR	MR
Narendra dhan	R	MR	R	MR						R	
Sonal	R				MR			R	MR	R	
Pratheeksha	R			MR	MR			MR		MR	
Luna sampad	R			MR	MR		MR	MR			MR
Luna suvarna	R			MR	MR		MR	MR			MR
Nua chinikamini	MR		R		MR		MR			R	
Phalguni	R		MR	MR	MR	MR	MR	MR	MR	R	R

Bold: Hybrids; R: Resistant; MR: Moderately resistant; Bl: Blast; BLB: Bacterial leaf blight; RTV: Rice tungro virus; ShBl: Sheath blight: BS: Brown spot: SB: Stem borer; BPH: Brown plant hopper; WBPH: White backed plant hopper; GM: Gall midge

Table 4: Quality characteristics of important export quality basmati varieties

Variety	*Kernel traits*		*Milling traits*			*Other quality traits*				
	Length (mm)	*Breadth (mm)*	*L/B ratio*	*Milling (%)*	*HRR*	*Alkali value*	*Amylose content (%)*	*KLAC (mm)*	*ER*	*Aroma*
Pusa Sugandh 5	8.01	1.90	4.21	71.0	62.9	7.0	24.46	14.2	1.76	Present
Pusa Basmati 6	7.78	1.70	4.40	66.5	56.8	7.0	22.17	16.7	2.23	Present
Improved Pusa Basmati 1	7.62	1.76	4.32	66	50.1	7.0	24.89	14.1	1.85	Present
Vasumati	7.27	1.81	4.00	68.4	53.7	4.0	25.05	13.31	1.81	Present
CSR 30	7.12	1.77	4.02	64.8	58.7	4.7	23.30	13.1	1.84	Present
Pusa RH 10	7.53	1.87	4.04	71.1	46.2	7.0	22.45	12.9	1.71	Present
Pusa Basmati 1121	7.9	1.92	4.11	63.1	52.9	7.0	20.70	17.8	2.25	Present
Checks										
Basmati 370	6.93	1.86	3.73	67.9	46.2	6.2	22.10	12.80	1.85	Present
Pusa Basmati 1	7.47	1.75	4.28	69.6	60.6	7.0	22.30	15.17	2.03	Present
Taroari Basmati	7.2	1.79	4.02	66.8	48.3	5.6	22.74	14.53	2.01	Present

HRR: Head rice recovery, KLAC: Kernel length after cooking (mm); ER: Elongation ratio

advantage over the best varietal check or over the best hybrid check) The grain quality is not compromised (HRT-MS trial).

The entries at different stages of testing are simultaneously included in the pest screening nurseries for assessing their reaction to major pests *viz.*, blast, bacterial leaf blight, sheath blight, rice tungro virus, sheath rot, sheath blight, false smut; brown plant hopper, white-backed plant hopper, gall midge, stem borer and leaf folder. At AVT-2 the entries are also evaluated for agronomic performance and grain quality. Entries that show consistent yield advantage over the check varieties/hybrids through three year and posess other desirable traits are identified for formal release for commercial planting in states/regions of the country, where they have been found promising. This systematic and stringent evaluation has helped identification of additional 31 hybrids for commercial cultivation during 2001-10 (Table 5). Among these, 17 are from the public sector institutions while 14 are from the private sector. Of the public institute bred hybrids, 11 have been released by the State Variety Release Committees, while 20 *viz.*, PA 6444, Pusa RH-10, RH 204, Ganga, Suruchi, DRRH-2, JKRH-401, PA 6129, Sahyadri-4, GK 5003, DRH 775, HRI 157, PAC 835, PAC 837, DRRH-3, US 312, Rajlaxmi, Indam 200-017,CRHR-32 and 27P11 by the CSCCSN & RV. Among the six centrally released hybrids, *viz.*, Pusa RH-10, DRRH-2, Sahyadri-4, DRRH-3, Rajlaxmi and CRHR 32 are from the public sector and the remaining are from the private sector (Table 5). Beside developing hybrids, the private sector has played a major role in production and marketing of hybrid seed. The private sector accounts for over 80% of hybrid seed production and marketing. Many companies are now strengthening their R&D component and are expected to contribute significantly in the coming years to the development of competitive hybrids, quality seed production, marketing and extensive adoption of hybrid rice technology.

Hybrids with improved grain quality

Quality preference in rice varies with the stakeholder. Millers see s milling quality in terms of milling percentage and head rice mill recovery while grain fineness, kernel colour and appearance are desired parameters of the trader. The consumer looks for acceptable cooking quality which varies from region to region; many early generation hybrids were pronounced poor quality because they were sticky when cooked. Excessive use of the low amylose male sterile line IR58025A as a parent of many hybrids was reason for poor cooking quality of the resulting hybrids. The seed harvested from the F_1 (*i.e.*, F_2 grain) segregates for physio-chemical properties of starch resulting in unacceptable quality. Thus a hybrid should be of high milling percentage, least grain breakage (head rice recovery) and preferably with medium slender (Length of 5.5–6.6m and L/B ration 2.5) to long slender (>6.6 mm and L/B ratio of 3.0) translucent grain. It should have intermediate gelatinization temperature (GT) and intermediate amylose content (~22.25%). In a prized basmati quality hybrid, the kernel should be extra long (>6 mm) have high L/B ratio (3.0) and excessive elongation ratio (~2.0) of cooked rice. It should also have the distinct basmati aroma. Because of conscious breeding for desirable grain quality, many hybrids not being commercialized are of medium slender type with enhanced level of amylose content. This has been due to the replacement of the low amylose A-line IR 58025 A in case of non basmati hybrids. For the only basmati quality hybrid, Pusa RH 10, it took many years to assemble A, B and R lines of basmati quality.

Hybrids with resistance to insect pests and diseases

For stable performance of hybrids across locations/seasons, it is necessary that hybrids possess resistance/tolerance to major insect pests and diseases. Hence, incorporation of resistance to major pests is one of the major objectives of hybrid rice breeding. In addition to planned development of parental lines with desired level of resistance to biotic stresses, hybrids in the coordinated trials are regularly screened for resistance to major insect pests and diseases through the national hybrid rice pest screening nurseries. Considering the fact that in any given region, rice suffers from more than one pest/ biotypes or pathovar and breakdown of single gene conferred resistance is common, breeding priority and strategy is to evolve multiple pest/biotype resistant hybrids/varieties. Gene deployment strategies like pyramiding of resistance genes to more than one pest and biotype/pathotype by marker assisted backcross breeding received special breeding attention. Parental lines (A and R) widely used for production of many hybrids have been introgressed with resistance to bacterial blight (*Xa21, Xa13, Xa5* etc.) and blast (*pikh,pil, piz* etc.).

Large scale adoption of hybrid rice depends on how profitable it is as compared to popular inbred varieties in any given rice ecology. The profitability in turn depends on the level of yield heterosis and price offered to the produce. The yield advantage may not be striking but value addition in terms of grain quality and resistance to major diseases and insect pests makes many of the latest hybrids advantageous. By developing and choosing appropriate parental lines,

Table 5: List of rice hybrids released in India (2001-2010)

Hybrid	*50% flowering duration (DFF)*	*Year of release*	*Notification no.*	*Date of notification*	*Developed by*	*Released for the states/ special feature*
PA 6444	105	2001	1134 (E)	5.11.2001	Bayer Bio-Science, Hyderabad	UP, Tripura, Orissa, AP, Karnataka, Maharashtra and Uttarakand
Pusa RH-10	95	2001	1134 (E)	15.11.2001	IARI, New Delhi	Haryana, Delhi, UP and Uttarakhand, Basmati quality.
Ganga	100	2001	599 (E)	25.4.2006	Paras extra growth seeds Ltd. Hyderabad	Uttarakhand, Punjab, Nagaland, Haryana, U.P., Orissa and Bihar
RH-204	95	2002	283 (E)	12.3.2003	Parry Monsanto seeds Ltd. Bengaluru	AP, Karnataka, TN, West Haryana, Uttarakhand and Rajasthan
Suruchi	103	2004	122 (E)	2.2.2005	Mahyco Ltd, Aurangabad	Haryana, Andhra Pradesh, Gujarat, Orissa, Chhattisgarh, Karnataka and Maharashtra.
Pant Sankar Dhan-3	92	2004	599 (E)	25.4.2006	GBPUA&T, Pantnagar	Uttarakhand
Narendra Usar Sankar Dhan-3	105	2005			NDUA&T, Faizabad	Saline and alkaline areas of U.P.
DRRH-2	86	2005	1566 (E)	5.11.2005	DRR, Hyderabad	Haryana, Uttaranchal, West Bengal and Tamil Nadu.
Rajlaxmi	98/128 (Boro)	2005	1572 (E)	20.9.2006	CRRI, Cuttack	Orissa and Boro areas of Assam and shallow lands in eastern India
Ajay	98	2005	1572 (E)	20.9.2006	CRRI, Cuttack	Orissa
Sahyadri-2	85	2005	122 (E)	6.2.2007	RARS, Karjat (BSKKV)	Maharashtra
Sahyadri-3	95	2005	122 (E)	6.2.2007	RARS, Karjat (BSKKV)	Maharashtra, Mahsuri grain type

Table 5: (*Contd....*)

Table 5: *(Contd....)*

Hybrid	*50% flowering duration (DFF)*	*Year of release*	*Notification no.*	*Date of notification*	*Developed by*	*Released for the states/ special feature*
HKRH-1	104	2006	122 (E)	6.2.2007	RARS, Kaul (CCSHAU)	Haryana
JKRH-401	110	2006	122 (E)	6.2.2007	JK Agri Genetics Ltd, Hyderabad	West Bengal, Bihar and Orissa, shallow lowlands
CORH-3	85	2006	1178 (E)	20.7.2007	TNAU, Coimbatore	Tamil Nadu
Indira Sona	98	2006	1178 (E)	20.7.2007	IGKVV, Raipur	Chhattisgarh
JRH-4	87	2007	1178 (E)	20.7.2007	JNKVV, Jabalpur	Madhya Pradesh
JRH-5	87	2007	1178 (E)	20.7.2007	JNKVV, Jabalpur	Madhya Pradesh
PA 6129	85	2007	1703 (E)	5.10.2007	Bayer Bio-Science, Hyderabad	Punjab, T.N. and Pondicherry.
GK 5003	88	2008	454 (E)	11.2.2009	Ganga Kaveri seeds Pvt Ltd, Hyderabad	Andhra Pradesh and Karnataka
Sahyadri 4	88	2008	454 (E)	11.2.2009	RARS, Karjat (BSKKV)	Maharashtra, UP and Punjab, Haryana and West Bengal
JRH-8	90	2008	449 (E)	11.2.2009	JNKVV, Jabalpur	Madhya Pradesh
DRH 775	96	2009	2187 (E)	27.8.2009	Metahelix Life Sciences Pvt. Ltd., Hyderabad	Jharkhand, Chhattishgarh and West Bengal
HRI-157	104	2009	2187 (E)	27.8.2009	Bayer Bio-Science, Hyderabad	UP, MP, Bihar, Jharkhand, Tripura, Chhattisgarh, Orissa, Maharashtra, Gujarat, Andhra Pradesh, Karnataka and Tamilnadu
PAC 835	102	2009	2187 (E)	27.8.2009	Advanta India Ltd. Hyderabad	Orissa and Gujarat

Table 5: *(Contd....)*

Table 5: *(Contd....)*

Hybrid	*50% flowering duration (DFF)*	*Year of release*	*Notification no.*	*Date of notification*	*Developed by*	*Released for the states/ special feature*
PAC 837	100	2009	2187 (E)	27.8.2009	Advanta India Ltd. Hyderabad	Western Gujarat, Eastern Chhattisgarh, Northwestern J&K, Andhra Pradesh and Karnataka
DRRH 3	101	2009	211 (E)	29.01.2009	DRR, Hyderabad	Andhra Pradesh, Orissa, Gujarat, Madhya Pradesh & Uttar Pradesh; Mahsuri grain type
US 312	98	2010	2137 (E)	31.8.2010	Seed Works International, Hyderabad	Tamil Nadu, Karnataka, Andhra Pradesh, Bihar, Uttar Pradesh and West Bengal
Indam 200-017	96	2010			Indo American Seeds, Hyd	Orissa, Chhattisgarh, Gujarat, Maharashtra and Andhra Pradesh
CRHR-32	112	2010			CRRI, Cuttack	Late–irrigated/shallow lowlands of Bihar and Gujarat
27P11	130	2011			PHI Seeds Private Ltd, Hyd	Karnataka and Maharashtra

Hybrids in bold font are released by CSCCSN & RV

it is now possible to progressively improve hybrids with higher yield, duration range for major rice ecologies, resistance to pests and grain quality according to consumer preference that varies from region to region.

Development of new CMS lines

Realising that excessive dependence on a single cytosterility source poses the danger of genetic vulnerability to sudden outbreak of any biotic stress, as happened by extensive use of WA cytosterility source, there has been search for alternate sources. But success has been limited to just two *viz.*, Kalinga cytoplasm by CRRI and Gambiaca cytoplasm by Karjat centre. However, based on WA source, many valuable male sterile lines relevant to rainfed shallow lowland ecologies and basmati quality have been developed and are in use.

Promising CMS lines developed by recurrent backcrossing are given in Table 6. Here also marker assisted backcross breeding has been of great help.

Table 6: Promising CMS lines developed

Centre	*Promising CMS lines developed*
DRR, Hyderabad	DRR 6A, 9A, 10A, 12A, 14A, 15A
IARI, New Delhi	Pusa 3A, 4A, 5A, 6A, 10A, 11[a]
PAU, Ludhiana	PMS 3A, 10A, 12A, 17A
RARS, Karjat	KJTCMS 1A, 2A, 3A, 4A
ARS, Ratnagiri	RTN 8A, 10A, 13A, 14A
APRRI, Maruteru	APMS 6A, 8A, 9A
CRRI, Cuttack	CRMS 32A, CRMS 31A

Genetic male sterility facilitated population improvement

Population improvement is a medium to long term activity and this breeding strategy is largely confined to the improvement of parental lines of hybrid rice. As against quick fixation of genes during selfing in recombination breeding, genetic male sterility facilitated intersubmiting and recurrent selection provide for continuous recombination, accumulation of favourable genes, broadening of genetic base and breaking of undesirable linkages. Through this method, composite populations in the background of maintainers and restorers are developed. Productive segregants with desired traits are handled by pedigree method, which ultimately results in new and improved parental lines.

Biotechnology Intervention for Rice Improvement

Progressive improvement of rice for enhanced productivity, stability and profitability has so far been through exploitation of easily accessible genetic variability by adopting conventional recombination breeding and phenotype based selection. Despite laudable successes achieved from this approach many issues still remain unaddressed because of limitation of genetic variability and phenotype-based selection procedures. Advances in cellular and molecular biology have provided three important tools to overcome these limitations *viz.*, anther culture, molecular marker assisted selection-breeding and recombinant DNA technology. They have been found increasingly rewarding in designing and tailoring genotypes of interest as briefly discussed below.

Doubled Haploid Breeding

Since the development and demonstration of the unique advantage of anther culture as a rapid method for obtaining homozygous recombinants from F_1 hybrids, it has been employed in rice at IRRI, China and India. During the 70s and 80s, many anther culture derived lines have been released as varieties largely in China. IRRI and India have also used doubled haploid(DH) strategy in breeding for tolerance to abiotic stresses characteristic to rainfed lowland/deepwater ecologies. Sathya Krishna (CR Dhan70), a 135 day long slender grain variety of high yield potential developed at the Central Rice Research Institute, Cuttack (India) was released in 2008 for general cultivation. Subsequently one more variety *viz.*, Phalgun (CRACC 2224) for rainfed upland ecology has been developed by DH approach at CRRI and released for general cultivation. Prior to this a variety developed by anther culture for tolerance to iron deficiency was released in Maharashtra (India). Some of the IRRI developed dihaploid lines tolerant to salinity have been found promising in some countries. Despite its advantage of shortened breeding cycle, DH breeding approach is not widely followed on account of low frequency of regeneration and very low probability of obtaining desired recombinants in small populations.

Despite its limitations, DH breeding is still in use in breeding for high yielding varieties for long duration and problem soil areas. The strategy was found highly rewarding for developing high yielding NERICA (New Rices for Africa) varieties for environmentally harsh conditions of Africa through interspecific crosses involving the Asian cultivar *O. sativa* subsp. *indica* and African cultivar *O. glaberrima*.

Another utility of DH strategy has been the development of immortal mapping populations for mapping and tagging of genes of interest to facilitate marker assisted selection (MAS) of complexly inherited traits and marker assisted backcross breeding. Significantly, the IRRI developed mapping populations *viz*. IR64 x Azucena has been extensively used world over to map many valuable genes including those for abiotic stress tolerance genes/QTLs.

Molecular Marker Aided Breeding

Like other genetic markers, DNA markers located in the neighbourhood of a trait (gene) of interest co-segregate with it in the segregating populations enabling their selection at genotypic level. Unlike phenotype-based selection, molecular marker-based genotypic selection of a trait is highly reliable. Marker technology is relevant and rewarding in instances, where natural variability exists for a given trait but phenotype-based selection is expensive, cumbersome or unreliable as happens for simply inherited resistance to biotic stresses as well as complexly inherited tolerance to abitoic stresses. The technology is also effective and useful in cases, where diverse gene sources conferring common phenotype are to be selectively staked in a single genome for pyramiding resistance genes against multiracial/biotype pests, rapidly recovering recurrent parent genome in backcross breeding and unfolding of novel QTLs of value from wild/weedy relatives.

Marker choice is determined following a set of criteria including its suitability for mapping a trait. Degree of polymorphism and capacity to differentiate homozygous and heterozygous allelic state are the principal criteria for choosing a marker. Microsatellite (SSR) markers are widely used because they are highly polymorphic, co-dominant, abundant and reliable. Rice is particularly a good crop for employing marker assisted selection (MAS) because it has over 20,000 microsatellite marker dense molecular linkage map (www.gramene.org). Recently, Narshimulu *et al.* (2010) identified evenly distributed hypervariable microsatellite markers for effective background selection. Many traits of economic value have already been mapped and tagged using a variety of markers and MAS in breeding and selection programmes. They broadly include (a) genetic yield enhancement (b) defending against biotic and abiotic stresses and (c) enrichment of nutritive quality.

Marker Aided Genetic Yield Enhancement

Ever since the development of dwarf high yielding varieties in the mid 1960 that breached the centuries long yield barrier, rice breeders have been looking for the second yield breakthrough. Of the various medium and long term strategies contemplated for raising the yield ceiling through molecular marker technology (a) enhancement of yield vigour through super hybrids (b) tailoring of new plant type varieties (c) genome-wide pooling of yield genes and (d) discovery and rational exploitation of novel yield genes of wild/weedy progenitor species are important.

Development of super hybrids

It goes to the credit of Prof. Longpin Yuan of China, who succeeded in raising the genetic ceiling to rice yield from 10 to 12.5t/ha through hybrid technology, the second landmark after the semidwarf varieties in the history of rice breeding. Unlike in China, where it was planted over 55% (18 million ha) of the rice area in just 12 years from the year of release, pace of adoption of the technology in India has left much to be desired. It is yet to cross two million hectares even after 15 years since the advent of hybrid technology. Of various reasons for the slow adoption of the technology, the major one is its non satisfactory yield advantage over the best inbred varieties, besides lower acceptable grain quality, vulnerability to pests and crop maturity not ideally suiting long wet season. A strategy that is rewarding for raising yield vigour and yield advantage is shift in breeding emphasis from intra subspecific (*indica/indica*) to intersubspecific (*indica/japonica*) hybrids, which are capable of yielding more than 12.5 tonnes per hectare. The major hurdle in exploiting the heterotic *indica/japonica* combinations, however, has been persistent hybrid semisterility. This sterility has been attributed to a series of duplicate recessive gametic development and duplicate fertility genes by one group (Oka and Doida, 1962) and to cryptic structural differences by the other (Henderson *et al.*, 1958; Sastry and Misra, 1961). It is now established, however, that simply inherited sterility-neutralizing genes, also known by Wide Compatibility Gene loci (WCG) (Ikehashi and Araki, 1984, 1986), can overcome this sterility. In the presence of the WC gene in either of the parents in an *indica/japonica* cross, the resultant hybrid would be fertile. Using early discovered WCG loci like $S^n{}_5$ from Dular and Keta Nanga, many heterotic hybrids such as Tiang You Paijiu, Eie You 9308 etc with yields surpassing that of intra-subspecific hybrids, have been developed. Recently, linking of all the major WCG loci to molecular markers has enabled breeders to selectively introgress them with ease to one or both parents. Taking advantage of comparative mapping as many as seven WCG loci *viz.*, $S^n{}_5$, $S^n{}_6$, $S^n{}_8$ mapped to chromosome 6, $S^n{}_7$ to chromosome 7 and $S^n{}_{15}$ and $S^n{}_{17}$ to chromosome 12 have been tagged to rice microsatellite markers. The loci $S^n{}_5$ and $S^n{}_6$ tightly linked to the flanking markers RM225 - RM14 and RM114 - RM141 respectively for instance, have become widely usable now in developing appropriate parents for evolving sterility-free inter-subspecific hybrids (Siddiq and Singh, 2005; Singh *et al.*, 2006).

As an alternative to the CMS-based 3-line hybrid breeding, Environment Sensitive Genic Male Sterility based 2-line breeding systems *viz.*, Photoperiod Sensitive Genic Male Sterility (PGMS) and Temperature Sensitive Genic Male Sterility (TGMS) has not only dispensed with the cumbersome and expensive 3-line hybrid seed production but also facilitated development of inter-subspecific hybrids with relative ease. TGMS sources ideally suited to tropical country like ours such as, *Annong* S_1 and *Norin 12* along with indigenously identified *SA2* and *F61* (Ali, 1993: Ali *et al.*, 1995) have helped further development of inter-subspecific hybrids, especially after tagging them to robust microsatellite markers. The two indigenous sources, *SA2* and *F61,* found non-allelic to all the previously known genes have been linked respectively to the flanking microsatellite markers RM257 - TS 200 on chromosome 9 and (SCAR marker) and RM 224 and RM21 on chromosome 11, respectively (Reddy *et al.*, 2000; Hussain *et al.*, 2011). Pyramiding of some of the TGMS gene loci has been reported to be more effective than use of single gene (Nas *et al.*, 2005).

Marker technology has also been exploited successfully in many other ways, including establishment of genetic purity, rational genetic identity, gene deployment etc. Maintenance of genetic purity of parental lines is very important for production of quality hybrids seed for realizing the expected performance of hybrids. Purity of A (seed parental) line depends on the purity of B (maintainer) line to ensure total and stable male sterility. Equally important is R line purity in terms of level of fertility restoration and combining ability. Impurity at any level would result in failure of the hybrid. Conventionally purity is assessed of the parental lines (A and R lines) at flowering time when off-types are removed. Now, molecular markers have become handy to ensure genetic purity of all the parental lines at early stages of hybrid development. Recently, parental lines of the Basmati quality hybrid Pusa RH 10 *viz.*, Pusa6B and PRR78 have been improved for resistance to BLB by introgressing two BLB resistance genes (*Xa21, Xa13*) by marker assisted selection (MAS) (Basavaraj *et al.*, 2010). Yet another piece of work that is sustaining the country's basmati export is the development of molecular marker based 'Basmati Verifier Kit', a CE-based multiplex microsatellite marker assay, which not only distinguishes the highly priced traditional basmati rices from the low priced look-alike non-basmati rices but also detects the level of adulteration of basmati rices with non-basmati types by unscrupulous traders. This assay multiplexes eight microsatellite loci *viz.*, RM1, RM171, RM202, RM44, RM241, RM55, RM348 and RM72 to generate variety-specific allele profiles that can detect adulteration from 1% upwards. The protocol also incorporates a quantitative-competitive PCR-based analysis for quantification of adulteration. Accuracy of quantification has been shown to be ± 1.5% (Archak *et al.*, 2007; Vemireddy *et al.*, 2007). Significantly, CDFD is the APEDA recognized nodal institute for certifying the authenticity and genetic purity of traditional basmati rice consignments meant for export.

Genome-wide pooling of yield genes

Broadening genetic base for complexly inherited traits like yield is the key not only to raise genetic yield level but also to incorporate stability against yield depressing stresses. Bill and Melinda Gates Foundation, Chinese Academy of Agricultural Sciences and IRRI jointly launched Green Super Rice Project is a massive breeding effort to this effect which places emphasis on developing ecologically and economically sustainable varieties in high yield backgrounds for resource poor rice farmers in Asia and Africa. The breeding strategy consists of developing advanced backcross populations involving as many as 46 popular high yielding varieties chosen from different countries as recurrent parents and 500 varieties with unique traits as donors, screening of segregating generation (BC_2F_2) under targeted stress conditions for transgressive segregants and pyramiding of promising complex trait-specific non-allelic QTLs derived from different donor sources into country specific popular varieties like IR64, BR11, BG300, Huang Hua Zhan etc. The recovery of promising lines in large numbers is attributed to harmonious complementation of genetic networks for complex traits, which otherwise are incomplete in both the parents (Ali and Santiaguel, 2011). While the breeding emphasis of GSR strategy is for tailoring eco-friendly farmer-friendly varieties/hybrids, the massive exercise will also facilitate pooling of yield enhancing genes/QTLs from different genomes involved in this strategy. Some of the so developed varieties and hybrids, now in advanced stages of testing in Asia and Africa, are reported to be higher yielding with high stability against biotic and abiotic stresses.

Unfolding of novel yield QTL in progenitor species

Tracing the geneology of modern high yielding varieties reveals that, very few accessions account for about 90% of their genetic constitution warranting accelerated search for still unfolded variability and its use in breeding for further yield enhancement. Wild progenitor species and primitive landraces are potential reserves of still unlocked sizeable variability. It was the successful mapping of two rare yield QTLs *viz.*, *yld* 1.1 and *yld* 2.1 in *O. rufipogon* by marker associated QTL approach by Cornell University (Xioa *et al.,* 1996), that aroused worldwide interest among rice geneticists to search for more and newer yield genes in wild/weedy species as well as primitive cultivars. The logic behind the strategy of going back to the progenitor species and primitive cultivars is that in the long process of evolution from origin to domestication and continued improvement, all alleles of yield related genes might not have been captured in the modern varieties and these alleles would still remain in the primitive and the wild. The search effort has led to the identification of many more yield related QTLs of promise in *O. rufipogon, O. nivara* and primitive cultivars. Among them *yld* 2.1 and *yld* 8.2 was mapped using backcross population involving an Indian accession of *O. rufipogon*, *yld* 2.2, *yld* 8.1 and *yld* 9.1 mapped in two accessions of *O. nivara* in the mapping populations involving them with the variety Swarna and many yield and yield component enhancing QTLs such as *gy* 12.4, *gy* 12.5, *npt* 8.2, *npt* 12.2, *npt* 12.4, *gw* 12.1, *bin* 8.2, *bin* 8.4 and *bin* 12.3 in the primitive landraces INR10192 and INRC10912 are quite promising (Marri *et al.,* 2005; Srividya *et al.,* 2011). Interestingly, many of the yield related QTLs have been found to map as clusters within flanking marker intervals of less than 20cM. While fine-mapping is in progress, marker assisted backcrossing to introgress and study their yield enhancing potential has yielded encouraging results. Several lines of KMR3 introgressed with *yld* 2.1, for instance, have shown yield increases upto 20% as compared to KMR3. Similarly, several of the Swarna lines introgressed with *yld* 2.2, *yld* 8.1 and *yld* 9.1 have shown 5-20% yield advantage over Swarna. Of the two introgression lines of KMR3 and Swarna now in advanced stages of testing in the All India Coordinated multi-location trials Swarna introgressed with still to be confirmed yield enhancing QTL of *O. nivara* has been identified by the varietal Identification Committee for formal release by the Central Variety Release and Notification Committee. This has been found to yield 15-20% higher than Swarna under rainfed upland conditions. This is the first variety ever developed by introgressing an yield QTL from a progenitor species. Equally promising KMR3 lines introgressed with *yld 2.2* and *yld 8.1* from *O. rufipogon* are soon to be released as varieties. Globally, over 20 genes underlying QTLs governing yield traits have so far been identified and functionally characterized. They include QTL affecting tillering (*MOC*1, *D*88 (*D*14), *OSTB*1), panicle development (*LAXI, SPA/MOCI*-3, *fZP, SPI, DEPI, AP01*), grain number (*Gn1a*, *Ghd*7, *Ghd*8), test grain weight (*GS3*, *GW*2, *qSW*5, *GW*5, *tgw* 6) and percentage grain filling (G_1F_1) (Xing and Zhang, 2010).

Yield enhancement through pleiotropic lodging resistance gene

Instances of lodging and loss of yield in short statured *indica* as well as *japonica* varieties grown under higher fertility conditions are not uncommon. The underlying reason is that dwarfness need not always be associated with culm strength. The gene *SCM2* (*Strong Culm 2*) recently identified by Prof. Yano's group (Ookawa *et al.,* 2010) is a pleiotropic QTL conferring resistance to lodging and productive panicle structure. Genotyping of the backcross population of the cross involving chromosome segmental substitution lines (CSSLS) (developed using the *indica* variety Habataki as the donor in the background of the lodging prone *japonica* variety Sasanishiki as the recurrent parent) and Sasanishiki and other lodging prone varieties has led to the identification of the QTL conferring enhanced culm strength designated as *Strong Culm* 2 (*SCM 2*). Several institutions in Japan, China, IRRI and CIAT are engaged in developing CSSLs using wild species as donors for mapping genes/QTLs for use in functional genomics/breeding. New genes are now being cloned from CSSLs through map-based cloning. A near isogenic line carrying *SCM2* showed not only enhanced culm strength but also vastly increased spikelet number possibly due to its pleiotropic effect on branching of main rachis. Positional cloning and expression of the QTL reveals *SCM2* to be identical to *Aberrant Panicle Organisation1 (APO1),* a gene previously reported to control panicle structure.

Identification of several QTLs directly and indirectly influencing yield has provided a fairly gene rich resource base for choosing QTLs of promise and combining the harmonious and synergistic ones for stepping up genetic yield level. The breeding strategy comprises stepwise development of QTL lines using QTLs of promise in a common varietal background, transfer of desired QTLs to the recurrent parent of choice by repeated backcrossing taking advantage of MAS and evaluation of lines introgressed with one or more yield QTLs for yield performance. Annotated genomic sequence of both the subspecies of *O. sativa,* has now provided us with highly valuable information on the rice genome.

Such information has become handy for researchers to isolate, functionally characterize and use many yield related QTLs/genes. Incidentally a serious research in a network mode is underway in India to exploit stable QTL by staking the harmonious ones as a strategy to raise the ceiling to genetic yield.

Yet another research strategy is to look for a novel allele of the widely used dwarfing gene *sd*1 that would help evolve a plant type with still higher yield potential. Differing from the traditional tall statured native varieties, wherein GA pathway remains intact as against the modern semi dwarf varieties, where GA pathway has been suppressed, looking for *sd*1-like allele or some other height reducing genetic mechanism that would least affect GA pathway is the strategy. According to Zhikang (personnel communication) this concept would help find a new yield threshold. Recently, some of the dwarf accessions have been found to be non-allelic to DGWG allele of *sd1* gene (Neeraja *et al.,* 2009).

Breeding for Broad Spectrum Resistance to Biotic Stresses

Resistance to diseases

Since the introduction of high yielding varieties, pest problem has assumed increasingly epiphytotic proportions causing serious crop losses year after year. Continuous cropping of rice facilitated by short maturity and photosensitivity, narrowing genetic base resulting in genetic uniformity and high selection pressure in single gene resistance breeding together with widening of pest spectrum horizontally and vertically has made the pest scenario increasingly complex and not easily manageable. There is a need now for an innovative breeding/selection techniques to cope with the challenge.

Molecular marker technology enabling genotype based screening/ selection has made breeding for resistance to major pests and their biotypes/races relatively easy and effective in rice. Of as many as 33 resistance genes for bacterial blight (BLB), many are now linked to robust/gene markers. Similarly, many race specific resistance genes have been linked to markers in case of blast, gall midge and hoppers. Such trait linked markers have been successfully used for selective improvement of popular high yielding quality rice varieties like BPT 5204 (Samba Mahsuri) and Pusa Basmati-1 for pest resistance by marker assisted advanced backcross breeding adopting foreground, recombinant and background selection strategies. Improved by such interventions, Pusa Basmati-1 introgressed with *Xa21* and *xa13* (Joseph *et al.,* 2004) and Improved Samba Mahsuri introgressed withh *Xa21, xa13* and *xa5* (Sundaram *et al.*, 2008) are more than 90% genetically closer to the respective parent varieties and practically no compromise on quality. The strategy of marker assisted introgression of dominant genes *(Xa4, Xa7, Xa21)* against BLB has been extended also to the parental lines of popular hybrids (Borines *et al.*, 2000; Zhang *et al.*, 2006).

Pyramiding diverse resistance genes against any pest is considered a good strategy for ensuring broad spectrum resistance and protection against frequent breakdown of resistance. As for the choice of gene sources, many of those accessed from wild/weedy germplasm appear to be of multi-racial/pathotype resistance. For instance, *Xa21,* which provides resistance to as many as 30 pathotypes of bacterial blight and shows synergism with country specific critical resistance genes, is from *O. longistaminata*, an African A-genome species, while *Xa30* and *Xa31* offering resistance to many pathotypes are from *O. nivara* and *Xa34* from *O. rufipogon*.

Resistance genes against blast equal the number of races. Since beginning breeders have exploited the host plant resistance gene sources against region/ecology specific race spectrum. For increasing the efficiency of disease containment molecular breeding approaches have been employed for mapping and linking of as many as 84 dominant resistance genes to closely placed markers. Of them *Pikh, Pitp (t), Pi38, Pi48* were mapped in India (Sharma *et al.*, 2005, 2011). Until now 14 genes including *Pikh* have been cloned and functionally characterised. This development has encouraged search for further variants of potential resistance genes by allele mining strategies in a wide range of landraces and wilds taking advantage of conserved synteny at genome level. Allele mining for *Pikh* for instance, has helped find many variables in primitive cultivars and wild progenitor species. As for marker assisted breeding for resistance to the pathogen, but for a few attempts to introgress a few resistance genes *(Pi1, Pi5, Piz, Pita)* by marker assisted breeding into varieties like Co39 (Hittalmani *et al.*, 2000), IR50 (Narayanan *et al.*, 2002) and Zhanshan 97A (Liu *et al.*, 2003), very little has been done. With the kind of gene sources and genomic tools now available, much needed stable broad spectrum resistance to this multi-racial pathogen could be achieved by gene pyramiding strategy.

Resistance gene sources for Rice Tungo Virus (RTV) in some cultivar accessions and only one gene source found in an Indian accession of *O. nivara* for Grassy Stunt Virus (GSV), have been of value in managing them by conventional breeding. As an alternative to host plant resistance-based recombination breeding, prospects of transgenic approach have been explored because no source of resistance against RTV *per se* and for resistance to GSV strain II could as yet be found.

Sheath blight (*Rhizoctonia solani*), that was once a minor disease is becoming a major devastative one todasy. In the absence of strong sources of resistance, two strategies are contemplated. They are (a) pooling moderate resistance genes found in diverse sources by marker assisted recurrent selection and (b) engineering with pathogenesis related antifungal protein genes. Work is underway at CRRI, IARI, DRR and ANGRAU towards identification of moderate resistance QTLs in diverse sources and tagging them to effective markers and pyramiding the harmonious and synergistic QTLs in good agronomic background. At CRRI a 'Foundation population' developed by male sterility facilitated random mating followed by repeated cycles of recurrent selection in a set of crosses involving 13 sources of moderate resistance is challenged with sheath blight inoculum for selecting plants with enhanced level of resistance. The strategy is found to be promising.

Resistance to insect pests

Rice hosts more than a dozen insect pests, of which stemborer, brown planthopper, white backed planthopper, leaf folder and gall midge cause serious yield losses. As a pest and vector of GSV, BPH has been the most devastating successively overcoming the introduced resistance genes. Outbreaks of gall midge at times result in very serious production losses as happened in 1985-86 in Andhra Pradesh. Confined to small endemic pockets till recent past it is increasingly becoming a serious pest causing sizeable crop losses even in non-endemic areas. This trend is attributed to emergence of newer biotypes differing vastly in their viruliference. As of today, there are as many as seven region specific biotypes.

Green leafhopper, which was never a major pest, has assumed BPH-like pest proportions in recent years. Host plant resistance available in abundance against all the major rice insect pests and their biotypes with the exception of stemborer and leaf folder has enabled breeders to manage them by sequentially releasing resistant varieties matching newly emerging biotypes. With continued emergence of newer and increasingly viruliferous biotypes and conventional breeding-approaches finding pest management difficult, innovative breeding-selection strategies have become inevitable. Molecular marker technology taking advantage of resistance gene sources available in the genepool and introgression of novel alien genes capable of providing resistance by recombinant DNA technology are employed for effective management of insect pest problem. Many of the host plant resistance genes have been mapped and linked to closely placed markers. Mapping and linking of as many as 5 BPH resistance genes (*Bph1, Bph2, Bph10 (t)* and *Bph8)* and four of gall midge resistance genes (*Gm1, Gm3, Gm4, Gm5,* and *Gm7*) to robust functional markers for instance, have made marker assisted breeding for insect pest resistance a successful strategy.

Nearly all the gall midge resistance genes (*Gm1, Gm2, Gm3, Gm4, Gm6, Gm7, Gm8* and *Gm11*) have already been linked to molecular markers. As for their pyramiding to realize broad spectrum resistance against more than one biotype. Rawat *et al.* (2012) are of the view that stacking of *Gm1* with any one of the genes in group II, which exclude *Gm4* and *Gm7* would confer resistance to five biotypes (*GMB1, GMB2, GMB3, GMB5* and *GMB6*). To cover all the biotypes, at least three genes preferably *gm1, gm2* and *gm4* would be required. Instead, it is desirable, if a local variety carrying the gene critical to the region is introgressed with two other appropriate genes.

TOLERANCE TO ABIOTIC STRESSES

Among physical stresses that severely depress productivity, drought, submergence and salinity are important. Plant's ability to withstand such stresses is derived from cumulative effect of a network of physiological and biochemical functions. Knowledge of genetics governing them is a prerequisite for finding solution to the problem. Unlike resistance to biotic stresses that largely follow Mendelian mode of inheritance, tolerance to abiotic stresses is genetically quite complex and polygenically controlled. Added to the inadequate knowledge on their genetics, lack of reliable and reproducible screening/selection techniques has made breeding all the more difficult, despite having good sources of resistance in the rice genepool. As a result, 55% of the rice area which is rainfed does not yet have high yielding varieties ideally suited to such ecologies. Since last 15 years advances in plant molecular biology have provided breeders with a variety of genomic tools and resources capable of finding genetic solution to such stresses.

Negative effects of abiotic stresses include stress imposed homeostasis imbalance, disruption of growth and metabolic activities and generation of cell damaging Reactive Oxygen Species (ROS). At molecular level, plants under abiotic stresses adapt to the situation through restoration of homoestasis (ion and osmatic gradient), control of damage and detoxification of ROS.

As for drought, several QTLs relating to different indices of tolerance such as osmotic adjustment, cell membrane stability, relative water content, root characteristics and stress recovery as well as yield *per se* under stress have been mapped and linked to robust markers. Most of the indices being polygenic, QTL in general are of small effect, though large effect QTL are not uncommon. Among those that co-map with yield under stress have been found appropriate for marker assisted breeding for drought tolerance. Some of the QTL of large and additive effect now being introgressed into high yielding varieties include one flanked by SSR markers RM316 and RM5219 on chromosome 9 affecting yield, biomass and harvest index (Yue *et al.,* 2005). Another, mapped in the Indonesian drought tolerant variety Way Rarem between the flanking markers RM28048 and RM511 on chromosome 12 for grain yield under the stress at flowering and grain filling phases, was found stable and quite promising (Bernier *et al.,* 2007). Intensive research is underway at IRRI and India to introgress the allele into popular varieties by marker assisted backcross breeding (MABB). Through MAS/ MABB, the first drought tolerant aerobic rice variety *i.e.*, MAS 946-1 released in India in 2007 was developed by crossing the deep-rooted upland *japonica* rice variety Azucena with the high yielding *indica* variety Kalinga III by MAS/MABB strategy. The newly bred variety requires 60% less water than traditional varieties (Steele *et al.*, 2006).

Area under stress is on the increase due to defective and inefficient irrigation-drainage management in general and irrigation command areas in particular. Though many sources of tolerance to the stress are available in the native germplasm such as SR26B, Nona Bokhra, Pokkali etc efforts to combine desired level of tolerance to the stress in high yield backgrounds by conventional breeding yielded no tangible results. Multiparameter based genetically complex nature of tolerance is attributable to low level breeding success. Internationally, as many as 80 QTLs of large effect relating to various tolerance indices have been mapped. Among them the QTL SALTOL relating to Na, K absorption and Na/K ratio at seedling and reproductive stages has been found to be of promise (Islam *et al.,* 2011). Intensive research at IRRI and India for identification of potential salt tolerance QTL in CSR27, Pokkali etc. has led to the discovery of two important QTL, *viz. Cl* ion concentration mapped in CSR17 on Chromosome 8 (Ammar *et al.,* 2004; Gautam *et al.,* 2011) and SALTOL detected in Pokkali on Chromosome 1. Fine mapping for narrowing down to the gene (s) *per se* underway at present might prove valuable for precise introgression and expression of these genes into popular high yielding varieties by marker assisted backcross breeding. As no single QTL/gene could provide desired level of tolerance, pyramiding of genes for tolerance relating to morpho-physiological features (Na and Cl exclusion, K uptake, Nα/ K ratio, tissue tolerance) and biochemical pathways (glyoxilase, abscisic acid, proline, glycine-betaine, polyamine) is being probed.

Flash floods caused submergence is the major yield depressing factor in rainfed shallow lowland rice ecology. Depending on growth stage and period under water, crop damage could vary from 10 to 100%. Over 13 million hectares are estimated to be prone to flash floods in the country and every year no less than 3 million ha are inundated. Unlike traditional tall varieties like FR13A and FR43B, which are known to adapt to submergence, short statured high yielding varieties suffer the most, even when the period of submergence is limited to only a few days. Hence, farmers in rainfed lowland ecologies prefer traditional varieties not minding their low productivity. Despite years of effort to breeding tangible submergence tolerance in high yield backgrounds using FR13A like donor sources, no progress could be made. This has been largely due to lack of precise understanding of genetics of submergence tolerance *per se* and of the morpho-physological parameters that govern it and non reliability of phenotype based selection under the stress condition. With the advent of molecular mapping techniques, genetics of this stress is better understood now. It is found to be controlled by a single major quantitative trait *locus* on chromosome 9 along with a number of small QTLs. The major QTL identified in the flood resistant landrace FR13A and designated as *sub1* explains 70% phenotypic variance. A joint international effort involving CRRI and NDUAT in India BRRI of Bangladesh, University of California, Davis, USA and IRRI was initiated for characterization and use of the QTL in marker assisted breeding for development of submergence tolerant varieties. Fine mapping and dissecting of the QTL revealed it to be a cluster of three genes encoding ethylene responsive factors designated as *sub1A, sub1B* and *sub1C*. Of them *sub1A* was identified as the key gene conferring resistance to submergence. Two alleles of *sub1A viz., sub1A-1* (tolerance specific) and *sub1A-2* (intolerance specific) were identified with a single nucleotide polymorphism changing proline to serine (Septiningsih *et al.*, 2008). In submergence tolerant genotypes, under the stress ethylene activates *sub1A-1* allele limiting ethylene response and promoting repressors of gibberellins resulting in reduced culm elongation and carbohydrate (energy) consumption.

Using marker assisted backcrossing *sub1A* was introgressed into the popular lowland variety Swarna, adopting foreground (for *sub1A gene*)-

recombinant (flanking marker)-background (Swarna genome) selection strategy. For background selection a large number of polymorphic SSR markers representing the genome of the recurrent parent (Swarna) were used. This is probably the first example of biotech product developed for abiotic stress condition by marker assisted breeding in rice. Swarna selectively introgressed with *sub1A* has been extensively evaluated under typical submergence condition all over south and southeast Asia and formally released for general cultivation as Swarna *sub1A* in India (Orissa, West Bengal, UP etc) Bangladesh, the Philippines and Indonesia. Adopting the same strategy of advanced backcrossing and foreground-recombinant-background selection, the submergence tolerance QTL has now been transferred successfully to many popular lowland varieties like Samba Mahsuri, Savithri, IR64 etc.

Genetic Engineering

Finding new yield thresholds

Cloning and introgression of alien genes of interest by recombinant DNA technology is a proven technology already deregulated by regulatory authorities for commercial planting in some countries. Several such genes are lying on laboratory shelves including in India awaiting approval for commercialization. Also, many are in the final stages of making which would help find new ways to meet still challenging problems in agriculture. Some of the success and likely success stories as well as research breakthroughs in rice are as under.

Manipulation of plant architecture

Believing that alteration of plant architecture as well as that of panicle might help further raise the genetic yield level, many have explored such possibilities at molecular level. One such effort that convinced us that plant architecture-yield relationship was through light regulated overexpression of *Arabidopsis phytochrome A* gene in rice was by Cornell University (Garg *et al.,* 2006). Phytochromes are a family of red/far red light absorbing photoreceptors, which control plant development and its metabolic activities. The Cornell group could demonstrate that by splicing *Arabidopsis PHY A* gene into rice by employing light regulated tissue transforming *rbc* promoter, plant stature could be altered by further reducing the height of an already semidwarf variety and increasing simultaneously number of productive tillers resulting in significantly higher yield as compared to the control variety.

Management of biotic stresses

Of many still not resolved problems that constrain production advance, stemborer, leaf folder and sheath blight, among the biotic stresses and drought, salinity and submergence among the abitoic stresses are important. In the case of the former, lack of source of resistance in the rice genepools and in the latter, inefficient selection/screening protocols, despite availability of genetic resources are attributed as reasons for no-breakthrough as yet. This situation has, therefore, warranted transgenic approach.

For engineering of resistance, many different insecticidal proteins and molecules known to be highly selective in their action against a given pest but causing no harm to non-target organisms are used. Among the widely used genes encoding insecticidal proteins/molecules against rice insect pests, endotoxin crystal proteins of *Bacillus thuringiensis*, digestive enzyme specific protease inhibitors, plant lectins, α amylase inhibitors, insect chitinases and insecticidal viruses are important. Of them *Bt* toxin genes (*cry IA, cry IB, cry IC*) and protease inhibitors (*cowpea serine P1*) against stemborer and lectin protein gene *(Gna, Asa Lectin*) against hoppers have been reported to be effective.

As regards diseases, Lin *et al.* (1995) have been the pioneers in coming up with Chitinase (*ChiII*) gene based transgenic rice against sheath blight. Nothing is available as yet to manage the fungal diseases. The transgenic strategy has been tried using pathogen derived resistance sources like coat protein gene against viral diseases.

In respect of insect pests, Fujimoto (1993) was the first to develop and demonstrate success of *Bt* rice against stemborer. Since then several workers have successfully engineered rice with different *Bt* genes (*cry IA (b), cry IA (c), cry IAb*) alone as well as in fusion forms against the pest (Tu *et al.*, 1996 a; Data *et al.,* 1996). Researchers in India recently succeded in transforming rice varieties with *Bt* gene. Biswal (2010) have reported transgenic IR64 developed using indigenously isolated *cry IAc*. Despite success stories elsewhere including in China, Spain, Pakistan etc *Bt* rice is yet to be deregulated for commercial planting. Unduly long delay in deregulation and reservation against GM rice are due to apprehended risk to human health though as yet no solid proof against such a notion could be advanced.

Nagadhara *et al.* (2003) are reported to have successfully engineered the variety Chaitanya with *gna lectin* protein and yet another variety with onion/garlic *lectin* (*asa lectin)* (Saha *et al.,* 2006) conferring resistance to rice sucking pests. In their effort to develop durable resistance against BPH, GLH and WBPH, *ASACI* and *GNA* protein genes have been pyramided by crossing single gene based transgenic lines. Lines thus developed have been found to surpass in their level of resistance to all the three hoppers as compared to single gene based parental transgenics (Ramesh *et al.*, 2004).

Management of Abiotic Stresses

Many genes conferring abiotic stresses have been identified through transgenic approach.

Tolerance to salinity

Engineering rice with relevant genes accessed from various sources, has been reported to confer tolerance. Cellular ion homeostasis following salt stress is mediated by either Na efflux across plasma membrane or compartmentalization of it into vacuoles. Many important genes involved in salt tolerance belong to salt overly sensitive (SOS) pathway. A vacuolar Na^+/H^+ antiporter is involved in the sequestration of Na^+ to the vacuole. In rice, tonoplast Na^+/H^+ antiporter NHY1 gene is induced by salt stress. Together with antioxidants like ascorbic acid and glutathione, ROS scavenging enzymes like super oxide dismutase, ascorbate peroxidase, catalase etc. provide cells with efficient machinery to detoxify ROS. Increased activity of these enzymes has been reported to confer enhanced-tolerance to salinity stress. Transgenic rice overexpressing yeast MnSOD resulting in increased level of ascorbate peroxidase and chloroplastic SOD has been found to grow well under saline conditions (Tanaka *et al.,* 1999).

Of several candidate and regulatory genes overexpressed in rice, the following have shown promise. Expression of two genes of glyoxalase pathway *viz*., *glyI* and *glyII,* preferably together have been shown to confer tolerance to salinity (Singla-Pareek *et al.,* 2006). Overexpression of the genes pyramided with the helicase gene *PDH 45* has been found to further enhance the tolerance level. In advanced stages of development the transgenic has now been transferred to a private sector Biotech company, under agreement for intensive testing for bioefficacy and biosafety under contained field conditions.

The study by the Cornell group reveals that trehalose accumulation confers high level of tolerance to salinity as well as drought (Garg *et al.,* 2002). Trehalose is a non-reducing disaccharide of glucose. Functioning as a compatible solute/sugar (osmoprotectant) it stabilizes biological structures from damage under salinity stress. Regulated overexpression of trehalose biosynthetic genes, *(Ots A and Ots B)* encoding trehalose-6-phosphate synthase and trehalose-6-phospate phosphotase accessed from *E. coli* and constructed as a fusion gene (TPSP), confers high level of tolerance to salinity in the variety Pusa Basmati-1. The transgenics exhibit sustained plant growth, less photooxidative damage and more favourable mineral balance under salt stress. The transgenic Pusa Basmati-1 is now used as a donor for transferring the gene by backcross breeding to many high yielding varieties popular in salt affected areas.

Constitutive overexpression of *Rab16A* (ABA inducible group II LEA protein) cloned from the salt tolerant variety Pokkali driven by endogenous + ORF promoter in the salt sensitive high yielding variety Khitish has been found to confer high level of tolerance to ORF1 salinity (Karabi Datta – personal communication).

Given availability of many novel genes conferring high level of tolerance to moisture stress, and efficient transformation-regeneration protocols in place, genetic engineering for drought tolerance is gaining relevance. The effective novel genes broadly belong to structural and regulatory categories. To the structural category belong the genes involved in biosynthetic pathways of enzymes that catalyse stress tolerance specific osmolytes, LEA proteins, water channel proteins, chaperones etc., The regulatory category are those that govern expression of upstream genes and include transcription factors and signal transduction components of receptor related proteins. The candidate genes tried with some success in rice are trehalose *(tps)* cloned from *Arabidopsis*, trehalose-6-phosphate synthase and trchalose-6-phosphate phosphatase *(tps1)* from yeast, pyrolline carboxylate synthase (*P5cb*) from *V. aconitifolia,* chlroplastic glutamine synthetase (*GS2*) from rice and choline oxidase (glycine-betaine synthesis) (*cod A*) from *Arthrobactor globiformis*. Among the regulatory genes tried out are DRE binding protein (*dreb1á*), calcium dependant protein kinase *(OsCDPK7)* and those involved with cellular level tolerance like helicases (*PDH45*) and APZ/ERF family DREB transcription factors are important. Significantly, many of the candidate as well as regulatory genes have been reported to confer resistance not only to drought but also to salinity.

Enrichment of Nutritive Value

Under- and mal- nutrition is a bane of many in the developing world. Whereas protein-energy malnutrition (PEM) affects growth and development of all but young children in particular, the numbers suffering from 'hidden hunger' is no less. Over 80% of women, especially pregnant and lactating mothers and 75% of children suffer from iron deficiency caused nutritional anaemia, about 50,000 children go blind every year due to Vitamin A deficiency and 30% of new born are of low birth weight. As a staple of majority in the world, rice deserves attention for nutritional upgradation. Attempts at nutritional enrichment of rice by conventional breeding have been unsuccessful but possibilities of achieving this objective through genetic engineering look brighter.

Among the nutritional limitations of rice, low protein content is the foremost. Transformed potato with albumin gene (*AmA1*) of *Amaranthus* sp was shown to have higher protein content and better quality protein (Subhra Chakrabarthy *et al.*, 2002). Genetic intervention for metabolically engineering beta carotene biosynthesis to produce the Golden Rice by the group lead by Ingo Potrykus has opened up unprecedented opportunities for nutritional amelioration of food crops. Sadly, however, though golden rice was developed almost 10 years back, it is still not moving out of laboratory because of regulatory hurdles. With accelerated efforts underway to transfer the trait to several popular varieties coordinated by IRRI, it is hoped that the product would be deregulated for commercial planting and marketing in coming years.

For mineral nutrients like iron and zinc, efforts on enrichment of rice are underway through conventional recombination in combination with DNA technology. IRRI in collaboration with selected rice growing countries including India is making steady progress in combining high iron and zinc in high yield backgrounds by recombination breeding. Simultaneously, genetic engineering is underway to transform rice with iron storage protein (ferritin) using the gene cloned from *Phaseolus vulgaris*. Iron deficiency is not only due to inherently low iron content but also due to presence of compounds in high contents that interfere with its absorption during digestion. To overcome this problem, a three pronged research has been mounted in Japan and Switzerland which includes (i) making iron storage protein gene cloned from *Phaseolus vulgaris* express in endosperm (ii) reducing phytic acid, the main inhibitor of iron absorption, by introducing phytase gene cloned from *Aspergillus niger* and (iii) introducing iron absosrption facilitator protein, the cysteine rich metallothione, cloned from rice. In assembling the ferritin and phytase genes, Kolkata University has made significant progress (Karabi Datta – personal communication).

Genetic engineering is a very powerful and potent technology with enormous potential for successfully meeting many existing and numerous emerging challenges relevant to productive, efficient, sustainable and safe agriculture. It is important that the technology is not kept a hostage to imagined reservations and fears. This calls for developing a shared goal by research scientists, policy planners and administrators and civil society leaders to arrive quickly at a policy and regulatory framework that will allow this technology to translate science concepts to products for good of the society.

REFERENCES

Ali, J. (1993). Studies on temperature sensitive genic male sterility and chemical induced male sterility towards development of two-line hybrids in rice (*Oryza sativa* L.). Ph.D. thesis submitted to the Indian Agricultural Research Institute, New Delhi, India. p. 138.

Ali, J. and Santiaguel, F.A. (2011). Greener rice. *Rice Today,* 10: 26–27.

Ali, J., Siddiq, E.A., Zaman, F.U., Abraham, M.J. and Ahmed, I. (1995). Identification and characterization of temperature sensitive genic male sterile sources in rice (*Oryza sativa* L.). *Indian J. Genet.*, 55: 243–259.

Ammar, M.H.M., Pandit, A., Singh, R.K., Sameena, S., Chauhan, M,S., Singh, A.K., Sharma, P.C., Gaikwad, K., Sharma, T.R., Mohapatra, T. and Singh, N.K. (2004). Mapping of QTLs controlling Na^+, K^+ and Cl^- ion concentrations in salt tolerant *indica* rice variety CSR27. *J. Plant Biochem. Biotechnol.,* 18: 139–150.

Archak, S., Lakshminarayanareddy. V. and Nagaraju, J. (2007). High-throughput multiplex microsatellite marker assay for detection and quantification of adulteration in Basmati rice (*Oryza sativa*). *Electrophoresis* 28: 2396–2405.

Basavaraj, S.H., Singh, V.K., Singh, A. Singh, A., Singh, A., Anand, D., Yadav, S., Gopala Krishnan, S., Nagarajan, M., Mohapatra, T., Prabhu, K.V. and Singh, A.K. (2010). Marker-assisted improvement of bacterial blight resistance in parental lines of Pusa RH10, a superfine grain aromatic rice hybrid. *Mol. Breed.,* 26: 239–305.

Bernier, J., Kumar, A., Venuprasad, R., Spaner, D. and Atlin, G. (2007). A large-effect QTL for grain yield under reproductive-stage drought stress in upland rice. *Crop Science*, 47: 507–518.

Biswal, A.K. (2010). Sub-cloning of synthetic Bt genes into binary vector and development of Bt transgenic rice resistant to yellow Stem Borer. Ph. D. thesis submitted to Jawaharlal Nehru Technological University, Hyderabad, India.

Borines, L.M., VeraCruz, C.M., Redona, E.D., Hernandez, J.F., Natural, M.P., Raymundo. A.D. and Leung, H. (2000). Marker-aided pyramiding of bacterial blight resistance genes in maintainer lines for hybrid rice production. *IRRI Conf. Abstr.* 4(2): 162.

Fujimoto, H., Itoh, K., Yamamoto, M., Kyozuka, J. and Shimamoto, K. (1993). Insect resistant rice generated by introduction of a modified d-endotoxin gene of *Bacillus thuringiensis*. *Bio / Technology*, 11: 1151–1155.

Garg, A.K., Kim, J.K., Owens, T.G., Ranwala, A.P., Choi, Y.D., Kochian, L.V. and Wu, R.J. (2002). Trehalose accumulation in rice plants confers high tolerance levels to different abiotic stresses. *Proc. Natl. Acad. Sci. USA*. 99: 15898–15903.

Garg, A.K., Sawers, R.J., Wang, H., Kim, J.K., Walker, J.M., Brutnell, T.P., Parthasarathy, M.V., Vierstra, R.D., and Wu, R.J. (2006). Light-regulated overexpression of an *Arabidopsis* phytochrome A gene in rice alters plant architecture and increases grain yield. *Planta* 223: 627–636.

Gupta, S.K., Rai, A.K., Kanwar, S.S., Chand, D., Singh, N.K. and Sharma, T.R. (2012). The single functional blast resistance gene *Pi54* activates a complex defence mechanism in rice. *J. Exp. Botany.*, 63: 757–772.

Henderson, M.T., Yeh, E.P. and Exner, B. (1959). Further evidence of structural differentiation in the chromosomes as a cause of sterility in intervarietal hybrids of rice. *Oryza sativa* L. *Cytologia,* 24: 415–422.

Hittalmani, S., Parco, A., Mew, T.W., Zeigler, R.S. and Huang, N. (2000). Fine mapping and DNA marker-assisted pyramiding of the three major genes for blast resistance in rice. *Theor. Appl. Genet.*, 100: 1121–1128.

Hussain, A.J., Ali, J., Siddiq, E.A., Gupta, V.S., Reddy, O.U.K. and Ranjekar, P.K. (2011). Mapping of *tms8* gene for temperature-sensitive genic male sterility (TGMS) in rice (*Oryza sativa* L.). *Plant Breeding.*, 131: 42–47.

Ikehashi, H. and Araki, H. (1984). Varietal screening of compatibility types revealed in F_1 hybrids of distant crosses in rice. *Jpn. J. Breed.*, 34: 304–313.

Ikehashi, H. and Araki, H. (1986). Genetics of F_1 sterility in remote crosses of rice. *In* Rice Genetics I, International Rice Research Institute, P.O. Box 933, Manila, Philippines, pp. 119–130.

Islam, M.R., Gregorio, G.B., Salam, M.A., Collard, B.C.Y., Tumimbang-Raiz, E., Adorada, D.L., Mendoza, R.D., Singh, R.K. and Hassan, L. (2011). Validation of a major QTL for salinity tolerance on chromosome 1 of rice in three different breeding populations, *Agrochemica*, 55: 355–366.

Lin, W., Anuratha, C.S., Datta, K., Potrykus, I., Muthukrishnan, S. and Datta, S.K. (1995). Genetic engineering of rice for resistance to sheath blight. *Bio/Technology* 13: 686–691.

Liu, S.P., Li, X., Wang, C.Y., Li, X.H. and He, Y.Q. (2003). Improvement of resistance to rice blast in Zhenshan 97 by molecular marker-aided selection. *Acta Botanica Sinica* 45: 1346–1350.

Marri, P.R., Sarla, N., Reddy, V.L.N. and Siddiq, E.A. (2005). Identification and mapping of yield and related QTLs from an Indian accession of *Oryza rufipogon*. *BMC Genet.*, 6: 33.

Nagadhara, D., Ramesh, S., Pasalu, I.C., Rao, Y.K., Krishnaiah, N.V., Sarma, N.P., Brown, D.P., Gatehouse, J.A., Reddy, V.D. and Rao, K.V. (2003). Transgenic indica rice resistant to sap-sucking insects. *Plant Biotechnol. J.,* 1: 231–240.

Narayanan, N.N., Baisakh, N., Vera Cruz N., Gnananmanickam, S.S., Datta, K. and Datta, S.K. (2002). Molecular breeding for the development of blast and bacterial blight resistance in rice cv. IR50. *Crop Sci.*, 42: 2072–2079.

Nas, T.M.S., Sanchez, D.L., Diaz, G.Q., Mendioro, M.S. and Virmani, S.S. (2005). Pyramiding of thermosensitive genetic male sterility (TGMS) genes and identification of a candidate *tms5* gene in rice. *Euphytica*. 145: 67–75.

Neeraja, C.N., Vemireddy, L.R., Malathi, S. and Siddiq, E.A. (2009). Identification of alternate dwarfing gene sources to widely used Dee-Gee-Woo-Gen allele of *sd1* gene by molecular and biochemical assays in rice (*Oryza sativa* L.). *Electronic J. Biotechnology* 12.

Oka, H.I. (1957). Phylogenetic differentiation of cultivated rice. 15. Complementary lethal genes in rice. *Japan. J. Genet.*, 32: 83–87.

Oka, H.I. and Doida, Y. (1962). Phylogenetic differentiation of cultivated rice, XX. Analysis of the genetic basis of hybrid breakdown in rice. *Jpn. J. Genet.*, 37: 24–35.

Ookawa, T., Hobo, T., Yano, M., Murata, K., Ando, T., Miura, H., Asano, K., Ochiai, Y., Ikeda, M., Nishitani, R., Ebitani, T., Ozaki, H., Angeles, E.R., Hirasawa, T. and Matsuoka, M. (2010). New approach for rice improvement using a pleiotropic QTL gene for lodging resistance and yield. *Nat. Commun.*, 1: 132.

Ramesh, S., Nagadhara, D., Reddy, V.D. and Rao, K.V. (2004). Production of transgenic *indica* rice resistant to yellow stem borer and sap-sucking insects, using super-binary vectors of *Agrobacterium tumefaciens*. *Plant Science,* 166: 1077–1085.

Rawat, N., Neeraja, C.N., Sundaram, R.M., Nair, S. and Bentur, J.S. (2012). A novel mechanism of gall midge resistance in the rice variety Kavya revealed by microarray analysis. *Functional and Integrative Genomics* 12: 249–264.

Reddy, O.U.K., Siddiq, E.A., Sarma, N.P., Ali, J., Hussain, A.J., Nimmakayala, P., Ramasamy, P., Pammi, S. and Reddy, A.S. (2000). Genetic analysis of temperature-sensitive male sterility in rice. *Theor. Appl. Genet*., 100: 794–801.

Sharma, T.R., Madhav, M.S., Singh, B.K., Shanker, P., Jana, T.K., Dalal, V., Pandit, A., Singh, A., Gaikwad, K., Upreti, H.C. and Singh, N.K. (2005). High resolution mapping, cloning and molecular characterization of the *Pikh* gene of rice, which confers resistance to *Magnaporthe grisea*. *Mol. Genet.* and *Genomics,* 274: 569–578.

Shastry, S.V.S. and Misra, R.N. (1961). Pachytene analysis in *Oryza*. II. Sterility in *japonica - indica* hybrids. *Chromosoma,* 12: 248–271.

Septiningsih, E. M., Pamplona, A.M., Sanchez, D.L., Neeraja, C.N., Vergara, G.V., Heuer, S., Ismail, A.M. and Mackill, D.J. (2008). Development of submergence tolerant rice cultivars: The Sub1 locus and beyond. *Ann. Bot.*, 103: 151–160.

Siddiq, E.A. and Singh, S.P. (2005). Wide-compatibility system for yield enhancement of tropical rice through inter-subspecific hybridization. *Adv. Agron.,* 87: 157–209.

Singh, S.P., Sundaram, R.M., Biradar, S.K., Viraktamath, B.C., Ahmed, M.I. and Siddiq, E.A. (2006). Identification of simple sequence repeat markers for utilizing wide-compatibility genes in inter-subspecific hybrids in rice. *Theor. Appl. Genet.,* 113: 509–517.

Singla-Pareek, S.L., Yadav, S.K., Pareek, A., Reddy, M.K. and Sopory, S.K. (2006). Transgenic tobacco overexpressing glyoxalase pathway enzymes grow and set viable seeds in zinc-spiked soils. *Plant Physiol.,* 140: 613–623.

Srividya, A., Vemireddy, L.R., Hariprasad, A.S., Jayaprada, M. and Sridhar, S. and Siddiq, E.A. (2010). Identification and mapping of landrace derived QTL associated with yield and its components in rice under different nitrogen levels and environments. *Int. J. Plant Breed. Genet.,* 4: 210–227.

Steele, K.A., Price, A.H., Shashidar, H.E. and Witcombe, J.R. (2006). Marker-assisted selection to introgress rice QTLs controlling root traits into an Indian upland rice variety. *Theor. Appl. Genet.,* 12: 208–221.

Sundaram, R.M., Vishnupriya, M.R., Biradar, S.K., Laha, G.S., Reddy, A.G., Rani, N.S., Sarma, N.P. and Sonti, R.V. (2008). Marker assisted introgression of bacterial blight resistance in Samba Mahsuri, an elite indica rice variety. *Euphytica,* 160: 411–422.

Tanaka, K., Hibino, T., Hayashi, Y., Tanaka, A., Kishitani, S., Takabe, T., Yokota, S. and Takabe, T. (1999). Salt tolerance of transgenic rice overexpression yeast mitochondrial Mn-SOD in chloroplasts. *Plant Sci.,* 148: 131–138.

Vemireddy, L.R., Archak, S. and Nagaraju, (2007). Capillary electrophoresis is essential for microsatellite marker based detection and quantification of adulteration of Basmati rice (*Oryza sativa*). *J. Agric. Food Chem.,* 55: 8112–8117.

Xing, Y. and Zhang, Q. (2010). Genetic and molecular bases of rice yield. *Annu. Rev. Plant Biol.,* 61: 421–442.

Yue, B., Xue, W., Xiong, L., Yu, X., Luo, L., Cui, K., Jin, D., Xing, Y. and Zhang, Q. (2005). Genetic basis of drought resistance at reproductive stage in rice: Separation of drought tolerance from drought avoidance. *Genetics,* 172: 1213–1228.

Zhang, J., Li, X., Jiang, G., Xu, Y. and He, Y. (2006). Pyramiding of *Xa7* and *Xa21* for the improvement of disease resistance to bacterial blight in hybrid rice. *Plant Breed.*, 125: 600–605.

3

Wheat

R. Tiwari, K. Venkatesh and R. Sendhil

Directorate of Wheat Research, Karnal, India

Rice, wheat and maize are the most widely grown crops of the world and largely supply the energy and nutritional requirements of the world population. Wheat is cultivated on 221 m ha globally and is the second most important food crop of the world with a production of about 698 m tonnes.

At the national level, wheat occupies an area of around 29 m ha (2012-13) and is the second highest contributor to the total food grain production of the country after Rice with the production of 92.96 m tonnes (Table 1). At the state level, Uttar Pradesh with 9.73 m ha of wheat area is the highest producer of wheat in the country with production of 30.3 m tonnes. Punjab and Haryana together called 'The Bread Basket of India' contribute 16.11 and 11.12 m tonnes, respectively to the total wheat produced in India with the record highest productivity at the national level (Table 1).

Wheat improvement in India has witnessed a glorious past by experiencing the Green Revolution and by having a constant increase in wheat production and productivity over the years (Fig. 1). The reasons collectively attributed to this glorious past are said to be (i) Identification and utilisation of semi-dwarf, input-responsive, high yielding varietal features from International Maize and Wheat Improvement Center's (CIMMYT) wheat breeding programme, which in turn replaced the native tall, photosensitive, low input responsive, lodging susceptible varieties

V.L. Chopra (Ed.): Breeding Field Crops II: Advances, Studium Press (LLC), USA.

of 50's. (ii) Identification and utilisation of source of major genes conferring resistance to major diseases (Vertical resistance) as well as adopting the recent concept of Adult Plant Resistance (APR) in breeding programmes. (iii) Adoption of environment friendly and time saving cultivation practices such as Conservation Tillage/Agriculture (CA).

Table 1: Area, production and productivity of wheat during 2011-12 and 2012-13 in major producing states along with coverage under irrigation (2009-10)

State/ Country	***2011-12*** Area (000'ha)	Production (000't)	Yield (kg/ha)	***2012-13**** Area (000'ha)	Production (000't)	Yield (kg/ha)	***Area under irrigation (%) in 2009-10***
Andhra Pradesh	8	11	1375	8	7	875	N.A.
Assam	53	60	1147	48	57	1188	N.A.
Bihar	2142	4725	2206	2215	5375	2427	92.10
Chhattisgarh	109	133	1227	101	141	1396	N.A.
Gujarat	1351	4072	3014	1050	3135	2986	89.80
Haryana	2522	12686	5030	2497	11117	4452	99.40
Himachal Pradesh	357	596	1671	359	543	1514	20.20
Jammu and Kashmir	296	500	1689	296	416	1404	30.00
Jharkhand	159	303	1908	156	267	1716	91.50
Karnataka	225	193	858	234	172	735	56.70
Madhya Pradesh	4889	11539	2360	5300	13133	2478	84.70
Maharashtra	843	1313	1558	594	875	1473	73.90
Odisha	1.46	2.40	1644	1.28	2.14	1672	N.A.
Punjab	3528	17280	4898	3522	16106	4573	98.70
Rajasthan	2935	9320	3175	2820	8954	3175	99.60
Uttar Pradesh	9731	30293	3113	9734	30302	3113	98.10
Uttarakhand	369	878	2379	358	838	2341	57.50
West Bengal	316	873	2765	322	907	2817	95.90
Others	92	177	1920	31	110	3514	N.A.
India	**29865**	**94882**	**3177**	**29647**	**92458**	**3119**	**91.70**

Note: *indicates the provisional data and N.A. indicates the non-availability of data.
Source: Directorate of Economics and Statistics, Government of India.

Today, with the 121 million population, demand for wheat in India is estimated to reach 100 m tonnes by 2030 (Chand *et al.*, 2011). Continuing the similar trend of constant increase in wheat production and productivity in future is a daunting task and its realization is being constantly challenged by many threats to wheat production world-wide in the form of climatic changes leading to temperature extremes, weather changes, erratic precipitation and changes in pest dynamics. Overall changes in the micro-environments are also expected to be more severe in the coming years coupled with challenges

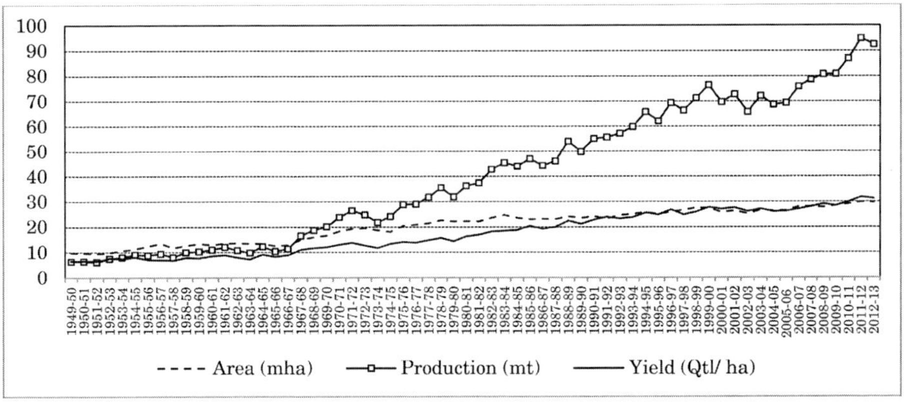

Data Source: Directorate of Economics and Statistics, Government of India

Fig. 1: Trends in area, production and productivity of wheat in India (1949-50 to 2012-13)

due to decrease in arable land caused by increasing urbanization, land degradation, limited water for irrigation. Similarly, the demand for high value products of wheat shall also have to be met (DWR Vision-2030, 2011).

Wheat breeding programme in India during the early decades of the 20th century started with the classification of mixtures on the basis of grain characters of different cultivars, called 'sorts'. Pure line selection methods were subsequently followed to improve the existing lots. During the last decade, several wheat varieties were released for commercial cultivation by the farmers. Some of the wheat varieties released for different zones and production conditions are given in Table 2.

Therefore, looking at the future challenges to wheat production and the urgent need for increasing production in the near future, latest developments and inventions in the related/other disciplines such as molecular markers, linkage maps, marker assisted selection and new generation genome sequencing technologies have come as a boon. These new tools and techniques are said to be the focal point of researchers world over because of their improved efficiency in selection of parents and progeny for both qualitatively and quantitatively inherited traits and also they are demonstrated to help in better understanding the underlying mechanisms in plant's response to stress and in overall growth and development of plant. The major areas which need consistent research efforts comprise the following.

Germplasm Characterization and Exchange

Further improvement in wheat requires a very strong germplasm development/exchange programme. In order to identify the potential

Table 2: Improved wheat varieties for different zones and production conditions

Agro-ecological zone	*Production conditions*			
	Irrigated timely sown	*Irrigated, late sown*	*Rainfed, timely sown*	*Others*
Northern Hills	VL 804, VL 738, HS 240	HS 490, VL 892, HS 420, HS 295	VL 804, VL 738, HS 240	High Altitude-HS 365, VL 832, SKW 196
North Western Plains	DPW 621-50, HD 2967, DBW 17, PBW 550, PBW 502, PBW 343, WH 542, UP 2338, HD 2687, PDW 291(d), PDW 233(d), WH 896 (d)	PBW 590, WH101, DBW 16, PBW 373, Raj 3765, UP 2425	WH 1080, PBW 396, PBW 175, Kundan	Timely sown, Restricted irrigation-HD 3043
North Eastern Plains	DBW 39, CBW 38, Raj 4120, K 0307, HD 2824, PBW 443, PBW 343, HD 2733, HUW 468, K 9107, HP 1769	HD 2985, HW 2045, DBW 14, NW 2036, HD 2643, NW 1014, HP 1744, Halna	HD 2888, MACS 6145, K 8027	Low fertility (LS, RF)-K 9465, K 8962
Central	HI 1544, GW 366, GW 322, GW 273, HI 8498 (d)	MP 1203, HD 2864, HD 2932, MP 4010, DL 788-2,	HI 1531, HI 1500, HW 2004, Sujata, HI 8627 (d), HD 4672 (d)	
Peninsular	Raj 4037, GW 332, HUW 510, DWR 162, HD 2189, MACS 2971 (Dic), DDK 1026 (Dic)	PBW 533, HD 2932, HD 2833, Raj 4083, NIAW 34	HD 2987, PBW 596, K 9644, HD 2781, NI 5439, AKDW 2997-16 (d)	
Southern Hills				HW 2044, HW 1085, COW (W)-1

(d)- Durum, (Dic.) - Dicoccum

attributes of the germplasm material, sets of nurseries/trials are planted at several locations world-wide under different programmes led by International Maize and Wheat Improvement Center (CIMMYT) and International Center for Agricultural Research in the Dry Areas (ICARDA) and at the national level by National Bureau of Plant Genetic Resources (NBPGR) and Directorate of Wheat Research (DWR). The data received from the centres are pooled and the genotypes having utility value are marked. Confirmed genetic stocks are registered and distributed to various breeding centers for their effective utilization as donor parents in breeding programmes. For instance, till the year 2010, a total of 138 genetic stocks of wheat were registered with NBPGR for various useful traits such as yield components, resistance to diseases, tolerance to drought, heat, salinity and quality traits. These stocks are available in public domain for utilization.

Germplasm Enhancement/Pre-breeding

In order to widen the genetic base of wheat, the emphasis is laid on introgressing genes from underutilized *buitre* types, synthetic hexaploids and sub compactoid eared germplasm. From these promising materials, genetic stocks have been developed. These genetic materials have more grains/spike, more spikelets, long spike and prolonged grain filling period. To widen the *de novo* genetic variability in the germplasm, the wild relatives of wheat (*Aegilops* spp.) are also evaluated for biotic, abiotic and yield traits and desirable accessions are utilized in wheat improvement programme. Interspecific crosses among tetraploid wheat with *T. turgidum, T. pyrimidale, T. persicum, T. carthlicum* and *T. dicoccoides* have also been made for durum improvement.

Exploiting Heterosis

Hybrid wheat development, both through Cytoplasmic Male Sterility (CMS) and Chemical Hybridizing Agent (CHA) approaches have been explored. The CMS lines have been developed by using cytoplasm from *Triticum timopheevii, T. araraticum, Ae. caudata* and *Ae. speltoides.* More than 80% fertility restoration was observed when the exotic stocks were crossed with *T. timopheevii* and *T. araraticum* based CMS lines. A few hybrids developed through CMS approach showed heterosis for yield components *viz.* spikelet number, spike length and tillers per plant. Through CHA approach, identified chemicals have been used which induced male sterility when sprayed at 10-15 mm of spike length at 50-60 days after germination. A large number of hybrids through CHA (and a few by CMS system) have been evaluated at wheat research centers. Based on the results and experiences gained from research on hybrid wheat, it is visualized that the CMS system, though cumbersome, provides useful and eco-friendly approach for hybrid seed production. However, CHA approach indicated certain inherent problems beyond control and, therefore, not found practically feasible to produce hybrid seed in wheat. The *timopheevii* based CMS system has been found most appropriate since the fertility restoration is highly satisfactory.

Simplifying the Complexity

Wheat breeding, similar to any other crop breeding programme, includes several agronomic and quality characters to be dealt with. However, breeding strategies and technologies keep becoming complex with attempts to prioritize these characters (yet retaining flexibility) and then combine improved levels of all these qualitative and quantitative traits in specific backgrounds. Total genetic progress becomes slow and incremental (Fig. 2). Yet, we can confidently say that plant breeding has been successful and essential to human welfare and it will continue to be important in future years as well. Several improvisations in the area of biotechnology are still required to sharpen it as a tool to plant breeding. This is supposed to be a safe and cost effective aid to plant breeding programs.

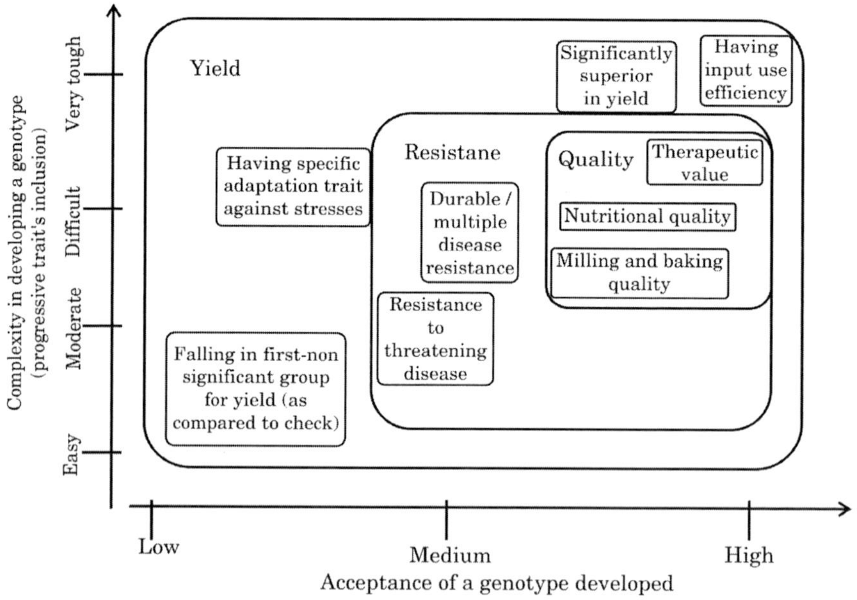

Fig. 2: Schematic presentation of wheat breeding priorities and relative efforts involved

Cytogenetics

Bread wheat was not amenable to early genetic studies mainly because of its hexaploid nature with presence of triplicate sets of genomes in a single nucleus, but assignment of genes to specific chromosomes became possible the availability of aneuploids, particularly the monosomics, (Sears, 1969). Construction of linkage maps with precise location of centromere became simple with the subsequent developments like availability of telocentrics of the 42 chromosome arms. The developments such as chromosome pairing studies contributed to understanding the mechanism of the evolution of polyploids involving diploidization. A variety of aneuploids such as monosomics, ditelosomics, double ditelosomics, and nullisomic–tetrasomics, and deletion lines developed later are of great importance and are used widely for both cytogenetic and genetic studies in wheat (Gupta *et al.*, 2005 and Zang *et al.*, 2007). The availability of ph mutants (*Ph1 and Ph2*) led to the development of alien addition/substitution lines and the introgression of alien chromosome segments into the wheat genome (Gupta *et al.*, 2005). The development of tools such as chromosome banding (Gill and Kimber, 1974a, 1974b) and fluorescence *in situ* hybridisation (FISH) (Rayburn and Gill, 1985) have allowed us to study the structure and function of chromosome landmarks such as heterochromatin (Bedbrook *et al.*, 1980; Appels *et al.*, 1981), rRNA genes (Mukai *et al.*, 1990; Jiang and Gill, 1994b), centromeres (Zhang *et al.*, 2001), subtelomeres (Zhang *et al.*, 2004a), and telomeres (Friebe *et al.*, 2001), and also the evolution of wheat and other species in the Triticeae family (Naranjo *et al.*, 1987; Jiang and Gill, 1994c; Schmidt and Heslop-Harrison, 1996). The more recent developments such as application of flow cytometry in carrying out systematic investigations on individual chromosomes have resulted in construction of chromosome-specific bacterial artificial chromosome (BAC) libraries, based on which the physical map of several wheat chromosomes are now established. This same chromosome sorting methodology is now being employed by the International Wheat Genome Sequencing Consortium (IWGSC) to establish the arm specific physical maps of the wheat chromosomes.

Exploiting the Power of Biotechnology

Biotechnology has provided us with new ways to generate genetic variation as well as methods for estimating it. Molecular markers allow the connection of phenotypic characters with chromosomal regions/ genomic loci responsible for them and are therefore an important tool for improvement. Conventional breeding programs work by generating segregating populations, then screening the phenotypes of plants for the presence of the desirable trait. This is followed by a time consuming costly process of repeated backcrosses or selfing and testing, for the accumulation (pyramiding) of genes encoding the same trait (*e.g.*, different resistance against same pathogen). The combination of recessive genes or combination of complex character encoded by multiple genes with additional effects Quantitative Trait Loci (QTL) is difficult to achieve with classical methods. Here comes the role of molecular markers in facilitating these processes. These approaches are to complement conventional methods of breeding.

While genetics revolutionized plant breeding at the beginning of the twentieth century, genomics is leading to a new plant breeding revolution at the beginning of the twenty-first century. This is because genomics allows the direct study of the genotype and its relationship with the phenotype for a high number of genes and traits (Tester and Langridge, 2010), and therefore provides new tools and resources that are leading to a paradigm shift in the breeding approaches (Jaime, 2011).

To accelerate the genetic progress for maintenance and enhancement of yield potential as well as to improve tolerance to biotic and abiotic stresses that are expected to increase in frequency and intensity, some of the new and innovative methods, that will help deliver improved varieties within shorter selection cycles are doubled-haploid production, marker-based selection, genomics selection, GM approaches etc.

Marker Assisted Selection (MAS) refers to selection of gene or combination of genes *via* linked markers (Chopra, 2000). This procedure is carried out at two levels *viz.*, foreground selection – selection for the trait of interest with the linked markers and background selection – in which the proportion of donor genome can be monitored during each backcross generation.

The breeding through the help of markers *i.e.*, Marker Assisted Breeding will greatly increase the efficiency of effectiveness in plant breeding compared to conventional breeding method. In the context of MAS, therefore, DNA-based markers can be effectively utilized for three basic purposes: (i) identification of suitable parental lines for improvement with respect to a specific target trait; (ii) tracing favorable allele(s) (dominant or recessive) across generations; and (iii) identifying the most suitable individual(s) among the segregating progeny, based on allelic composition across a part or the entire genome. Pyramiding of effective genes against rust diseases in desirable genotypes having a better potential of getting recognized as a cultivar, is the safest and economically viable solution to avoid any yield losses by these diseases. Gene pyramiding often requires improvement of a specific cultivar simultaneously for several characters. There are two options for transferring multiple characters: (a) transfer the genes simultaneously in one backcrossing programme or (b) transfer the genes in independent backcrossing programmes and combine them into one individual at the end. The problem with transferring several genes simultaneously is the large number of backcross seeds that must be obtained to be certain that a genotype with the entire desired genes is present. The stepwise transfer approach involves transfer of genes from different donor parents to a common recurrent parent in independent backcrossing programmes but simultaneously to combine them into one individual at the end. This approach is fast and at the same time requires lesser population size. Simulations on backcross breeding strategies have indicated that MAS should enable the recovery of desired target genotype after three cycles of backcrossing compared with a minimum of six cycles with traditional approaches (Tanksley *et al.*, 1989).

Multiplexing of markers and efficient data reading and analysis systems will be areas where improvements are possible through intensive efforts. To use molecular tools effectively, breeding programmes will require advances in automated genotyping, data management and bio-computing. To enhance the efficiency of MAS, DNA sequence (of the gene) enabled designing of perfect or diagnostic markers will be key area. Since such markers are located within the actual gene sequence, it eliminates the possibility of recombination between marker and the targeted gene (Ellis *et al.*, 2002; Ogbonnanya *et al.*, 2001). Such an effort will also help to discover and mark high value traits (Quantitative Trait Loci – QTL) and genes, which may produce higher quality, disease-resistant wheat. Knowing the location of these traits and genes on the wheat genome will allow breeders more precision in selecting trait combinations for specific varieties.

Doubled Haploid Prodcution

In doubled haploid production using Maize x Wheat crossing approach, F_1 generation wheat plants are crossed with maize. Subsequently maize chromosomes get eliminated and haploid condition is observed. Colchicine treatment then results in two copies of each chromosome which are essentially a clone of the gamete used. The haploid genome of the gamete, when doubled, results in a plant with a complete genome with two identical copies of every gene. Doubled haploids are homozygous at every locus achieved in a single generation as against six or more generations in conventional inbreeding programs. This helps wheat breeders in trying to identify and isolate valuable new traits for future varieties. Doubled haploid breeding also speeds up the identification of new genes since traits in different lines can be compared by using genetic markers and therefore in developing the lines quicker. The doubled haploid process is in the beginning time consuming and more delicate than traditional breeding. However, subsequently, it helps farmers and end users, by developing the genotypes five and six years sooner than traditional breeding. It also works well with the genomic selection. Breeders can use a synergy of the methods to integrate individual traits into wheat varieties they develop in the future without the time of selfing and backcrossing.

Breeding by design

Breeding by design is a concept that aims to control all allelic variation for all genes of agronomic importance through the information on molecular markers linked to them and then predicting the outcome of a set of crosses for reconstitution of a desired genotype containing all favorable genes coming from various parents and crosses (Peleman and van der Voort, 2003).

Genomic selection (GS)

Genomic selection (GS), which is based on the simultaneous estimation of effects on the phenotype of all available loci, haplotypes, and markers without a previous selection of markers with effects on the phenotype (Jannink *et al.*, 2010). The GS requires availability of phenotypic and genotypic data and the establishment of a genetic model, so that the differences in the phenotype are explained by the markers analyzed (Lorenz *et al.*, 2011) (Adopted from Jaime, 2011).

Association mapping

Marker/trait associations so far have mainly been made through the application of genetic markers by way of traditional bi-parental mapping in plants. However a large gap between identifying important QTL and applying marker assisted selection to breeding material has been there. To close this gap association mapping (AM), also known as association analysis or linkage disequilibrium mapping is in offing. Bi-parental populations usually don't have phenotypic information and were needed to be analyzed in the greenhouse or field after the populations is created. By making germplasm selections from lines that existed in advanced stages of the breeding process in National / International Research programmes, association mapping can potentially utilize historical phenotypic data already collected, reducing costs and time. As compared to other experimental designs that require sampling within families, AM offers the important advantage of sampling unrelated individuals in the population in the study of genetics of complex traits (Risch, 2000).

This approach has been tried in recent past. For example, in one such study, analysis indicated identification of six agronomic traits on chromosome 4A and *Stagonospora nodorum* blotch resistance in wheat using SSR markers. (Liu *et al.,* 2010; Tommasini *et al.*, 2007). Significant SSR markers associated with kernel size, Russian wheat aphid have also been reported (Breseghello and Sorrells, 2006; Peng *et al.*, 2009). Association mapping was used to identify QTLs influencing the agronomic performance of durum wheat grown across a broad range of soil moisture conditions and the results were further analyzed in order to ascertain the presence of different marker-trait association patterns across widely differing environments as water availability and, consequently, grain yield (Maccaferri *et al.*, 2010). This indicates that there is a good possibility of utilizing this approach in identification and application of the marker for genetic enhancement of specific traits in wheat.

Next generation sequencing (NGS)

With the advent of Next Generation Sequencing (NGS), ability to characterize the genomes of crops, including wheat, has been revolutionized. Identification, characterization and validation of SNPs from the sequence data holds promise of removing several bottlenecks of marker identification and marker assisted selection. In this procedure, full genome sequencing and genome-wide molecular markers are utilized to effectively address various genomic and environmental factors through a representative or complete set of genetic resources and breeding materials. These strategies are now increasingly based on understanding of specific genomic regions, genes/alleles, haplotypes, linkage disequilibrium (LD) block(s), gene networks and their contribution to specific phenotypes. Large-scale and high density genotyping and genome-wide selection are two important components of these strategies. As components of whole-genome strategies, molecular breeding platforms and methodologies should be backed up by high throughput and precision phenotyping and e-typing (environmental assay) with strong support systems such as breeding informatics and decision support tools (Xu *et al.*, 2012).

Transgenics

Process of transformation in wheat was successfully carried out in spring wheat, mainly in the Bob White variety. Internationally, scientists kept working with a known anti-fungal gene (glucanase) from alfalfa and one from rice (chitinase) to help make wheat more leaf rust resistant. This could occur by making the plant's cells less prone to fungal attack. Experiments involving a bacterium thought to influence drought tolerance *via* its manufacture of a sugar alcohol, mannitol was also carried out. Mannitol is known to protect cellular membranes under drought stress. Potential value of any kind of transgenic wheat, and its profitability for either public sector and/or companies is highly debated. Recently, a process of laying down specific guidelines for generating information from confined field trials for environmental safety and efficacy of specific GE plants, including wheat, has been in progress jointly by the ICAR and DBT in India.

In spite of the fact that there seems to be less probability of transgenic wheat varieties being planted at farmer's fields in the near future, yet there appears to be a tremendous potential for wheat improvement, as transgenesis could be used to add herbicide resistance, drought tolerance, and disease resistance, including viral disease resistance etc. Therefore, a core competence on transformation technique and containment facilities for evaluation of the transgenic wheat should be available with the wheat network programmes.

FUTURE BREEDING GOALS

Plant breeding, and hence wheat breeding, is a continuous process, more so ever due to changing climatic situations. A need will always be there to identify better adapted genotypes suited to the humankind for specific needs and sustainability. Some of the priority areas for contemporary wheat breeding are:

1. Grain yield potential, stability, and wide adaptation
2. Durable resistance to diseases such as stem, leaf, and yellow (*Puccinia striiformis* West.) rusts and other emerging diseases.
3. Improvement of Radiation Use Efficiency (RUE)
4. Improvement of Nutrient Use Efficiency (NUE)
5. Developing product specific wheat varieties to meet industrial demand
6. Developing varieties which can perform well in warmer temperatures
7. Developing varieties with improved water-use efficiency or drought tolerance to suit to changing climatic conditions

REFERENCES

Appels, R., Dennis, E.S., Smyth, D.R. and Peacock, W.J. (1981). Two repeated DNA sequences from the heterochromatic regions of rye (*Secale cereale*) chromosomes. *Chromosoma*, 84: 265–277.

Bedbrook, J.R., Jones, J., O'Dell, M., Thompson, R.D. and Flavell, R.B. (1980). A molecular description of telomeric heterochromatin in *Secale* species. *Cell*, 19: 545–560.

Breseghello, F. and Sorrells, M.S. (2006). Association mapping of kernel size and milling quality in wheat (*Triticum aestivum* L.) cultivars. *Genetics*, 172: 1165–1177.

Chand, R., Kumar, P. and Kumar, S. (2011). Total factor productivity and contribution of research investment to agricultural growth in India. Policy Paper 25, National Centre for Agricultural Economics and Policy Research, New Delhi, India.

DWR Vision, 2030. 2011. Directorate of Wheat Research, Karnal-132001, India.

Ellis, M.H., Spielmeyer, W., Gale, K.R., Rebetzke, G.J. and Richards, R.A. (2002). Perfect markers for the Rht-B1b and Rht-D1b dwarfing genes in wheat. *Theor. Appl. Genet.*, 111: 423–430.

Friebe, B., Kynast, R.G., Zhang, P., Qi, L., Dhar, M. and Gill, B.S. (2001). Chromosome healing by addition of telomeric repeats in wheat occurs during the first mitotic divisions of the sporophyte and is a gradual process. *Chromosome Res.*, 9: 137–146.

Gill B.S. and Kimber G. (1974a). The giemsa C-banded karyotype of rye. *Proceedings of the Natl. Acad. Sci. U.S.A.*, 71: 1247–1249.

Gill, B.S. and Kimber, G. (1974b). Giemsa C-banding and the evolution of wheat. *Proc. Natl. Acad. Sci. U.S.A.*, 71: 4086–4090.

Gupta, P.K., Kulwal, P.L. and Rustgi, S. (2005). Wheat cytogenetics in the genomics era and its relevance to breeding. *Cytogenet. Genome Res.*, 109: 315–327.

Jaime, P. (2011). Plant breeding: A success story to be continued thanks to the advances in genomics. *Front. Pl. Sci.*, 2: 1–2.

Jiang, J. and Gill, B.S. (1994). Different species-specific chromosome translocations in *Triticum timopheevii* and *T. turgidum* support the diphyletic origin of polyploid wheats. *Chromosome Res.*, 2: 59–64.

Liu, L., Wang, L., Yao, J., Zheng, Y. and Zhao, C. (2010). Association mapping of six agronomic traits on chromosome 4A of wheat (*Triticum aestivum* L.). *Mol. Pl. Breed.*, 1: 1–10.

Maccaferri, M., Sanguineti, M.C., Demontis, A., Ahmed, A.E., Moral, L.G., Maalou, F., Nachit, M., Nserallah, N., Ouabbou, H., Rhouma, S., Royo, C., Villegas, D. and Tuberosa, R. (2010). Association mapping in durum wheat grown across a broad range of water regimes. *J. Exp. Bot.*, 1: 1–30.

Mukai, Y., Endo, T.R. and Gill, B.S. (1990). Physical mapping of the 5S rRNA multigene family in common wheat. *J. Hered.*, 81: 290–295.

Naranjo, T., Roca, A., Goicoechea, P.G. and Giraldez, R. (1987). Arm homoeology of wheat and rye chromosomes. *Genome*, 29: 873–882.

Ogbonnanya, F.C., Subrahmanyam, N.C., Moullet, O., de Majnik, J., Eagles, H.A., Brown, J.S., Eastwood, R.F., Kollmorgen, J., Appels, R. and Lagudah, E.S. (2001). Diagnostic DNA markers for cereal cyst nematode resistance in bread wheat. *Aust. J. Agric. Res.*, 52: 1367–1374.

Peng, J.H., Bai, Y., Haley, S.D. and Lapital, N.L.V. (2009). Microsatellite-based molecular diversity of bread wheat germplasm and association mapping of wheat resistance to the Russian wheat aphid. *Genetics*, 135: 95–122.

Rayburn, A.L. and Gill, B.S. (1985). Use of biotin-labeled probes to map specific DNA sequences on wheat chromosomes. *J. Hered.*, 76: 78–81.

Risch, N.J. (2000). Searching for genetic determinants in the new millennium. *Nature*, 405: 847–856.

Schmidt, T. and Heslop-Harrison, J.S. (1996). High resolution mapping of repetitive DNA by *in situ* hybridization: Molecular and chromosomal features of prominent dispersed and discretely localized DNA families from the wild beet species *Beta procumbens*. *Plant Mol. Biol.,* 30: 1099–1114.

Sears, E.R. (1969). Wheat cytogenetics. *Annu. Rev. Genet.,* 3: 451–468.

Tommasini, L., Schnurbusch, T., Fossati, D., Mascher, F. and Keller, B. (2007). Association mapping of *Stagonospora nodorum* blotch resistance in modern European winter wheat varieties. *Theor. Appl. Genet.,* 115: 697–708.

Xu, Y., Lu, Y., Xie, C., Gao, S., Wan, J. and Prasanna, B.M. (2012). Whole-genome strategies for marker-assisted plant breeding. *Mol. Breeding*, 29: 833–854.

Zhang, P., Friebe, B., Lukaszewski, A.J. and Gill, B.S. (2001). The centromere structure in Robertsonian wheat-rye translocation chromosomes indicates that centric breakage-fusion can occur at different positions within the primary constriction. *Chromosoma*, 110: 335–344.

Zhang P., Li, W., Fellers, J., Friebe, B. and Gill, B.S. (2004). BAC-FISH in wheat identifies chromosome landmarks consisting of different types of transposable elements. *Chromosoma*, 112: 288–299.

Zhang, P., Friebe, B., Gill, B. and Park, R.F. (2007). Cytogenetics in the age of molecular genetics. *Aust. J. Agr. Res.,* 58: 498–506.

4

Maize

B.M. Prasanna[1] **and B.S. Dhillon**[2]

[1]CIMMYT (International Maize and Wheat Improvement Center), Nairobi, Kenya
[2]Punjab Agricultural University, Ludhiana, India

Maize (*Zea mays* L.) is currently produced on nearly 100 million hectares in 125 developing countries and is among the three most widely grown crops in 75 countries of that category (FAOSTAT, 2010). The crop provides over 20% of total calories in human diets in 21 countries, and over 30% in 12 countries that are home to over 310 million people (Shiferaw *et al.*, 2011). Globally, 765 million metric tons of maize was harvested in 2010 from less than 153 million hectares. About 73% of this area was located in the developing world. For 900 million farmers and consumers in low- and middle-income countries, maize is a preferred crop or food.

During 2003-08, maize production increased annually by 6.0% in Asia, as compared to 5.0% in Latin America, and 2.3% in Sub-Saharan Africa (FAOSTAT, 2010). However, between now and 2050, the demand for maize in the developing world will double and, by 2025, it will have become the crop with the greatest production globally and in the developing world (Rosegrant *et al.*, 2008). The growth in demand for human consumption of maize in the developing world is predicted to be 1.3% per annum until 2020. Moreover, rising incomes are expected to result in a doubling of consumption of meat across the developing world (Naylor *et al.*, 2005), leading to a predicted growth in demand for feed maize of 2.9% per annum. The importance of improving maize production

V.L. Chopra (Ed.): Breeding Field Crops II: Advances, Studium Press (LLC), USA.

and productivity in the developing world could be gauged from the fact that one-third of all malnourished children are found in systems where maize is among the top three crops (Hyman *et al.*, 2008).

The challenges being imposed by the global climate changes are tremendous and real. The maize fields in many countries, especially in Sub-Saharan Africa and South Asia, are now increasingly experiencing rising temperature, more frequent droughts, excess rainfall/flooding, as well as new and evolving pathogens and insect-pests. Although maize researchers in CIMMYT, in partnership with the national programs, have been successful in developing drought, low nitrogen or waterlogging stress tolerant maize germplasm suitable for diverse agro-ecologies (Cairns *et al.*, 2012), there is a need to further intensify and accelerate the efforts toward development and deployment of multiple stress resilient maize in the developing world.

TRENDS IN AREA, PRODUCTION AND PRODUCTIVITY

The average yields in several tropical, rainfed maize-growing regions of the world are still low, due mainly to poor soil fertility, frequent occurrence of droughts, high incidence of insect-pests, diseases and weeds, farmers' limited access to fertilizer, and lack of access to improved maize seed. India is among countries with sizable maize area but with less than 3 t/ha yield (Prasanna, 2011).

Both the area and production of maize in India have grown significantly in the past few decades. Maize grain production has increased from about 7 million tons in 1980/81 to about 21 million tons in 2010/11. The impressive growth of maize in India has been largely driven by the increasing demand for maize grain as feed for the rapidly expanding poultry industry (Hellin and Erenstein, 2009). The adoption of maize in non-traditional areas, the strong role of the private sector in the maize seed industry, and the development and delivery of higher-yielding, single-cross hybrids, are some of the major factors behind this. Annual growth rates for production have been recorded as 1.9%, 3.3% and 5.3% for the decades 1980–1990, 1990–2000, 2000–2010, respectively. While growth in the first two decades (1980-2000) was driven mainly by the yield increases due to improved adoption of high-yielding cultivars, area expansion has constituted more than half the growth over the past decade (DAC, 2010).

Simultaneous with these trends, maize prices have more than doubled over the past 10 years. With increasing globalization of agriculture, maize prices in a region may be greatly influenced by the developments in the rest of the world.

TECHNOLOGICAL ADVANCES AND EMERGING OPPORTUNITIES

The ability to develop and deliver maize germplasm with abiotic and biotic stress tolerance, coupled with improved nutritional quality and higher yields, will be critical for the resilience as well as diversification of cropping systems. While conventional breeding has been successful in improving maize yields over the decades, rapid advances in producton and productivity for keeping pace with the significantly rising demand for maize are possible by combining field-based and molecular breeding approaches. This integration entails increased penetration of genetic analyses, high-throughput genotyping and phenotyping, together with prediction models and decision support tools.

Next-Generation Sequencers and High Throughput Genotyping

Advances in genomics led to the identification of numerous DNA markers in maize during the last two decades, including thousands of mapped microsatellite or simple sequence repeat (SSR) markers, and more recently, single nucleotide polymorphisms (SNPs) markers. Besides molecular marker tagging of genes, a large number of genes controlling various aspects of plant development, biotic and abiotic stress resistance, quality characters, etc. have been cloned and characterized in maize, which are excellent assets for molecular breeding.

SSRs have been the most widely used markers by the maize researchers till recently due to their availability in large numbers in the public domain (Maize GDB; http://www.maizegdb.org), simplicity and effectiveness. There are now excellent opportunities for undertaking high throughput genotyping in maize using the SNP markers, as SNPs are highly amenable for automation, and offer significant advantages for genetic analysis and breeding purposes. Compared with the genomes of other cultivated plant species, SNP frequency in maize is high, with one SNP being found every 28–124 bp (*e.g.*, Vroh Bi *et al.*, 2006). A database and resource for SNP discovery and trait dissection has been established for maize in which genotype, phenotype and polymorphism data can be accessed for diverse maize inbreds and populations (http://www.panzea.org). Nearly one million maize SNPs are available in public databases (http://www.panzea.org), and several high throughput genotyping platforms have been developed for commercial use.

The genome sequencing of B73, a very popular inbred line in the US Corn Belt (Schnable *et al.*, 2009), and Palomero, an important popcorn landrace in Mexico (Vielle-Calzada *et al.*, 2010) are significant landmarks in maize research. With next-generation DNA sequencing technology (Shendure and Ji, 2008), it will be feasible to sequence any maize genotype of interest or undertake high-density genotyping (Metzker, 2010). Genotyping-by-sequencing (GBS) strategies are now feasible for maize. GBS procedure based on next generation sequencing in conjunction with genome complexity reduction has been applied on maize intermated B 73 x Mo 17 (IBM) population using 200,000 sequence tags (Elshire *et al.*, 2011).

As whole genome molecular information builds up, new insights are being obtained into the maize genome organization and evolution, as well as strategies to utilize the rapidly expanding genomic information for maize improvement (Xu *et al.*, 2012). Whole-genome array based comparative genomic hybridization has revealed a level of structural diversity between the inbred lines B73 and Mo17 that is unprecedented among higher eukaryotes (Springer *et al.*, 2009). Several thousand DNA sequences, including genes are present in B73 but not in Mo17. The copy number variation and presence/absence variation in genome content of inbred lines is an unexpected finding with implications for phenotypic diversity and heterosis in maize.

Trait Dissection using Molecular Markers

Maize researchers worldwide have generated numerous reports of molecular markers tagging genes/QTLs for diverse traits of agronomic and scientific interest. QTLs for several important traits affecting maize have been mapped, including resistance to several diseases (*e.g.* downy mildews, Northern corn leaf blight/Turcicum leaf blight, common smut, *Fusarium moniliforme* ear rot, Banded leaf and sheath blight (BLSB), aflatoxins, etc.), abiotic stresses (*e.g.*, drought, waterlogging, low nitrogen stress, etc.) and specialty traits (*e.g.*, high-oil content, etc.). Such studies have contributed to a greater understanding of the genetic architecture of various traits, particularly disease resistance and drought tolerance in maize (Prasanna *et al.*, 2010).

QTL analysis is also providing greater insights into genetic basis of heterosis for grain yield in maize hybrids (Tang *et al.*, 2010). In one of the studies, QTL for heterosis of grain yield has been reported to be located in bins adjacent to the centromere and their positions congruent across three hybrids (Schon *et al.*, 2010). Molecular markers, particularly in absence of pedigree and combining ability information have been found to be useful for prediction of hybrid performance in maize (Schrag *et al.*, 2010).

Association mapping through Linkage Disequilibrium (LD) analysis has led to identification of many genes controlling several simply inherited traits in various plant species (Zhu *et al.*, 2008). This approach is now being applied to dissect complex traits and identify superior alleles contributing to improved phenotypes in crops like maize. Association mapping seeks to identify a statistically significant genetic association between a change in the DNA sequence and a change in a trait of interest using a large population of diverse individuals, to remove circumstantial correlations. The approach provides excellent mapping resolution and the ability to investigate many alleles at the same time.

Nested association mapping (NAM) has been pioneered in maize as a means of combining the advantages of linkage analysis and association mapping (Yu *et al.*, 2008). NAM uses both historic and recent recombination events to have high mapping resolution while accommodating greater allelic diversity. Twenty five diverse maize lines were crossed with inbred line B73 (line used for the first maize genome sequencing) to generate 200 recombinant inbred lines (RILs) from each cross resulting in a total of 5000 RILs in the NAM population. Each RIL has been genotyped with a common set of 1106 markers besides being subjected to genome sequencing. Genome wide association studies using NAM population are helping to probe deeper into the genetic architecture of key agronomic traits (Tian *et al.*, 2011).

Doubled Haploid Technology in Maize Breeding

Cost- and time-effective development of homozygous lines is an important component of maize breeding. Traditionally, homozygous lines are obtained by repeated selfings of heterozygous material for 6-7 generations, which is a time-consuming and expensive process. The induction and subsequent doubling of maternal haploids is an efficient alternative to generate homozygous lines in quick time frame (two generations). CIMMYT has adapted and optimized the doubled-haploid (DH) technology, in collaboration with the University of Hohenheim (Germany), to accelerate the development of homozygous lines in diverse, elite, highly adapted genetic backgrounds (Prasanna *et al.*, 2012). Modern maize inducers have haploid induction rates of about 8% on average. Further, genetic variability has been reported for haploid induction ability (Kebede *et al.*, 2011; Prigge *et al.*, 2011; Prigge and Melchinger, 2012) which can be exploited to improve the haploid induction rate.

The DH technology is a powerful means to accelerate the introgression of novel germplasm into elite maize breeding lines. It enhances "forward breeding" and provides an opportunity to have an earlier look at the potential of new lines, greater knowledge about their environmental adaptability before they are fully tested, and further used as parental lines for hybrid development and commercial cultivation. Use of DH technology can potentially enhance the efficiency of recurrent selection or genomic selection based schemes for traits with low heritability, particularly for breeding programmes without access to offseason nurseries (Bouchez and Gallais, 2000). Furthermore, the DH technology enables shifting of resources from the labour-intensive task of repeated inbreeding to generate inbred lines, and spending more time on evaluation of the DH lines for yield and other adaptive traits, and using the identified lines for producing hybrids and synthetics. Elaborate schemes for hybrid maize breeding with doubled haploids have been examined (Wagenast *et al.*, 2010; Melchinger *et al.*, 2011).

CIMMYT's research on DH technology, at present, is focused towards development of tropicalized haploid inducer lines (in both yellow and white kernel backgrounds) to meet the needs of the CIMMYT and NARS breeding programs in Africa, Latin America and Asia, besides optimizing agronomic management protocols for efficient, reliable and high throughput DH line development.

High Throughput and Precision Phenotyping

High throughput and precision phenotyping protocols are now being developed and applied for an array of traits in crops like maize; this is extremely important in accelerated breeding for stress tolerance and improved nutritional quality. High throughput phenotyping platforms (HTPP) could be particularly useful for obtaining detailed measurements of plant characteristics that collectively provide reliable estimates of trait phenotypes. These platforms are also useful in modelling (especially taking into account 'hidden variables') for predicting genotypic performance in different climate scenarios (under controlled experimental conditions). In recent years, there has been an increasing interest in establishing HTPPs not only by the major private sector institutions, which have been pioneering such endeavour, but also by some of the public research institutions worldwide. This is the case, for example, of the "The Australian Plant Phenomics Facility" which includes, up to now, the "High Resolution Plant Phenomics Centre", placed in Canberra, and "The Plant Accelerator" at the University of Adelaide (http://www.plantphenomics.org.au) (Finkel, 2009). There are also companies that are actively engaged in developing such facilities (*e.g.* LemnaTec, http://www.lemnatec.com) at both hard- and software levels for plant phenomics and high-throughput phenotyping (Prasanna *et al.*, 2013).

There is also an increasing interest to develop relatively low-cost, field phenotyping platforms to overcome the limitation of greenhouse or growth chamber phenotyping like that of the High Resolution Plant Phenomics, Centre (http://www.plantphenomics.org/HRPPC). Such field platforms include placement of the remote sensing cameras in aerial platforms such as balloons, zeppelins or remote-controlled airplanes or "polycopters" (*e.g.* http://www.mikrokopter.de/ucwiki/en/MikroKopter) or using light curtains and spectral reflectance sensors mounted on a tractor for evaluating crop performance under field conditions (Montes *et al.*, 2011).

In addition to above, improving throughput and precision of field-based phenotyping, using low cost easy-to-handle tools, is an area of research that is being intensively pursued, since many of the national agricultural research systems (NARS) and small- and medium-sized seed companies from developing countries cannot afford to establish and maintain expensive high throughout phenotyping platforms. CIMMYT Global Maize Program is making intensive efforts for characterizing field variability at the key phenotyping sites worldwide, and for improving field-based phenotyping. This includes approaches like non-destructive estimation of biomass using NDVI (Normalized Differential Vegetation Index), monitoring soil moisture using neutron probes/TDR, chlorophyll content using SPAD meter, canopy behavior using infra-red thermography, etc (Prasanna *et al.*, 2012).

MARKER-ASSISTED BREEDING FOR MAIZE IMPROVEMENT

Significant progress has been made worldwide in optimizing molecular marker-assisted selection (MAS) for improvement of both qualitatively and quantitatively inherited traits using maize as a model system. An important fact that emerges is that there is no single "recipe" for application of MAS for improving diverse and important traits in maize. The genetic architecture of the trait should decide the way in which the MAS-based interventions are to be applied in breeding programmes.

MAS for Simply Inherited Traits

Simply inherited traits are those that are largely controlled by a few major genes, with high heritability. One of the successful examples of

MAS for maize improvement, and of particular use to the developing world, is the utilization of *opaque2*-specific SSR markers in conversion of maize lines into Quality Protein Maize (QPM) lines with enhanced nutritional quality (Babu *et al.*, 2005; Gupta *et al.*, 2009; Prasanna *et al.*, 2010). A MAS-derived QPM hybrid, 'Vivek QPM Hybrid 9' has been released by the Vivekananda Parvatiya Krishi Anusandhan Sansthan (Vivekananda Research Institute for Hill Area Agriculture; VPKAS) at Almora, India. This QPM hybrid was developed through marker-assisted transfer of the *o2* gene and phenotypic selection for endosperm modifiers in the parental lines (CM145 and CM212) of Vivek Hybrid 9. This strategy was used to develop QPM versions of several elite, early maturing inbred lines adapted to the hill regions of India (Gupta *et al.*, 2009), as well as QPM versions of elite inbred lines, which are the parents of three single-cross hybrids, at the Indian Agricultural Research Institute (IARI) at New Delhi (Prasanna *et al.*, 2010).

Another potential application of MAS in maize could be for improving the provitamin A content of grain. Quantifying the provitamin A carotenoid content of maize samples is difficult, time-consuming and expensive, and breeding programmes will therefore benefit greatly from use of MAS to reduce the need for phenotypic assays. Collaborative research work under Harvest Plus led to the detection of important allelic variation and development of useful markers for favourable alleles of two critical genes in the carotenoid biosynthetic pathway, namely *LycE* (*Lycopene epsilon cyclase*) and *CrtRB1* (*Carotene beta-hydroxylase 1*) (Harjes *et al.*, 2008; Yan *et al.*, 2010). Significant success has also been achieved by CIMMYT under Maize Harvest Plus for molecular marker-assisted development of provitamin A-enriched maize, that included phenotypic selection for deep orange ears coupled with MAS for favourable alleles in the two critical genes (*LycE* and *CrtRB1*) using seed DNA-based genotyping (Prasanna *et al.*, 2010; Babu and Prasanna, 2013).

MAS for Polygenic Traits

Several of the disease resistance traits in maize follow polygenic inheritance, governed by a few to many genes/QTL in each case, with low G x E, and reasonably high heritability. For such traits, the best possible strategy could be detection and validation of marker-trait associations, fine mapping for identifying breeder-friendly molecular markers (preferably SNPs), and pyramiding of the favourable alleles using such markers in the desired genetic backgrounds. CIMMYT is presently following this strategy for some of the important maize diseases, especially Maize Streak Virus (MSV), Gray Leaf Spot (GLS), and Turcicum leaf blight, as these diseases have significant impact on maize production and food security in several developing countries.

The genome-wide selection or genomic selection (GS) is a potential strategy for enhancing genetic gains in breeding for complex traits (Heffner *et al.*, 2009). This approach could help in effectively avoiding issues pertaining to the number of QTL controlling a trait, the distribution of effects of QTL alleles, and epistatic effects due to genetic background (Bernardo and Yu, 2007). GS also relies on MAS and is under evaluation for the feasibility of incorporating desirable alleles at many loci that have small genetic effect when used individually. In this approach, breeding values can be predicted for individual lines in a test population based on phenotyping and whole-genome marker screens. These values can then be applied to progeny in a breeding population based on marker data only, without the need for phenotypic evaluation. Modeling studies indicate that this method can lead to considerable increases in the rates of genetic gain by accelerating the breeding cycles (Heffner *et al.*, 2009).

New breeding and selection strategies like GS rely on the availability of cheap, robust and reliable marker systems. Pilot projects on the implementation of rapid-cycling GS using much higher marker densities are being initiated by CIMMYT on new platforms based on next generation sequencing technologies, with the ultimate aim of its routine application across the CIMMYT and NARS maize breeding programs in Sub-Saharan Africa, Latin America and Asia.

TRANSGENIC TECHNOLOGY AND PUBLIC-PRIVATE PARTNERSHIPS

Among the transgenic or genetically modified (GM) crops that are being grown worldwide, maize has an important place along with soybean, cotton and canola. GM crops occupied an area of 160 million hectares in 2011. Of the 29 Biotech/GM crop-growing countries, 16 have GM maize (James, 2012). Several transgenic maize products have been developed and released in USA, mostly by the multinational concerns. Also, an array of transgenic maize products are in pipeline for release in many developing countries, including the *Bt* maize, herbicide resistant maize, maize with resistance to Maize Streak Virus (MSV), and those with stacked traits, mainly herbicide resistance + Insect-resistance (*Bt*).

Despite the impressive progress, GM crops are not a "magic bullet" for providing solutions to all the problems of agriculture. However, the developing world should have access to all cutting-edge technologies, including GM, to improve agricultural production. Research institutions need to judiciously utilize the GM technology in breeding improved maize genotypes for tackling some intractable problems, especially biotic and abiotic stress tolerance.

Public-private partnerships are also important for effective deployment of improved cultivars in the developing world, and to leverage cutting-edge technologies for development of novel products for the benefit of farmers and consumers. Considering the high cost of molecular research platforms, the need for extensive and precise phenotyping, the increasing complexity of bioinformatics tools to manage and interpret data, and the ever-growing intellectual property rights restrictions to germplasm exchange, such partnerships offer a synergistic way for accessing cutting-edge technologies for maize improvement as well as effective sharing of scientific and infrastructure capacities.

REFERENCES

Babu, R., Nair, S.K. and Kumar, A. (2005). Two-generation marker-aided backcrossing for rapid conversion of normal maize lines to quality protein maize (QPM). *Theoretical and Applied Genetics*, 111: 888–897.

Babu, R. and Prasanna, B.M. (2013). Molecular marker-assisted breeding for high quality protein maize (QPM): Status and prospects. *In*: Advances in Genomics of Plant Genetic Resources (*Eds*. Tuberosa, R. and Varshney, R.K.). Springer-Verlag (In Press).

Bernardo, R. and Yu, J. (2007). Prospects for genome-wide selection for quantitative traits in maize. *Crop Science*, 47: 1082–1090.

Bouchez, A. and Gallais, A. (2000). Efficiency of the use of doubled haploids in recurrent selection for combining ability. *Crop Science*, 40: 23–29.

Cairns, J.E., Sonder, K., Zaidi, P.H., Verhulst, N., Mahuku, G., Babu, R., Nair, S.K., Das, B., Govaerts, B., Vinayan, M.T., Rashid, Z., Noor, J.J., Devi, P., San Vicente, F. and Prasanna, B.M. (2012). Maize production in a changing climate: Impacts, adaptation and mitigation strategies. *Advances in Agronomy,* 114: 1–58.

DAC, (2010). Agricultural statistics at a glance – 2010. Department of Agriculture and Cooperation, Government of India, New Delhi.

FAOSTAT, (2010). Statistical databases and data-sets of the Food and Agriculture Organization of the United Nations. http://faostat.fao.org/default.aspx (accessed April 2010).

Finkel, E. (2009). With 'phenomics,' plant scientists hope to shift breeding into overdrive. *Science,* 325: 381–381.

Gupta, H.S., Agrawal, P.K. and Mahajan, V. (2009). Quality protein maize for nutritional security: Rapid development of short duration hybrids through molecular marker assisted breeding. *Current Science*, 96: 230–237.

Harjes, C.E., Rocheford, T.R. and Bai, L. (2008). Natural genetic variation in *Lycopene epsilon* cyclase tapped for maize biofortification. *Science,* 319: 330–333.

Heffner, E.L., Sorrells, M.E. and Jannink, J.L. (2009). Genomic selection for crop improvement. *Crop Science*, 49: 1–12.

Hellin, J. and Erenstein, O. (2009). Maize-poultry value chains in India: Implications for research and development. *Journal of New Seeds*, 10: 245–263.

Hyman, G., Fujisaka, S., Jones, P., Wood, S., de Vicente, M.C. and Dixon, J. (2008). Strategic approaches to targeting technology generation: Assessing the coincidence of poverty and drought-prone crop production. *Agricultural Systems*, 98: 50–61.

James, C. (2012) Global status of commercialized biotech/GM crops: 2011. Brief 43. ISAAA (http://www.isaaa.org).

Kebede, A.Z., Dhillon, B.S., Schipprack Wolfgang, Araus, J.L., Banziger, M., Semagn Kassa, Alvarado, G. and Melchinger, A.E. (2011). Effect of source germplasm and season on the *in vivo* haploid induction rate in tropical maize. *Euphytica,* 180: 219–226.

Melchinger, A.E., Technow, F. and Dhillon, B.S. (2011). Gene stacking strategies with doubled haploids derived from biparental crosses: Theory and simulations assuming a finite number of loci. *Theoretical and Applied Genetics,* 123: 1269–1279.

Metzker, M.L. (2010). Sequencing technologies — the next generation. *Nature Reviews Genetics*, 11: 31–46.

Montes, J.M., Technow, F., Dhillon, B.S., Mauch, F. and Melchinger, A.E. (2011). High throughput non-destructive biomass determination during early plant development in maize under field conditions. *Field Crops Research*, 121: 268–273.

Naylor, R., Steinfeld, H. and Falcon, W. (2005). Losing the links between livestock and land. *Science,* 310: 1621–1622.

Prasanna, B.M. (2011). Maize in Asia – trends, challenges and opportunities. *In*: Addressing Climate Change Effects and Meeting Maize Demand for Asia. Book of Extended Summaries of the 11th Asian Maze Conference (*Eds*. Zaidi, P.H. *et al.*), Nanning, China, 7-11 November 2011. CIMMYT, Mexico, D.F., pp. 3–6.

Prasanna, B.M., Araus, J.L., Crossa, J., Cairns, J.E., Palacios, N., Das, B. and Magorokosho, C. (2013). High-throughput and precision phenotyping for cereal breeding programs. *In*: Cereal Genomics-II (*Eds*. Gupta, P.K. and Varshney, R.K.). Springer-Verlag.

Prasanna, B.M., Pixley, K.V., Warburton, M. and Xie, C. (2010). Molecular marker-assisted breeding for maize improvement in Asia. *Molecular Breeding*, 26: 339–356.

Prasanna, B.M., Chaikam, V. and Mahuku, G. (2012). Doubled haploid technology in maize breeding: theory and practice. Mexico D.F.: CIMMYT, p. 50.

Prigge, V. and Melchinger, A.E. (2012). Production of haploids and doubled haploids in maize. *In*: Plant Cell Culture Protocols (*Eds*. Víctor, M.L. and Neftalí, O.), pp. 161–172, Springer Link.

Prigge, V., Sánchez, C., Dhillon, B.S., Schipprack, W., Araus, J.L., Bänziger, M. and Melchinger, A.E. (2011). Doubled haploids in tropical maize: I. Effects of inducers and source germplasm on *in vivo* haploid induction rates. *Crop Science*, 51: 1498–1506.

Rosegrant, M.W., Msangi, S., Ringler, C., Sulser, T.B., Zhu, T. and Cline, S.A. (2008). International model for policy analysis of agricultural commodities and trade (IMPACT): Model Description. International Food Policy Research Institute: Washington, D.C. http://www.ifpri.org/themes/impact/impactwater.pdf (accessed March 10, 2010).

Schnable, P.S., Ware, D. and Fulton, R.S. (2009). The B73 genome: Complexity, diversity, and dynamics. *Science,* 326: 1112–1115.

Schon, C.C., Dhillon, B.S., Utz, H.F. and Melchinger, A.E. (2010). High congruency for QTL positions for heterosis of grain yield in three crosses of maize. *Theoretical and Applied Genetics*, 120: 321–332.

Schrag, T.A., Mohring, J., Melchinger, A.E., Kusterer, B., Dhillon, B.S., Piepho, H. and Frisch, M. (2010). Prediction of hybrid performance in maize using molecular markers and joint analyses of hybrids and parental lines. *Theoretical and Applied Genetics*, 120: 451–461.

Shendure, J. and Ji, H. (2008). Next-generation DNA sequencing. *Nature Biotechnol.,* 26: 1135–1145.

Shiferaw, B., Prasanna, B., Hellin, J. and Banziger, M. (2011). Crops that feed the world 6. Past successes and future challenges to the role played by maize in global food security. *Food Security*, 3: 307–327.

Tang, J., Yan, J., Ma, X., Teng, W., Wu, W., Dai, j., Dhillon, B.S., Melchinger, A.E. and Li, J. (2010). Dissection of genetic basis of heterosis in an elite maize hybrid by QTL mapping in an immortalized F_2 population. *Theoretical and Applied Genetics*, 120: 333–340.

Tian, F., Bradbury, P.J., Brown, P.J., Hung, H., Sun, Q., Flint-Garcia, S., Rocheford, T.R., McMullen, M.D., Holland, J.B. and Buckler, E.D. (2011). Genome-wide association study of leaf architecture in the maize nested association mapping population. *Nature Genetics*, 43: 159–162.

Vielle-Calzada, J.P., de la Vega, O.M. and Hernández-Guzmán, G. (2010). The Palomero genome suggests metal effects on maize domestication. *Science,* 326: 1078.

Vroh Bi, I., McMullen, M.D. and Villeda, H.S. (2006). Single nucleotide polymorphisms and insertion-deletions for genetic markers and anchoring the maize fingerprint contig physical map. *Crop Science*, 46: 12–21.

Wagenast, T., Utz, F.H., Friedrich, C., Login, H., Maurer, H.P., Dhillon, B.S. and Melchinger, A.E. (2010). Hybrid maize breeding with doubled haploids: V. Selection strategies for test cross performance with variable sizes of crosses and S1 families. *Theoretical and Applied Genetics*, 120: 699–708.

Xu, Y., Lu, Y., Gao, S., Wan, J. and Prasanna, B.M. (2012). Whole genome strategies for molecular marker-assisted plant breeding. *Molecular Breeding*, 29: 833–854.

Yan, J., Kandianis, C.B. and Harjes, C.E. (2010). Rare genetic variation at *Zea mays crtRB1* increases β-carotene in maize grain. *Nature Genetics*, 42: 322–327.

Yu, J., Holland, J.B., McMullen, M.D. and Buckler, E.S. (2008). Genetic design and statistical power of nested association mapping in maize. *Genetics,* 178: 539–551.

Zhu, C., Gore, M., Buckler, E.S. and Yu, J. (2008). Status and prospects of association mapping in plants. *The Plant Genome*, 1: 5–20.

5

Sorghum

SUJAY RAKSHIT AND J.V. PATIL

Directorate of Sorghum Research, Hyderabad, India

PRODUCTION TREND OF SORGHUM

Globally sorghum is cultivated in 40.51 m ha with maximum area in India (7.67 m ha) followed by Sudan (5.61 m ha), Nigeria (4.74 m ha), Niger (3.32 m ha) and other countries in the semi-arid tropics (SAT) (Rakshit *et al.*, 2012a). India ranks second to USA in terms of sorghum production. However, the productivity of sorghum in India and other SAT is much lower than that in USA. In India sorghum is losing area to other competitive crops like soybean, cotton and maize; it is now only 42.5% of what it was in 1960s (Fig. 1). However, production of sorghum has reduced only by little over 25% during this period. This has been due to increase in the productivity by over 80% of the average of 1960s (511 kg/ha) to a current five yearly average of 923 kg/ha. The increased productivity is more prominent in *kharif* season due to release of 54 cultivars at the national level and over 60 cultivars at state levels over last fifty years. Year-wise releases of the cultivars are given in Table 1. Out of these 54 released cultivars, 34 are of *kharif* adaptation (17 hybrids and 17 varieties), while 14 are *rabi* cultivars (8 varieties and 6 hybrids) besides two being forage hybrids, one forage variety, two sweet sorghum varieties and one sweet sorghum hybrid.

V.L. Chopra (Ed.): Breeding Field Crops II: Advances, Studium Press (LLC), USA.

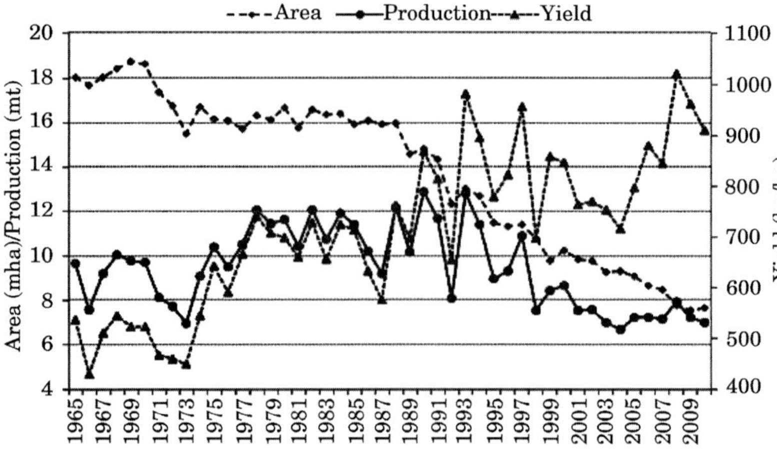

Fig. 1: Area, production and productivity trend of sorghum in India

Table 1: Hybrids and varieties released through AICSIP at National Level

Sl. no.	*Hybrid*	*Year of release*	*Pedigree of the hybrid*	*Grain yield (kg/ha)*	*Fodder yield (kg/ha)*
Kharif hybrids					
1	CSH 1	1964	CK 60A × IS 84	3000-3500	7500
2	CSH 2	1965	CK 60A × IS 3691	3000-3500	
3	CSH 3	1970	2219A × IS 3691	3500-3800	
4	CSH 4	1973	1036A × Swarna	3500-3800	
5	CSH 5	1975	2077A × CS 3541	3800-4000	9300
6	CSH 6	1977	2219A × CS 3541	3376	8100
7	CSH 9	1983	296A × CS 3541	4000-4200	9800
8	CSH 10	1984	296A × SB 1085	3633	12000
9	CSH 11	1986	296A × MR 750	4172	9180
10	CSH 14	1992	AKMS 14A × AKR 150	3840	8800
11	CSH 13 K&R	1995	296A × RS 29	3924	14400
12	CSH 16	1997	27A × C43	4308	9676
13	CSH 17	1998	AKMS 14A × RS 673	4186	10400
14	CSH 18	1999	IM 9A × Indore 12	4336	13100
15	CSH 21	2005	ML SA 848 × MLR 34		
16	CSH 23	2005	MS 7A × RS 627	4100	8700
17	CSH 25	2008	PMS 28 A × C 43	4370	12710
Kharif varieties					
1	CSV 1	1968	Sel. from IS 3924	3000-3500	
2	CSV 2	1974	IS 3922 × Karad local	3000-3500	
3	CSV 3	1974	IS 2954 × BP 53	3500-4000	
4	CSV 4	1974	IS3675 × IS 3541	3000-3500	
5	CSV 5	1974	IS 3687 × Aispuri	3000-3500	
6	CSV 6	1974	IS 3922 × Aispuri	3200-3500	

Table 1: (*Contd....*)

Table 1: (*Contd....*)

Sl. no.	*Hybrid*	*Year of release*	*Pedigree of the hybrid*	*Grain yield (kg/ha)*	*Fodder yield (kg/ha)*
7	CSV 7	1982	CS 3541 (Tall mutant)		
8	CSV 9	1983	CS 3541 × Tall Mutant	3000-3500	8940
9	CSV 10	1983	SB 1066 × CS 3541	3000-3500	9010
10	CSV 11	1985	(SC 108-3 × CS 3541)-	3250	9600
11	CSV 12 (Co 26)	1985	11-1 (IS 2947 × SPV232) × 1022	2500-4000	3000-6000
12	CSV 13	1988	(IS 12622 × 555) × (IS 3612 × E 35-1-52)	3525	9700
13	CSV 15	1996	SPV 475 × SPV462	3621	12100
14	CSV 17	2002	SPV 946 × SPV 772	2500	6800
15	CSV 20	2006	SPV 946 × Kh 89-246	3100	13300
16	CSV 22	2007	SPV 1359 × RSP 2	2300	7100
17	CSV 23	2007	SPV 861 × SU 248	2500-3000	15500
Rabi varieties					
1	M35-1	1969	Landrace selection from local maldandi bulk	2080	6110
2	CSV 7R	1974	IS 2950 × M 35-1	2000-2500	
3	CSV 8R	1979	R 24 × R 16	2500-3000	
4	Swati	1984	SPV 86 × M 35-1	2157	5277
5	CSV 14R	1992	(M 35-1 × (CS 2947 × CS 2644) × M 35-1	2270	5537
6	Sel.3	1995	Sel. from Bidar *Rabi* local	500-600	1500-1800
7	CSV 216R	2000	Land race Sel. from *rabi* germplasm Dhulia	2000-2500	7000-8000
8	CSV 18	2005	A selection from cross (CR 4 × IS 18370)	3500-3800	8700-9000
Rabi hybrids					
1	CSH 7R	1977	36A × 168	2500-3000	
2	CSH 8R	1977	36A × PD3-1-11	2500-3000	3675
3	CSH 12R	1986	296A × M 148-138	2564	4665
4	CSH 13R	1991	296A × RS 29	3924	14400
5	CSH 15R	1995	104A × RS 585	3194	5570
6	CSH 19R	2000	104A × R354	2953	5838
Forage hybrids and varieties					
1	CSH 20 MF	2005	2219 A × UPMC 503	87219	24278
2	CSV 21 F	2006	GSSV 148 × SR 897	37840	11380
3	CSH 24 MF	2008	ICSA 467 × Pant chari	91394	23165
Sweet sorghum hybrids and varieties					
1	CSV 19SS	2005	RSSV 2 × SPV 462	800-1000	40000-45000
2	CSH 22 SS	2005	ICSA 38 × SSV 84	1750	42000-45000
3	CSV 24SS	2011	NSS 1005A × (SSV 84 × 401B)	1270	39100

Sorghum Research Domains (SRDs)

As in India, sorghum is cultivated the world over under different ecologies, growing conditions and for different end uses. Based on utilization pattern, sorghum may be grouped as grain sorghum, forage sorghum and sweet sorghum, wherein grain sorghum may further be classified into *kharif* and *rabi* based on season of cultivation. Taking the needs of different sorghum types as well as growing conditions into consideration, Bantilan *et al.* (2004) have grouped sorghum research activities into eight sorghum research domains (SRD) representing homogenous eco-regions in terms of soil and climatic conditions. These domains are: wide adaptability (SRD I), dual purpose with specific adaptability (SRD II), dual purpose with fodder emphasis (SRD III), forage sorghum (SRD IV), early-sown post-rainy sorghum (SRD V), late-sown post-rainy sorghum (SRD VI), irrigated sorghum (SRD VII) and extreme altitude sorghum (SRD VIII). Out of these eight SRDs first six are of relevance to India with SRD IV and V exclusively specific in Indian sub-continent (Kumar *et al.*, 2011). SRD I to IV represent *khairf* season, while SRD V and VI encompass *rabi* season. Principal breeding procedures do not change drastically for respective research domains. However, research objectives and genetic resource employed change depending on research goal. Shoot fly (*Atherigona soccata* Rondani) remains a major constraint in first six SRDs, stem borer (*Chilo partellus* Swine) is important in SRD II – IV. Grain mold is important in SRD I and II, while leaf diseases in SRD IV. Drought is a major concern in SRD II and VI.

Sorghum Genetic Resources and their Utilization

Irrespective of the SRD, genetic resources play a very important role. Principal repository of global sorghum germplasm collection is International Crop Research Institute for Semi-Arid Tropics (ICRISAT), India with a collection of 36,774 accessions from 90 countries across the world (Reddy *et al.*, 2008). Towards best utilization of the variability in Gene Bank, establishment of core collection (~10% of entire collection), representing over 70% of the genetic variation present in the entire collection with 95% certainty is very essential (Brown, 1989). Towards this direction, Rao and Rao (1995) first attempted to establish a core collection of 3475 accessions. However, more systematic approach of establishing core collection in sorghum was made by Grenier *et al.* (2001) who employed three sampling procedures – constant, logarithmic, and proportional – to establish three subsets of the core collection, each possessing 2247 accessions. Still, the size of this collection was quite large for multi-location evaluation. To address this issue Upadhyaya and Ortiz (2001) suggested the mini-core collection approach, which is a core of a core collection (1% of the entire collection) representing the species diversity. Following this strategy, a mini-core collection with 242 accessions has been established at ICRISAT (Upadhyaya *et al.*, 2009). Of late, massive explosion in genomic resources has brought out association mapping as a potential tool for fine mapping of quantitative trait loci (QTL). In this regard, a diverse germplasm collection is of immense importance. Recently, a germplasm panel for association mapping comprising of 377 genotypes has also been reported by Casa *et al.* (2008). These collections of diverse germplasm lines are potential resources for breeding and genomics study. Nevertheless, their adaptability and utility in a given SRD needs to be ascertained before deploying them in breeding programme.

Sorghum Breeding in India

The progress in sorghum breeding particularly in last three decades is discussed here. Hybrid breeding has predominated the *kharif* breeding programme, while varieties rule *rabi* breeding initiatives. Grain mold has remained a major constraint during *kharif* season, while shoot fly is a challenge in both seasons, particularly under late sown conditions. Among abiotic stresses, post-flowering moisture stress is a major bottleneck, particularly for *rabi* sorghum as it is grown mostly under residual and receding soil moisture.

Grain sorghum breeding

In *kharif* grain sorghum breeding, major emphasis is on hybrid development. Hybrid sorghum breeding is based on cytoplasmic male sterility (CMS). Till now several sources of CMS, *viz.*, A_1, A_2, A_3, A_4, A_5, A_6, M 35-1, VZM etc. have been reported in sorghum. Fertility restoration of A_1 or milo-cytoplasm is most simple and complete among all CMS systems available. Thus, till now most of released sorghum hybrids across the world are based on A_1 or milo cytoplasm (Reddy and Stenhouse, 1994). Exclusive use of milo cytoplasm in hybrid sorghum has brought concern about vulnerability of sorghum hybrids as happened with southern leaf blight on hybrid maize carrying T-cytoplasm. Further it restricts use of more germplasm accessions as A-lines or R-lines. Hence, need was felt to diversify the seed parents as well as restorers. During the initial period of seed parent diversification, attention was given towards grain and fodder yield and quality alone along with geographic adaptation (such as maturity). Soon it was felt necessary to incorporate resistance to biotic and abiotic constraints into hybrid parents. Such efforts have led to identification of promising lines in advanced generations, which were subsequently converted to CMS background. However, in recent past major emphasis is on race-specific and alternative (non-milo) CMS system-specific diversification of A-lines. Towards this direction A_1 and A_2 cytoplasms are being given greater emphasis as A_2 CMS system is as efficient as A_1 with a slight edge over A_1 for commercial exploitation (Kumar *et al.*, 2008). Based on A_2 CMS system first sorghum hybrid, Zinza No. 2 has been released in China, which occupied one sixth of the total sorghum area in China (Liu *et al.*, 2000). Harnessing the potential of A_2 cytoplasm in India is yet to be achieved on a commercial scale. Towards further diversification of CMS, different races as well as donors for various traits are being used. Till now only *caudatum* and *durra* races have been extensively exploited. However, other races like *guinea* and *kafir* possess genes for yield and other traits. Such races may potentially be exploited to enhance heterosis levels in commercial hybrids (Umakanth, 2008).

Seed parent diversification needs to be complemented with diversification of restorers as well. Presently, R-line breeding is rested mainly on varietal development programme, wherein different germplasm donor lines are being utilized. One early success of restorer diversification programme is derivation of CS 3541 from temperate-tropical cross, IS 3541 (African Zerazera) × IS 3675, which was used as male parent of CSH 5, CSH 6 and CSH 9. In the process of developing restorers, in addition to yield potential, plant height is given equal importance. Early hybrids CSH 2 and CSH 3, though promising, yet could not become popular due to shorter stature of the male parent leading to seed production problem (Umakanth, 2008).

Heterosis in *rabi* sorghum has not been exploited on commercial scale mainly due to the narrow genetic base of the *rabi* genotypes (Sajjanar *et al.*, 2011). Thus, *rabi* grain sorghum breeding mainly focuses on varietal development. The main challenge in *rabi* breeding programme is to replace M 35-1, a selection from Maldandi, which is dominating the *rabi* tract over last 75 years. Rakshit *et al.* (2012b) have reported wide variability at morphological and molecular level among the Maldandi accessions in the Gene Bank. This has opened up scope for using promising selections from Maldandi accessions in *rabi* breeding programme. *Rabi* sorghum is cultivated in different soil depths with receding soil moisture. In this regard, plant ideotype plays a very important role (Donald, 1968). Reddy *et al.* (2012) have proposed a crop ideotype suitable particularly for medium soil depth as follows: higher grain yield and per day grain productivity (at least double that of M 35-1); higher biomass and per day fodder productivity (50% higher biomass than M 35-1), harvest index greater than 30; greater earhead exertion (5-10% more than M 35-1), bolder seeds (5% bolder than M 35-1), panicle dry matter constituting 50% of total dry matter and higher relative water content (5% higher than M 35-1). Towards widening of *rabi* gene pool, use of *rabi* × *kharif* crosses have been attempted with limited success as segregants from such crosses have adverse grain quality and adaptability problem. *Rabi* landraces possess excellent grain quality and adaptive traits to mitigate terminal drought, shoot fly attack and charcoal rot. Efforts are being made to use these adapted landraces as restorers for *caudatum*-based A-lines derived from exotic germplasm. This strategy has shown good potential in the initial studies at ICRISAT (Umakanth, 2008).

Diversification of hybrid parents has led to identification of a series of CMS as well as restorer lines over years, which have contributed to release of eight *kharif* and three *rabi* hybrids since 1990 (Table 1). Among different CMS lines, 296A is found among the parentage of five hybrids (three *kharif* and two *rabi*). Till mid-nineties, after release of CSH 13 and CSH 14, though there was no significant grain yield improvement yet diversification for early maturity and fodder yield could be achieved (Umakanth, 2008). RS 29, the R-line for CSH 13 was developed from a cross between American elite line, SC 108 and SPV 126 (a tall mutant of CS 3541). The R-line of CSH 15R (RS 585) is a product of *rabi* × *kharif* introgression programme. *Kharif* hybrid, CSH 16 with grain mold tolerance genes from Ethiopian germplasm line IS 23549 was released in 1997 (Audilakshmi *et al.*, 2003). R line of this hybrid (C 43) has high grain yield potential and has been used as parent in the latest hybrid CSH 25 as well. Additional hybrids, CSH 17, CSH 18, CSH 19R, CSH 21 and CSH 23 have been released in the intervening period. However, none of these gained popularity except CSH 23, which matures in 103 days and is tolerant to terminal drought. The male parent of the hybrid, RS627 has been derived from *rabi* × *kharif* introgression programme. Among all the released hybrids, CSH 16 is most widely cultivated. Over last twenty years, five *kharif* and four *rabi* varieties have been released at national level. These are predominantly produced through direct selection from elite land races (CSV 216R) or through pedigree selection from crosses between elite lines, land races and temperate germplasm (CSV 15, CSV 16, CSV 17, CSV 18, CSV 20, CSV 22 and CSV 23). Many of *rabi* varieties have M 35-1 as a progenitor in their pedigree. This is mainly because consumers prefer grains having lustre and boldness like Maldandi.

In recent past, bio-fortification (increasing the grain Fe and Zn status through genetic means) of sorghum is gaining importance to address hidden malnutrition in Sub-Saharan Africa and South Asia. Towards this objective, breeding target has been fixed at 70 ppm Fe and 40 ppm Zn in grain (Kumar *et al.*, 2011). Through screening of a total of 2267 core germplasm accession under HarvestPlus Challenge Programme at ICRISAT, wide variability for Fe and Zn content has been reported leading to identification of potential donors (Kumar *et al.*, 2009). These donors are being used in breeding programme for developing mapping population to identify QTL for use in bio-fortification programme.

Forage sorghum breeding

In forage sorghum breeding, biomass production and its quality play the most important role. Cell wall composition and structure, which is determined by the content and composition of lignin, cellulose and hemicellulose, determines biomass quality, whereas biomass yield is determined by agronomic traits like plant height and stem diameter. Intake potential and energy availability of forage crops is determined by cell-wall concentration and digestibility. Forage evaluation can be done biochemically (*in vitro*) or biologically (*in vivo*). Forage quality indicators to dairy farmers include crude protein (CP), neutral detergent fiber (NDF), acid detergent fiber (ADF), energy, total soluble sugar (TSS) and presence of anti-nutritional factors. CP is easily measurable and directly proportional to forage quality. Plant fiber consists of three components, *viz.*, cellulose, hemi-cellulose and lignin. Fiber contents are inversely proportional to its energy content. NDF measures all fiber found in forage, while ADF measures the cellulose and lignin contents. Both NDF and ADF are partially digestible and low ADF is desirable. Lignin is the key element that limits cell-wall digestibility and affects saccharification efficiency of forage (Oliver *et al.*, 2005). However, lower lignin reduces biomass yield and survival and also causes lodging (Pedersen *et al.*, 2005). Brown midrib (*bmr*) mutations, originally a naturally occurring mutant identified in corn and subsequently reported in sorghum and millet, are reported to alter lignin subunit composition and reduce lignin content. They are characterized by having brown pigmentation in the leaf midrib (distinct by 4 to 6 leaf stage), stem, tassel, cob and roots. Reduced lignin, thus, improves forage quality. There are examples of reduced lignin without obvious phenotypic changes. In addition to lignin content cellulose and hemicelluloses also play a very crucial role in determining forage quality (Wang *et al.*, 2008). The availability of cellulose and hemicellulose as a source of energy, however, depends on the overall structural properties of the cell wall. In this regard also lignin impedes access of hydrolytic enzymes to the cell wall polysaccharides, in turn quality of forage. TSS increases sweetness of forage and in turn its palatability, which increases milk yield of animals. *In vitro* dry matter digestibility (IVDMD) is a laboratory technique and most reliably used to study the forage quality. Among the anti-nutritional factors HCN or dhurin content is very critical in sorghum forage. HCN content above 500 ppm on fresh weight basis is fatal to animals and this tends to remain high during early stages of crop growth (up to 45-50 days) and also tends to increase under drought situation. Plants that are grown on soils high in nitrogen and low in phosphorus and potassium tend to have high levels of cyanide. Regeneration ability of cultivars is an additional trait for consideration, particularly while breeding for multi-cut sorghum.

Focused forage research has led to release of over 30 cultivars across different states of which two hybrids, CSH 20MF and CSH 24MF and one variety CSV 21F have been released in this millennium. Sorghum cultivars are dominated by varieties with high green foliage. Recently released varieties, *viz*., MP Chari, UP Chari 1, UP Chari 2, Rajasthan Chari 1, Rajasthan Chari 2, Pusa Chari 23, HC 260, HC 308, CSV 21MF have improvement in terms of resistance to leaf spot diseases and stem borer (Bhat, 2008). New dual purpose varieties, CSV 15, CSV 20 and CSV 23 have also immense prospect particularly in North West India. In spite of release of these cultivars, forage sorghum area of North India is dominated by SSG 59-3 (Meethi Sudan), which has potential of up to four cuts (Bhat, 2008). The recent variety, CSV 21F has been developed by pedigree selection from a cross between GSSV 148 and SR 897. Recent hybrid, CSH 24MF is a product of a cross between ICRISAT CMS line, ICSA 467 and a popular forage variety, Pant Chari. In forage hybrid seed production, the main limitation is the cost of hybrid seeds as the seed yield in forage sorghum is low. In this regard, use of three-way cross hybrid is suggested to be an alternative. Forage sorghum breeding mainly focuses on reducing lignin content by using *bmr* mutant lines. Towards improvement of forage sorghum, biotechnological tools have been suggested to be a viable solution. Recently, QTL for lignin and cell wall components in corn have been successfully identified (Murray *et al*., 2009; Shiringani and Friedt, 2011). Similarly, QTL for plant biomass traits also has been reported (Brown *et al*., 2008; Murray *et al*., 2008). This has opened up possibilities of marker-assisted breeding in forage sorghum. To reduce HCN content, anti-sense RNA technology using the genes from HCN pathway is being used. Initial study has suggested reduction in HCN content below toxic limit.

Sweet sorghum breeding

Sweet sorghum is a new generation bioenergy crop having a potential to accumulate large amounts of sugars (10-20%) in its stalks (Hunter and Anderson, 1997). Sweet sorghum stalk juice can be used successfully for production of syrup, fuel-grade ethanol, glycols, specialty and bulk organic chemicals, industrial alcohol etc., without compromising grain production. To produce ethanol from plant biomass, cellulose/ hemicelluloses in the cell wall need to be saccharified. *i.e.*, hydrolyzed into fermentable sugars (glucose and xylose). Lignin content of cell walls determines, among other factors, sugar release and thus the efficiency of the ethanol extraction process (Lorenz *et al.*, 2009). Thus, like forage breeding the main sweet sorghum breeding goal is to develop cultivars with high biomass yield as well as cellulose and hemicelluloses content with reduced lignin content. In the selection process, biomass and stem brix values (>18%) are given emphesis. In India, sweet sorghum breeding has received importance since 1980s. This has led to release of two sweet sorghum varieties, CSV 19SS, CSV 24SS and a hybrid CSH 22SS. CSV 19SS is a selection from a cross between a sweet sorghum donor, RSSV 2 and a normal grain cultivar, SPV 462. CSV 24SS is a cross between NSS1005 and a derivative of (SSV84 x 401 B). CSH 22SS has been developed by crossing an ICRISAT CMS line, ICSA 38 to a nationally developed sweet sorghum donor, SSV 84. Recent release, CSV 24SS, is a cross between NSS1005 and a derivative of SSV84 and promising grain sorghum parent, 401 B.

Resistance breeding

IS 18551 is a reliable source of resistance against shoot fly. However, it has many undesirable traits limiting its utility as a potential donor in regular breeding programme. IS 2312 is another donor which has been exploited for shootfly resistance breeding. Several sources of resistance in desirable agronomic background are now available (ICSV 702, ICSV 705, ICSV 708, PS 21318, PS 30715-1, PS 35805, IS 923, IS 1057, IS 1071, IS 1082, IS 1096, IS 2394, IS 4663, IS 5072, IS 18369, IS 4664, IS 5470 AND IS 5636). At ICRISAT, through trait-based pedigree breeding, a large number of shoot fly resistant parents for both *kharif* season (ICSA-/B-409 to 436) and *rabi* season (ICSA-/B-437-463) have been reported (Kumar *et al.*, 2011). Genetic analyses have demonstrated that shoot fly resistance in sorghum is complex, polygenically inherited with high influence of environment (Singh *et al.*, 2004; Aruna and Padmaja, 2009). Studies on shoot fly resistance mechanisms suggested that dead hearts, oviposition, leaf glossiness, trichomes on the abaxial surface of the leaf can be used as marker traits to select for resistance to shoot fly besides use of biochemical parameters like silica and sugar content (Dhillon *et al.*, 2005; Gomashe *et al.*, 2012). In recent past DNA markers have been deployed to identify QTLs associated with shoot fly resistance contributing traits. Major QTLs influencing leaf glossiness and trichome density have been reported on chromosome V and X (Satish *et al.*, 2009). Aruna *et al.* (2011) identified as many as 25 QTLs influencing different shoot fly resistance attributing traits, many of which were in common with as reported by Satish *et al.* (2009). Currently, these QTLs are being deployed into promising parental lines for shoot fly tolerance through MAS.

M 35-1 (IS 1054) and BP 53 (IS 1055) show reasonable amount of resistance against stem borer. Resistance against stem borer is polygenically inherited (Prem Kishore, 1986a). Similar is the case with midge resistance. The quantitative nature of inheritance of resistance to these pest makes the breeding task difficult. Both pedigree and population breeding methods have been used to incorporate resistance into good agronomic backgrounds (Rana *et al.*, 1985; Prem Kishore, 1986b). Some moderately stem borer resistant lines derived through breeding approaches are E 302, CSV 8R, SPV 103 and SPV 104, while those against midge are PM 6751, PM 7061, PM 7318-2, PM 7322, PM 7493 etc. At ICRISAT, a new sorghum variety, SPV 1015 with resistance to shoot fly and stem borer has been developed (Prem Kishore, 1992). Though some progress has been made towards insect resistance breeding, concerted efforts are needed to incorporate multiple insect resistances among the breeding population.

Screening 7132 germplasm lines from world collection, Bandyopadhyay *et al.* (1988) identified 156 genetically diverse colored-grain lines with high level of resistance to grain mold. 14 out of these 156 lines were without testa layer and were low in tannin content. This indicates that resistance in colored-grain sorghum is not always associated with high level of tannin in testa and there are possibilities to get high level of resistance in red-grain sorghum which lacks testa (Das *et al.*, 2012). However, in recent past, moderate to high level of mold resistance has been identified in elite white-grain sorghum lines as well (Audilakshmi *et al.*, 1999; Thakur *et al.*, 2003). The resistance against grain mold has been reported to be complex. Both major and minor genes with additive and epistatic effects have been reported to influence grain mold, which show significant genotype by environment interactions (GEI) (Rodriguez- Herrera *et al.*, 2000). Involvement of 2-3 dominant major genes in governing grain mold resistance in colored grain sorghum have been reported by Audilakshmi *et al.* (2000). However, polygenic inheritance with significant additive × additive gene interaction has been reported in white-grain (Audilakshmi *et al.*, 2005). Presence of different mechanisms of inheritance, involvement of many pathogenic fungi and high GEI make the genetics of mold resistance very complex (Audilakshmi *et al.*, 2011). QTL analysis of grain mold tolerance has led to identification of 5 QTL located on 5 different linkage groups (Klein *et al.*, 2001). High GEI influences the QTL analysis significantly. To reduce these environmental effects, Audilakshmi *et al.* (2011) suggested that resistance against pre-physiological maturity is genetic and breeding research should focus on this rather than post-maturity grain mold, in which many saprophytic fungi are involved.

Screening of germplasm for stalk rot tolerance has led to identification of some sources, which are associated with stay green traits. In recent past, much progress has been made towards identification of major QTLs for stay green traits. This and other genomics tools have opened up the scope for marker assisted breeding for drought tolerance. Cold tolerance is another abiotic stress, which has not received much attention. Under Indian situation, low temperature at anthesis causes much damage to *rabi* sorghum, particularly in seed production plots. Concerted attention towards this direction is very much needed.

Sorghum biotechnology

Over last two decades much progress has been made in the field of sorghum biotechnology both towards transgenic research and DNA marker studies including genomics. Compared to some other crop plants, sorghum is recalcitrant to tissue culture. However, in recent past the problem has been addressed and several successful transformation events have been reported mainly through particle bombardment (reviewed by Visarada, 2008). Stem borer resistant transgenic sorghum carrying *cry1Ac* has been reported from ICRISAT (Girijashankar *et al.*, 2005). Maheswari *et al.* (2010) developed drought tolerant sorghum lines using *pCAM mtlD*. *Agrobacterium* mediated transformation of M 35-1 with *Cry1Ab* has been reported by Balakrishna *et al.* (2008). Two of these events are under field evaluation at the Directorate of Sorghum Research. Though good progress has been made against stem borer through transgenic approach, no progress has been achieved against shoot fly.

In sorghum, various molecular techniques like restriction fragment length polymorphism (RFLP), random amplified polymorphic DNA

(RAPD), amplified fragment length polymorphism (AFLP), simple sequence repeat (SSR) and diversity array technology (DArT) markers are employed for wide range of applications including diversity studies (Anas and Yoshida, 2004; Mace *et al.*, 2008; Mutegi *et al.*, 2011; Rakshit *et al.*, 2012b), construction of genetic linkage maps (Bhattramakki *et al.*, 2000; Hausmann *et al.*, 2002), mapping of economically important genes (Satish *et al.*, 2009; Srinivas *et al.*, 2009; Aruna *et al.*, 2011; Mace *et al.*, 2011), population genetic studies (Casa *et al.*, 2008) and marker assisted selection (Knoll and Ejeta, 2008). Among the various DNA markers, SSR markers are most preferred. Ramu *et al.* (2009) exploited rice–sorghum synteny to develop large number of EST-SSRs in sorghum. Conducting an extensive study on diversity across diverse panel of sorghum accessions under Generation Challenge Programme a set of 48 SSR markers bas been identified as Sorghum Diversity analysis kit (http://sat.cirad.fr/sat/sorghum_SSR_kit/). Rakshit *et al.* (2012b) have demonstrated utility of SSR markers in combination with morphological markers in effectively studying diversity even among closely related accessions like Maldandi landraces. DNA markers have brought significant change in the way quantitative traits are being handled through QTL analysis. Large number of QTL influencing majority of traits have been identified in various studies (reviewed by Hash and Senthilvel, 2008; Mace and Jordan, 2010).

Genome sequencing of American variety, BT×623 by Paterson *et al.* (2009) has brought a paradigm shift in genomic studies, which has wide implications in sorghum breeding. With genome sequence in hand, Mace *et al.* (2009) have established sorghum consensus genetic map by integrating multiple component maps and DArT markers. Ramu *et al.* (2010) have aligned large number of publicly available DNA markers (5950 SSR markers, 125 CISP markers and 100 gene/ gene products) to the sequence-based sorghum physical map and the consensus map. They could map important genes and demonstrated the utility of the sequence-based markers in marker assisted backcross breeding. Mace and Jordan (2010) integrated 35 major effect genes into the consensus map using published information. Recently, Nelson *et al.* (2011) have reported 96000 potential SNPs by subjecting 8 diverse sorghum accessions to short-read genome sequencing and aligning the sequences with reference genome sequence. All these resources bring the potential to initiate association mapping and other QTL analysis in sorghum. Mannai *et al.* (2011) through association mapping identified 11 QTL controlling flowering time and photoperiod sensitivity using 107 accessions of sorghum grown under natural condition and 45 accessions grown under controlled conditions. Wang *et al.* (2011), identified several new SSR markers associated with height using pool-based genome-wide association mapping in sorghum. These studies have wide implication not only on sorghum breeding but on other sister crops as well.

Future Thrusts

Under changing climatic scenario and decreased water regime, sorghum assumes more importance. Loss in area needs to be compensated with increased productivity. Genetic and cytoplasmic diversification of seed parents should receive top priority followed by incorporation of stress tolerance among restorers. Efforts are needed to breed for dual purpose sorghum to meet the need of both grain and fodder. New research effort is needed to focus on bio-ethanol production from lingo-cellulose rather than use of stem juice alone. In spite of long effort, success towards shoot fly and grain mold tolerance is not encouraging. New genomics tools are available in sorghum, which need to be combined with conventional breeding tools in form of genomics-assisted breeding to address the challenges effectively. Value addition is receiving much attention towards diversification of the uses of sorghum. Towards this direction new breeding initiative is needed to breed cultivars suitable for specific end uses.

REFERENCES

Anas and Yoshida, T. (2004). Sorghum diversity evaluated by simple sequence repeat (SSR) markers and phenotypic performance. *Plant Prod. Sci.,* 7: 301–308.

Aruna, C. and Padmaja, P.G. (2009). Evaluation of genetic potential of shoot fly resistant sources in sorghum (*Sorghum bicolor* (L.)). *Journal of Agricultural Sciences*, 147: 71–80.

Aruna, C., Bhagwat, V.R., Madhusudhana, R., Sharma, V., Hussain, T., Ghorade, R.B., Khandalkar, H.G., Audilakshmi, S. and Seetharama, N. (2011). Identification and validation of genomic regions that affect shoot fly resistance in sorghum [*Sorghum bicolor* (L.) Moench]. *Theor. Appl. Genet.,* 122: 1617–1630.

Audilakshmi S., Aruna, C. and Kiran Kumar V.S.S. (2003). Utilization of germplasm for improvement of varieties in India. *In*: XXXIII Annual Sorghum Group Meeting, 1–3 May 2003, Surat.

Audilakshmi, S., Aruna, C., Garud, T.B., Nayakar, N.Y., Atale, S.B., Veerabadhiran, P., Rao, B.D., Ratnavati, C.V. and Idira, S. (2005). A technique to enhance the quality and market value of rainy season sorghum grain. *Crop Protection*, 24: 251–258.

Audilakshmi, S., Das, I.K., Ghorade, R.B., Mane, P.N., Kamatar, M.Y., Nrayana, Y.D. and Seetharama, N. (2011). Genetic improvement of sorghum for grain mold resistance. I. Performance of sorghum recombinant inbred lines for grain mold reaction across environments. *Crop Protection*, 30: 753–758.

Audilakshmi, S., Stenhouse, J.W. and Reddy, T.P. (2000). Genetic analysis of grain mold resistance in coloured sorghum genotypes. *Euphytica*, 116: 95–103.

Audilakshmi, S., Stenhouse, J.W., Reddy, T.P. and Prasad, M.V.R. (1999). Grain mold resistance and associated characters of sorghum genotypes. *Euphytica*, 107: 91–103.

Balakrishna D., Padmaja, P.G. and Seetharama, N. (2008). Sorghum *Bt* transgenic plants for stemborer resistance. *Jowar Samachar*, 5: 11–12.

Bandyopadhyay, R., Mughogho and Prasad Rao, K.E. (1988). Sources of resistance to sorghum grain molds. *Plant Disease*, 72: 504–508.

Bantilan, M.C.S., Deb, U.K., Gowda, C.L.L., Reddy, B.V.S., Obilana, A.B. and Evenson, R E. (2004). Sorghum genetic enhancement research process, dissemination and impacts. International Crops Research Institute for Semi–Arid Tropics, Patancheru, India. p. 320.

Bhat, B.V. (2008). Genetic improvement of forage sorghum. *In*: Reddy, B.V.S., Ramesh, S., Kumar, A.A. and Gowda, C.L.L. (*Eds*.) Sorghum Improvement in the New Millennium, International Crops Research Institute for Semi–Arid Tropics, Patancheru, pp. 124–131.

Bhattramakki, D., Dong, J., Chhabra, A.K. and Hart, G. (2000). An integrated SSR and RFLP linkage map of *Sorghum bicolor* (L.). *Moench. Genome*, 43: 988–1002.

Brown, A.H.D. (1989). Core collection: A practical approach to genetic resources management. *Genome*, 31: 818–824.

Brown, P.J., Rooney, W.L., Franks, C. and Kresovich, S. (2008). Efficient mapping of plant height quantitative trait loci in a sorghum association population with introgressed dwarfing genes. *Genetics*, 180: 629–636.

Casa, A.M., Pressoir, G., Brown, P.J., Mitchell, S.E., Rooney, W.L., Tuinstra, M.R., Franks, C.D. and Kresovich, S. (2008). Community resources and strategies for association mapping in sorghum. *Crop Science*, 48: 30–40.

Das, I.K., Audilakshmi, A. and Patil, J.V. (2012). Fusarium grain mold: The major component of grain mold disease complex in sorghum (*Sorghum bicolour* L. Moench). *The European Journal of Plant Science and Biotechnology*, p. 6.

Dhillon, M.K., Sharma, H.C., Reddy, B.V.S., Singh, Ram, Naresh, J.S. and Kai, Zhu. (2005). Relative susceptibility of different male–sterile cytoplasms in sorghum to shoot fly, *Atherigona soccata*. *Euphytica*, 144: 275–283.

Donald, C.M. (1968). The breeding of crop ideotypes. *Euphytica*, 17: 385–403.

Girijashankar, V., Sharma, H.C., Sharma, K.K., Swathisree, V., Prasad, L.S., Bhat, B.V., Royer, M., San Secundo, B., Narasu, M.L., Altosaar, I. and Seetharama, N. (2005). Development of transgenic sorghum for insect resistance against the spotted stem borer (*Chilo partellus*). *Plant Cell Reports*, 24: 513–522.

Gomashe, S.S., Misal, M.B., Mehtre, S.P., Rakshit, S. and Ganapathy, K.N. (2012). Assessing parental lines and crosses for shoot fly resistance in sorghum [*Sorghum bicolor* (L.) Moench]. *Indian J. Genet.,* 72: 31–37

Grenier, C., Hamon, P. and Bramel–Cox, P.J. (2001). Core collection of sorghum: II. Comparison of three random sampling strategies. *Crop Science*, 41: 241–246.

Hausmann, B.I.G., Hess, D.E., Seetharama, N., Welz, H.G. and Geiger H.H. (2002). Construction of a combined linkage map from two recombinant inbred population using AFLP, SSR, RFLP and RAPD markers, and comparison with other sorghum maps. *Theor. Appl. Genet.,* 105: 629–637.

Hunter, E.L. and Anderson, I.C. (1997). Sweet sorghum. *Horticultural Reviews*, 21: 73–104.

Klein, R.R., Rodriguez-Herrera, R., Schlueter, J.A., Klein, P.E., Yu, Z.H. and Rooney, W.L. (2001) Identification of genomic regions that affect grain-mould incidence and other traits of agronomic importance in sorghum. *Theor. Appl. Genet.,* 102: 307–319.

Knoll, J.E. and Ejeta, G. 2008. Marker-assisted selection for early season cold tolerance in sorghum: QTL validation across populations and environments. *Theor. Appl. Genet.* 116: 541–553.

Kumar, A.A., Reddy, B.V.S., Sharma, H.C., Hash, C.T., Rao, P.S., Ramaiah, B. and Reddy, P.S. (2011). Recent advances in sorghum genetic enhancement research at ICRISAT. *American J. Plant Sciences*, 2: 589–600.

Kumar, A.A., Reddy, B.V.S. and Kaul, S.L. (2008). Alternative cytoplasmic male sterility systems in sorghum and their utilization *In*: Reddy, B.V.S., Ramesh, S., Kumar, A.A. and Gowda, C.L.L. *Eds.* Sorghum Improvement in the New Millennium, International Crops Research Institute for Semi–Arid Tropics, Patancheru, pp. 132–144.

Kumar, A.A., Reddy, B.V.S., Ramaiah, B., Reddy, P.S., Sahrawat, K.L. and Upadhyaya, H.D. (2009). Genetic variability and plant character association of grain Fe and Zn in selected core collections of sorghum germplasm and breeding lines. *E–Journal of SAT Agricultural Research*, 7: 1

Liu, Q.S., Ping, J.A., Li, T.Y. and Zhang, F.Y. (2000). New grain sorghum cytoplasmic male sterile line A_2V_4A and F_1 hybrid Jinza No, 12 for Northwest China. *Int. Sorghum and Millets Newsletter*, 41: 31–32.

Lorenz, A.J., Anex, R.P., Isci, A., Coors, J.G., de Leon, N. and Weimer, P. (2009). Forage quality and composition measurements as predictors of ethanol yield from maize (*Zea mays* L.) stover. *Biotechnol. Biofuels,* 2: 5.

Mace, E.S. and Jordan, D.R. (2010). Location of major effect genes in sorghum (*Sorghum bicolour* (L.) Moench). *Theor. Appl. Genet.,* 121: 1339–1356.

Mace, E.S., Rami, J.F., Bouchet, S., Klein, P.E., Klein, R.R., Kilian, A., Wenzl, P., Xia, L., Halloran, K. and Jordan, D.R. (2009). A consensus genetic map of sorghum that integrates multiple component maps and high–throughput Diversity Array Technology (DArT) markers. *BMC Plant Biol.,* 9: 13.

Mace, E.S., Xia, L., Jordan, D.R., Halloran, K., Parth, D.K., Huttner, E., Wenzl, P. and kilian, A. (2008). DArT markers: Diversity analyses and mapping in *Sorghum bicolor*. *BMC Genomics*, 9: 26.

Mace, E.S., Singh, V., Van Oosterom, E.J., Hammer, G.L., Hunt, C.H. and Jordan, D.R. (2011). QTL for nodal root angle in sorghum (*Sorghum bicolor* L. Moench) co–locate with QTL for traits associated with drought adaptation. *Theor. Appl. Genet.,* 124: 97–109.

Maheswari, M., Varalaxmi, Y., Vijayalakshmi, A., Yadav, S.K., Sharmila, P., Venkateswarlu, B., Vanaja, M. and Saradhi, P.P. (2010). Metabolic engineering using mtlD gene enhances tolerance to water deficit and salinity in sorghum. *Biologia Plantarum*, 54: 647–652.

Mannai, Y.E., Shehzad, T. and Okuno, K. (2011). Variation in flowering time in sorghum core collection and mapping of QTLs controlling flowering time by association analysis. *Genet. Res. Crop Evol.,* 58: 983–989.

Murray, S.C., Rooney, W.L., Hamblin, M.T., Mitchell, S.E. and Kresovich, S. (2009). Sweet sorghum genetic diversity and association mapping for brix and height. *The Plant Genome*, 2: 48–62.

Murray, S.C., Sharma, A., Rooney, W.L., Klein, P.E., Mullet, J.E., Mitchel, S.E. and Kresovich, S. (2008). Genetic improvement of sorghum as a biofuel feedstock II: Quantitative loci for stem and leaf structural carbohydrates. *Crop Science*, 48: 2180–2193.

Mutegi, E., Sagnard, F., Semagn, K., Deu, M., Muraya, M., Kanyenji, B., deVilliers, S., Kiambi, D., Herselman, L. and Labuschagne, M. (2011). Genetic structure and relationships within and between cultivated and wild sorghum (*Sorghum bicolor* (L.) Moench) in Kenya as revealed by microsatellite markers. *Theor. Appl. Genet.,* 122: 989–1004.

Nelson, J.C., Wang, S., Wu, Y., Li, X., Antony, G., White, F.F. and Yu, J. (2011). Single–nucleotide polymorphism discovery by high–throughput sequencing in sorghum. *BMC Genomics*, 12: 352.

Oliver, A.L., Klopfenstein, T.J., Grant, R.J. and Pedersen, J.F. (2005). Comparative effects of the sorghum bmr–6 and bmr–12 genes. I. Forage sorghum yield and quality. *Crop Science*, 45: 2234–2239.

Paterson, A.H., Bowers, J.E., Bruggmann, R., Dubchak, I., Grimwood, J., Gundlach, H., Haberer, G., Hellsten, U., Mitros, T., Poliakov, A., Schmutz, J., Spannagl, M., Tang, H., Wang, X., Wicker, T., Bharti, A.K., Chapman, J., Feltus, F.A., Gowik, U., Grigoriev, I.V., Lyons, E., Maher, C.A., Martis, M., Narechania, A., Otillar, R.P., Penning, B.W., Salamov, A.A.,Wang, Y., Zhang, L., Carpita, N.C., Freeling, M., Gingle, A.R., Hash, C.T., Keller, B., Klein, P., Kresovich, S., McCann, M.C., Ming, R., Peterson, D.G., Mehboob-ur-Rahman, Ware, D., Westhoff, P., Mayer, K.F., Messing, J. and Rokhsar, D.S. (2009). The *Sorghum bicolour* genome and the diversification of grasses. *Nature*, 457: 551–556.

Pedersen, J.F., Funnell, D.L. and Vogel, K.P. (2005). Impact of reduced lignin on plant fitness. *Crop Science*, 45: 812–819.

Prem Kishore. (1986a). Host plant resistance in sorghum to shoot fly and stem borer. *Ann. Pl. Resit. Insects Newsl.* 13: 50–51.

Prem Kishore, (1986b). Breeding for resistance to shoot fly and stem borer in sorghum. *Insect Sci. Applic.* 7: 117–121.

Prem Kishore, (1992). SPV 1015 (PGS-1) A new sorghum variety endowed with multiple resistance to shoot fly, *Atherigona soccata* Rondani and stem borer. *Chilo partellus* (Swinhoe). *J. Ent. Res.* 16: 321–323.

Rakshit, S., Ganapathy, K.N., Gomashe, S.S., Rathore, A., Ghorade, R.B., Nagesh Kumar, M.V., Ganesmurthy, K., Jain, S.K., Kamtar, M.Y., Sachan, J.S., Ambekar, S.S., Ranwa, B.R., Kanawade, D.G., Balusamy, M., Kadam, D., Sarkar, A., Tonapi, V.A. and Patil, J.V. (2012a). GGE biplot analysis to evaluate genotype, environment and their interactions in sorghum multilocation data. *Euphytica*, DOI 10.1007/ s10681–012–0648–6

Rakshit, S., Gomashe, S.S., Ganapathy, K.N., Elangovan, M., Ratnavathi, C.V., Seetharama, N. and Patil, J.V. (2012b). Morphological and molecular diversity reveal wide variability among sorghum Maldandi landraces from India. *J Plant Biochem Biotechnol*, DOI: 10.1007/s13562–011–0079–9

Ramu, P., Deshpande, S.P., Senthilvel, S., Jayashree, B., Billot, C., Deu, M., Ananda Reddy, L. and Hash, C.T. (2010). *In silico* mapping of important genes and markers available in the public domain for efficient sorghum breeding. *Mol. Breeding*, 26: 409–418.

Ramu, P., Kassahun, B., Senthilvel, S., Ashok Kumar, C., Jayasree, B., Folkertsma, R.T., Ananda Reddy, L., Kuruvinashetti, M.S., Haussmann, B.I.G. and Hash, C.T. (2009). Exploiting rice–sorghum synteny for targeted development of EST–SSRs to enrich the sorghum genetic linkage map. *Theor. Appl. Genet.,* 119: 1193–1204.

Rana, B.S., Singh, B.U. and Rao, N.G.P. (1985). Breeding for shoot fly and stem borer resistance in sorghum. *In*: Proc. Int. Sorghum Ent. Workshop. July 15-21, 1984. Texas A and M Univ. College station, Texas, USA, ICRISAT, Patancheru, AP, India, pp. 347–360.

Rao, K.E.P. and Rao, V.R. (1995). The use of characterization data in developing a core collection of sorghum. *In*: Core collection of plant genetic resources (*Ed*. Hodgkin, T. *et al.*). Wiley–Sayee, Chichester, UK. pp. 109–116.

Reddy, B.V.S. and Stenhouse, J.W. (1994). Sorghum improvement for the semi–arid tropic region: Past, current and future research thrust in Asia. *Punjabrao Krishi Vidyapeeth Research Journal*, 18: 155–170.

Reddy. B.V.S., Kumar. A.A., Reddy. P.S. and Elangovan. M. (2008). Sorghum germplasm: Diversity and utilization. *In*: Reddy, B.V.S., Ramesh, S., Kumar, A.A. and Gowda, C.L.L. *Eds*. Sorghum Improvement in the New Millennium, International Crops Research Institute for Semi–Arid Tropics, Patancheru, pp. 153–169.

Reddy, P.S., Patil, J.V., Nirmal, S.V. and Gadakh, S.R. (2012). Improving post–rainy season sorghum productivity in medium soils: Does ideotype breeding hold a clue? *Current Science*, 102: 904–908.

Rodriguez– Herrera, R., Roony, W.L., Rosenow, D.T. and Frederiksen, R.A. (2000). Inheritance of grain mold resistance in grain sorghum without a pigmented testa. *Crop Science*, 40: 1573–1578.

Sajjanar, G.M., Biradar, B.D. and Biradar, S.S. (2011). Evaluation of crosses involving rabi landraces of sorghum for productivity traits. *Karnatak J Agric Sci,,* 24: 227–229.

Satish, K., Srinivas, G., Madhusudhana, R., Padmaja, P.G., Nagaraja Reddy, R., Murali Mohan, S. and Seetharama, N. (2009). Identification of quantitative trait loci for resistance to shoot fly in sorghum [*Sorghum bicolor* (L.) Moench]. *Theor. Appl. Genet*, 119: 1425–1439.

Shiringani, A.L. and Friedt, W. (2011). QTL for fibre related traits in grain x sweet sorghum as a tool for the enhancement of sorghum as a biomass crop. *Theor. Appl. Genet.,* 123: 999–1011.

Singh, B.U., Padmaja, P.G. and Seetharama, N. (2004). Stability of biochemical constituents and their relationship with resistance to shoot fly *Atherigona soccata* (Rond.) in seedling sorghum. *Euphytica*, 136: 279–289.

Srinivas, G., Satish, K., Madhusudhana, R., Nagaraja Reddy, R., Murali Mohan, S. and Seetharama, N. (2009). Identification of quantitative trait loci for agronomically important traits and their association with genic–microsatellite markers in sorghum. *Theor. Appl. Genet.,* 118: 1439–1454.

Thakur, R.P., Reddy, B.V.S., Indira, S., Rao, V.P., Navi, S.S., Yang, X.B. and Ramesh, S. (2003). Sorghum grain mold. Information Bulletin No. 72, International Crops Research Institute for Semi–Arid Tropics, Patancheru, India. p. 32.

Umakanth, A.V. (2008). Development of sorghum restorers and varieties. *In*: Reddy, B.V.S., Ramesh, S., Kumar, A.A. and Gowda, C.L.L. (*Eds.*) Sorghum Improvement in the New Millennium, International Crops Research Institute for Semi–Arid Tropics, Patancheru, pp. 79–92.

Upadhyaya, H.D. and Ortiz, R. (2001). A mini core collection for capturing diversity and promoting utilization of chickpea genetic resources in crop improvement. *Theor. Appl. Genet.,* 102: 1292–1298.

Upadhyaya, H.D., Pundir, R.P.S., Dwivedi, S.L., Gowda, C.L.L., Reddy, V.G. and Singh, S. (2009). Developing a mini core collection of sorghum for diversified utilization of germplasm. *Crop Science*, 49: 1769–1780.

Visarada, K.B.R.S. (2008). Development and testing of transgenics in sorghum. *In*: Reddy, B.V.S., Ramesh, S., Kumar, A.A. and Gowda, C.L.L. (*Eds.*) Sorghum Improvement in the New Millennium, International Crops Research Institute for Semi–Arid Tropics, Patancheru, pp. 202–205.

Wang, Y.H., Bible, P., Loganantharaj, R. and Upadhyaya, H.D. (2011). Identification of SSR markers associated with height using pool–based genome–wide association mapping in sorghum. *Molecular Breeding*, DOI: 10.1007/s11032–011–9617–3

Wang, J., Elliott, J.E. and Williamson, R.E. (2008). Features of the primary wall CESA complex in wild type and cellulose–deficient mutants of *Arabidopsis thaliana*. *J. Exp. Bot.,* 59: 2627–2637.

6

Pearl Millet

O. P. Govila

Formerly Division of Genetics, Indian Agricultural Research Institute, New Delhi, India

Both production and productivity of pearl millet since 1960s in India has been on the rise. The area planted under pearl millet declined during 1980s and 1990s but currently it is stabilizing around 9-10 m ha. The major pearl millet growing states are Rajasthan, Maharashtra, Gujarat, Uttar Pradesh and Haryana which account for more than 90% of pearl millet acreage in the country. The grain production has increased from

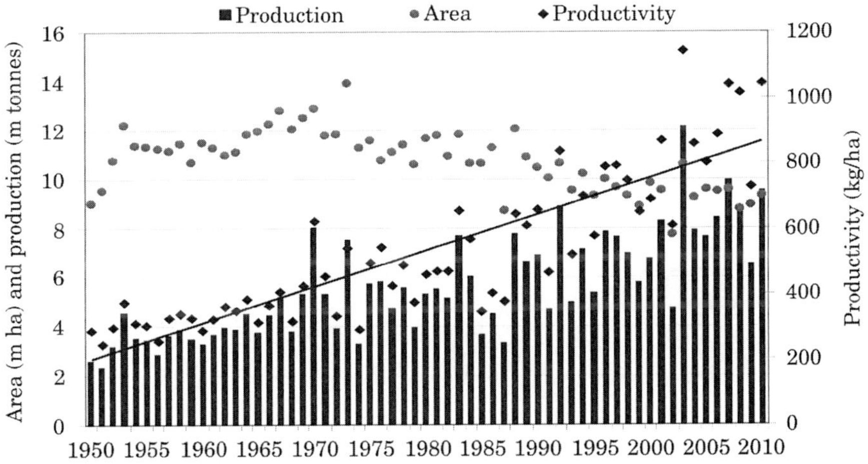

Fig. 1: Area, production and productivity of pearl millet in India (Yadav *et al.*, 2011)

V.L. Chopra (Ed.): Breeding Field Crops II: Advances, Studium Press (LLC), USA.

3.5 to 9.5 m tons due to continuous increase in productivity during the last six decades (Fig. 1).

The increase in yield has been due to adoption of high-yielding cultivars and suitable agro-production technologies (Khairwal and Yadav, 2005). The increase in yield in Rajasthan which accounts for half of the area in India under pearl millet is low. In arid Rajasthan western region productivity is still lower. However, even in Rajasthan and arid Rajasthan there is consistent improvement in yield (Fig. 2).

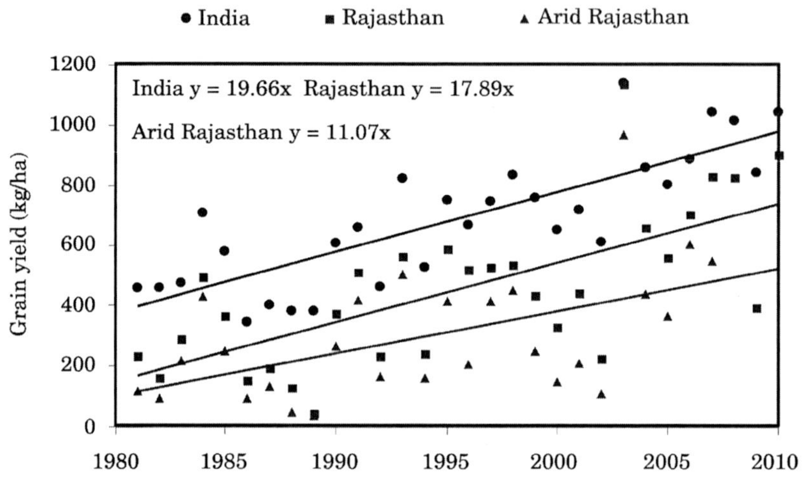

Fig. 2: Trends in pearl millet productivity in India, Rajasthan and arid regions of Rajasthan since 1981 (values inside figure indicate rate of improvement in grain yield in kg/ha/year) (Yadav *et al.*, 2011)

There have been four conspicuous phases of genetic improvement of pearl millet in India in different regions.

(i) *Pre-hybrid phase (1959-65)*: Traditional cultivars dominate with a few open pollinated varieties

(ii) *Initial hybrid phase (1966-80)*: A few hybrids with very narrow genetic base dominate, downy mildew epidemics common

(iii) *Improved hybrids phase (1981-95)*: Hybrids based on genetically diverse parental lines developed, downy mildew was largely contained

(iv) *Advanced hybrid phase (1996-2011)*: A large number of highly diverse CMS lines and pollinator parents used in heterosis breeding targeting niche adaptation in different zones

Hybrid development continues to be the main focus in this phase of pearl millet improvement. As a result, grain productivity has been increased to about 20 kg/ha/year (Fig. 1) Principles and strategies adopted for genetic enhancement have been in the following direction (Rai *et al.* see ref.).

1. CYTOPLASMIC MALE STERILITY SEARCH AND EVALUATION FOR STABILITY

The entire hybrid development program has been based mostly on the single CMS system (A_1 system) till 1998. This large-scale cytoplasmic uniformity exposed the hybrid seed industry to potential disease and insect-pest epidemics. Thus, efforts were made to assemble CMS lines, and characterize these for stability of male sterility and their maintainer frequency. Sources assembled at ICRISAT (A_1 and A_4 from USA, A_2 and A_3 from India, and Av from France) and two CMS sources identified at ICRISAT *Aegp* and A_5 were developed in each of the three diverse genetic backgrounds (81B, 5054B and ICMB 88004). Initial evaluation of A-lines for pollen shedders had shown that the A_2 and A_3 sources were most unstable. A-lines with *Aegp* cytoplasm and those with the A_4 cytoplasm showed very low frequency of pollen shedders (0.0-0.3%) were more stable than those with the A_1 cytoplasm (0.0-2.5% pollen shedders). The A-lines with the A_5 cytoplasm had no pollen shedders, implying that A_5 CMS system was the most stable. Evaluation of male sterility of topcross hybrids developed from crosses between isonuclear A-lines with the A_1, A_4 and A_5 cytoplasm and several improved populations from Asia and Africa regions showed that the frequency of maintainers was highest for the A_5 system (85-98%), followed by A_4 system (mostly 25-50%), and it was the lowest for the A_1 CMS system (mostly 10-40%), thus providing the prospects of greatest genetic diversification of A-lines with the A_5 system, and then with the A_4 and the A_1 systems, in that order.

2. CYTOPLASMIC DIVERSIFICATION OF SEED PARENT

Although the seed parent development using the A_1 CMS system continues as most of the parental lines in the public and private sector breeding programs are not restorers of the A_4 and A_5 cytoplasm, there has been an increasing emphasis in the national programme on the utilization of the A_4 and A_5 CMS systems in seed parents developments, with a parallel effort on developing their restorers. For instance, of the 99 designated A-lines developed till 2004, 62 are based on the A_1 cytoplasm, 33 on the A_4 cytoplasm, and only 4 on the A_5 cytoplasm, while of the 245 A/B pairs at the initial backcross stages, 83 backcross progenies had A_1 cytoplasm, 100 had A_4 cytoplasm, and 52 had A_5 cytoplasm. In addition, maintainers of the morphologically diverse 34 A_1-system A-lines were converted into A_4-system A-lines as well as into A5-system A-lines. This provides a wide range of genetically diverse A-lines of these two new and the most stable CMS systems. These two new and highly stable CMS sources also provide opportunities to develop stable male sterile lines that can be used in breeding forage hybrids. All three CMS systems (A_1, A_4 and A_5) are being used for seed parent development leading to the the higher yields.

3. GENETIC DIVERSIFICATION OF CMS LINES

Seed parents require stability for complete male sterility across seasons and sites, high grain yield and high levels of DM resistance with high degree of lodging resistance. These are the basic requirements for all hybrid parents, especially seed parents. Breeders' requirements for other traits such as maturity, tillering ability, plant height, panicle size, and grain size and color, however, vary considerably, depending on the agro-ecological conditions and farmers' preference. Thus, a trait-based approach in breeding seed parents with high grain yield potential and high levels of DM resistance has been followed, resulting in the development of large diversity in seed parents. For instance, the flowering of the 99 of the designated A-lines ranges from 39 to71 days, with 22 of these flowering in 39-45 days (useful for breeding early-maturing hybrids), and 73 flowering in 46-55 days (useful for breeding dual-purpose hybrids). The panicle length of these lines varies from 10 to 40 cm, panicle diameter from 15 to 40 mm, no. of panicles plant-1 from 1-5 and 100-seed mass from 7 to 16 g. This large morphological diversity stems from focused selection for these traits. Germplasm introgression for specific traits generated new variability. For instance, introgressed lines with large seed size (1000-seed mass of 18-20 g compared to mostly 10-12 g in commercial parental lines) and long panicles (45-60 cm compared to mostly 20-25 cm in commercial parental lines) have been produced. This breeding strategy has led to the development of several A-lines, most of which are resistant to 3-4 pathotypes, with some resistant to downy mildew (Rai *et al.*).

4. RESTORER PARENT DEVELOPMENT

Very few restorers were utilized in breeding programme before 2000. An approach was followed as in seed parent, with respect to trait-based breeding for morphological diversity, high grain yield potential and high levels of DM resistance. For the first time in 2006, nine R-lines (5 A_1 and 4 A_4 restorers) were developed that flowered in 46-60 days, had 11-30 cm of panicle length, 130-200 cm plant height and 5.5-13.5 g of 1000-seed weight. These were highly resistant to at least two of the pathotypes, with eight of these resistant to four of the five pathotypes of dowry mildew. R-lines with greater diversity are being developed and disseminated in the future as a large number of advanced breeding line.

In A_5 system also a pedigree breeding program from the planned crosses has been initiated, and early-generations breeding lines have been developed. A backcross breeding procedure to breed restorer lines from the promising A_1 system restorers (otherwise maintainers of the A_4 and A_5 CMS systems) was developed. Using this breeding scheme, 4 lines from ICRISAT and 9 lines from the NARS were converted into their A_4 restorer versions, while 29 lines from ICRISAT and 10 lines from NARS were converted into their A_5 restorer versions. The backcross breeding produced restorers of these alternative CMS systems (Rai *et al.*).

5. POTENTIAL PARENTAL LINES OF FORAGE HYBRIDS

Some of the male-sterile lines with dual-purpose characteristics (bred for grain hybrid) have been tested for the forage potential of their hybrids. The initial attempts led to the identification of ICMA 00999, which produced a high-yielding experimental top-cross hybrid (ICMA 00999 x IP 17315). This hybrid gave 15-30% more dry forage yield than a released hybrid Proagro 1 at 80-d harvest across several trials. Subsequent experiments identified a forage type OPV that yielded as well as this experimental hybrid (13 t ha^{-1}), and another OPV that out yielded it by 14%. A topcross hybrid trial identified three hybrids on ICMA 89111 and four hybrids on ICMA 00999 that had 12.5-13.8 t ha^{-1} of dry forage yield (6-17% more than the ICMA 00999 × IP 17315). Based on these results, a hybridization program among B-lines has been initiated to breed forage type A-lines, and inbreeding and selection will be initiated in pollinator populations of high-yielding hybrids to develop male parents.

6. EXPLOITATION OF HETEROSIS IN ARID ENVIRONMENTS

Pearl millet is mostly grown in arid environments (Western Rajasthan, part of Gujarat, Haryana). It is generally believed that heterosis is best exploited under favourable crop growing conditions. While this could largely be true (Fig. 3), several studies conducted in arid environments have demonstrated that heterosis can also be effectively exploited to improve both grain yield and stover yield in pearl millet under marginal conditions of arid zone.

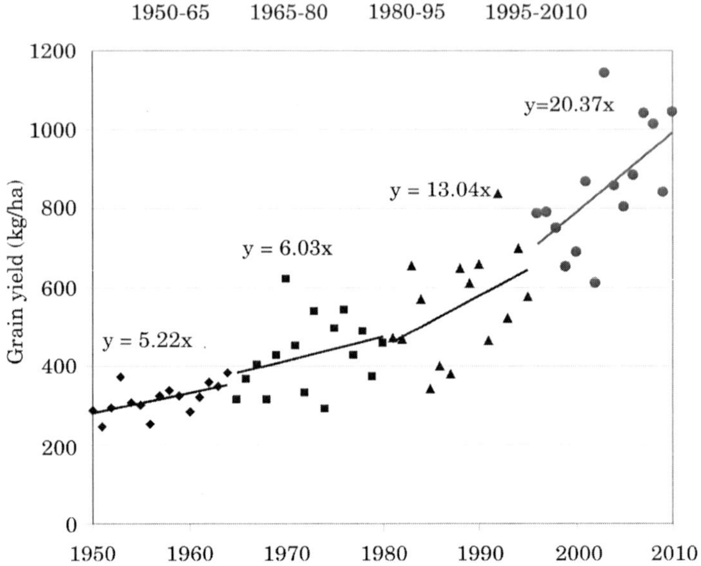

Fig. 3: Trends in pearl millet productivity during 1950-65, 1965-80, 1980-95 and 1995-2009 in India (values inside figure indicate rate of improvement in grain yield in kg/ha/year -during four phases (Yadav *et al.*, 2011)

Another requirement under arid zone conditions is that the improvement in grain yield should not be at the cost of stover yield. Yadav *et al.* (2000) reported an average of 15% heterosis in growth rate in topcross hybrids based on unimproved landrace pollinators, which translated to a positive biomass heterosis where duration of the hybrid was similar to that of the pollinator.

Research has shown that variation in biomass heterosis is the major determinant of both grain-yield heterosis and stover-yield heterosis (Bidinger and Yadav, 2009). The contribution of harvest-index heterosis to grain and stover yield heterosis has been observed to be of a comprising nature suggesting that strategy for grain yield by harvest index will not result in the desired outcome in the marginal arid environments where stover yield is also important. There are exploitable genetic differences among the male-sterile lines and restorers for their ability to produce heterotic crosses for biomass under arid environments. (Bidinger *et al.,* 2003; Bidinger and Yadav, 2009: Yadav *et al.,* 2009).

7. ADVANCES IN UNDERSTANDING DOWNY MILDEW

Characterization of the DM pathogen populations for virulence showed that there are at least five major pathotypes in India. While the collection of pathogen populations from new areas in India and their characterization for virulence diversity is a continuing process, evaluation of breeding lines at various stages of inbreeding and selection for two diverse and major pathotypes (one from Durgapura in the northern India, and the other one from Jalna in the central India), is followed by evaluation of candidate lines (lines having potential for designation and dissemination) for resistance to all the five pathotypes. All this screening is done through seedling inoculation under high disease pressure (generally >90% disease incidence in the susceptible checks) in the greenhouse. Those found highly resistant (<10% disease incidence) to at least two of the pathotypes are designated and disseminated as resistant line seed (Thakur *et al.*, 2006).

8. GENETIC ENHANCEMENT FOR STRESS TOLERANCE FOR SUMMER CULTIVATION

Drought and soil salinity are the two widespread abiotic production constraints in pearl millet in most of the traditional pearl millet growing areas. Pearl millet is the most drought and salinity tolerant warm-season cereal. Recently, pearl millet has emerged as an important crop for summer cultivations in parts of Gujarat state of India where air temperatures during flowering can go as high as 46-48°C. Most of the hybrids fail to set any seed at that high temperatures, but a few set excellent seed, and have wider farmers acceptance. Conventional approach to breeding for drought tolerance in pearl millet has resulted in the release of hybrids for summer cultivation (see Table 1).

9. IMPACTS ON YIELD IMPROVEMENT DUE TO DIVERSIFICATION OF PARENTAL LINES

The impact on yield due to parental line diversification and niche based hybrids and downy mildew resistance has been shown in Figs. 1 to 3. This has made remarkable contribution to on-field genetic diversity (largest number of hybrids on-farm in pearl millet as compared to any coarse grain cereal),and enhanced DM resistance (no DM epidemics in

Table 1: Parentage, year of release and major salient features of pearl millet hybrids released* and notified for cultivation

Sl. no.	*Hybrids*	*Source*	*Year of release/ notification*	*Parentage (AxR)*	*Areas of adoption*	*Maturity group /salient features*
1.	Nandi 32	Nandi seeds	1999	NMS 7A x NMP 24	Raj., Har, Guj, MP., UP, Delhi	78 days, long ear heads
2.	RHB 90	RAU, Durgapura	2000	81 A x RIB3135-18	Raj.	85 days
3.	HHB 94	HAU, Hissar	2000	ICMA 89111 x G73 – 107	Haryana	73-76 days, medium tall,
4.	CoHCu 8	TNAU, Coimbatore	2001	732 A x PT 4450	Tamil Nadu	80-85 days, tall
5.	Nandi 35	Nandi seeds	2001	NMS 11 A x NMP 42	Maha, Kar, AP, TN	80 days
6.	RHB 121	RAU, Durgapura	2001	ICMA 89111 x RIB 3135	Raj, Har, MP, Guj, UP, Delhi	85 days, bristled
7.	PB 112 (Proagro 9445)	Proagro seeds	2001	PSP 35 x PP 1	Raj, Har, MP, Guj, UP, Delhi	80-88 days, Good tillering
8.	PB 106 (Proagro 9443)	Proagro seeds	2001	PSP 41 x PP6	All India	76-80 days
9.	7688	Pioneer seeds	2001	PH 03 x PH 05	All India	78-80 days
10.	PB 172	Proagro seeds	2003	PSP 35 x PP 27	Guj- summer cultivation	84 days
11.	GHB 526	GAU, Jamnagar	2003	ICMA 95222 x J2372	All India summer cultivation	85 days
12.	GHB 558	GAU, Jamnagar	2003	ICMA 94555 x J 2290	All India	75-80 days
13.	HHB 146	HAU, Hissar/ ICRISAT	2003	ICMA 95222 x HTP 94/54	Raj, Ha, MP, Guj, UP, Delhi	75-80 days
14.	HHB 117	HAU Hissar	2004	HMS 7 A x H77/29-2	Haryana	74-78 days, stay green
15.	PB 180	Proagro seeds	2004	PSP 41 x PP 29	Raj,Guj. & other states summer cultivation	85 days, tall high tillering
16.	GHB 577	GAU Jamnagar	2004	JMS A 101x J 205	Raj., Ha, Guj, MP., UP, Del	80-85 days
17.	HHB 67 (Improved) through MAS	HAU Hissar	2005	ICMA 843-22 x H77/833-2-202	Western Raj. and drier parts of Guj & Ha,	Early maturity, highly resistant to moisture stress & downy mildew, medium maturity and lodging

Table 1: (*Contd...*)

Table 1: *(Contd...)*

Sl. no.	*Hybrids*	*Source*	*Year of release/ notification*	*Parentage (AxR)*	*Areas of adoption*	*Maturity group /salient features*
18.	B 2301	Zuari seeds	2007	B0009A x B 5103 R	Mah, Kar, AP, TN,	Late maturity
19.	NMH 68	Nandi seeds	2007	ICMA 97444 x NMP 48	Raj, Guj, Pun, UP, MP, Del	Late maturity
20.	GHB 719	GAU Jamnagar	2007	ICMA 95222 x J 2454	Western Raj, and drier parts of Guj and Har,	70-75 days
21.	GHB 757	GAU Jamnagar	2008	ICMA 92777 x J2467	Western Raj, and drier parts of Guj and Har	Early maturity
22.	GHB 744	GAU Jamnagar	2008	ICMA 98444 x J 2340	Raj, Guj, Har, Pun, UP, MP, Del	Medium maturity
23.	GHB 732	GAU Jamnagar	2008	ICMA 96222 x J2340	Raj, Guj, Har, Pun, UP, MP, Del	Medium maturity
24.	HHB 197	HAU Hissar	2008	ICMA 97111x HBL 11	Raj, Guj, Har, Pun, UP, MP, Del	Early maturity
25.	PHB 2168	PAU Ludhiana	2008	ICMA 92333 xPIB 686	Raj, Guj, Ha, Pun, UP, MP, Del	Early maturity
26.	PB 727 (Proagro 9555)	Bayer Bio Science	2008	PSP 51 x PP38	Summer growing areas of Guj., Mah, Raj, & TN	Late maturity
27.	B 2095 (MH1 257)	Zuari seeds	2009	B0009A x B 5220 R	Raj, Guj, UP, MP, Del	Medium maturity
28.	GK 1051	Ganga Kaveri seeds	2009	PM 678-II x PM 1081 R-I	Mah, Kar, AP, TN	Medium maturity
29.	JKBH 676	JK Agri Genetics	2009	JKMS 20 A x JKR6136	Raj, Guj, Ha, Pun, UP, MP, Del	Medium maturity, stay green
30.	RHB 154	RAU, Durgapura	2009	ICMA 95444 x RIB 57 S/05	Western Raj. and drier parts of Guj. and Har.	Early maturity
31.	Nandi 64	Nandi seeds	2010	NMS 2 11A x NMP 4-1	Summer growing areas of Guj., Mah, Raj, & TN	Late maturity
32.	HHB 216 (MN 121)	HAU Hissar	2010	HMS 37 A x HTP 3/13	Western Raj. and drier parts of Guj. and Har	Medium maturity.

Table 1: *(Contd...)*

Table 1: *(Contd...)*

Sl. no.	*Hybrids*	*Source*	*Year of release/ notification*	*Parentage (AxR)*	*Areas of adoption*	*Maturity group /salient features*
33.	HHB 223 (MH 1468)	HAU Hissar	2010	ICMA 94555 x HBL 11	Raj, Guj, Har, Pun, UP, MP, Del	Medium maturity.
34.	Nandi 61 (MH 1548)	Nandi seeds	2010	NMS 24A x NMP 64	Raj, Guj, Har, Pun, UP, MP, Del	Late maturity.
35.	Nandi 61 (MH 1549)	Nandi seeds	2010	NMS 24A x NMP 75	Raj, Guj, Har, Pun, UP, MP, Del	Late maturity.
36.	RHRBH 9808	MPKV Dhule	2010	RHRB 13A x RHRBI 1314	Mah,	Medium maturity
37.	*HHB-234 (MH 1561)	HAU, Hisar	2011	HMS 7 A x H-77/ 833-2-202	Drier parts of Raj, Har, Guj	Early maturity
38.	*VBBH-3040 (MH 1578)	Vibha Agrotech	2011	VBBA-310089 x VBBR 330585	Mah, Kar, AP and TN	Medium maturity
39.	*86 M 66 (MH 1617)	Pioneer Overseas Corporation	2011	MO96 F/M 118 R	Raj, Harana, UP, MP, Pun and Del	Late maturity
40.	*MP-7877 (MH-1610)	Metahelix Life Sciences Pvt. Ltd.	2011	M002 A/M004R	Raj, Harana, UP, MP, Pun and Del	Late maturity
41.	*MP-7792 (MH-1609)	Metahelix Life Sciences Pvt. Ltd.	2011	M001A/M004R	Raj, Harana, UP, MP, Pun and Del	Late maturity

*Identified by the AICPMIP

the last 15 years) and yield stability, and increased yield. The Table 1 shows the hybrids released from 2000 to 2011 in the national programme by public and private sector. Similarly, very large number of hybrids have been developed by private sector and sold to the farmers as truthfully labeled seed. The hybrids cover about 50% of total area, which is the highest among coarse cereal crops. Area under HYVs is the highest in Gujarat where almost whole area (>90%) has come under high hybrid coverage. As mentioned earlier, Rajasthan has the highest area under pearl millet, adoption of HYVs in this state has been very low (25-30%).

The Table 1 shows that 41 hybrids have been identified/released through All India Coordinated Pearl Millet Improvement Project (ICAR) during 2000-2011. These hybrids include 25 different CMS lines and 40 restorers. Nine hybrids have been release for low rainfall area (drier parts of Rajasthan, Haryana and Gujarat); seventeen hybrids for good rainfall areas of Rajasthan, Gujarat, Uttar Pradesh, Madhya Pradesh, Haryana, Punjab and Delhi; 5 for summer cultivation in different parts of India; 6 for southern states like Maharashtra, Karnataka, Andhra Pradesh and Tamil Nadu: one hybrid through marker assisted selection for dowry mildew and three for all India cultivation. The table clearly brings out the genetic variability utilization in the breeding programme during the last ten years in CMS and restorer lines. It also brings out the focus on cultivation in different niche areas. This has resulted in an overall increase in yield in India (Figs. 1-3). The genetic variability utilized from 1965-1999, was very limited as compared to the utilization from 2000-2011.

10. ROLE OF BIOTECHNOLOGY

Because of a history of limited success in breeding for drought adaptation by conventional approaches (Blum, 1988; Baker *et al*., 2005), this field has become a prime focus for molecular marker assisted breeding (MAB). Mapping of Quantitative Trait Loci (QTL) underlying drought tolerance of arid zone adapted land races is underway. Research at ICRISAT has identified QTL which has significant effects on pearl millet yield in drought stress environment (Yadav *et al*., 2002; Yadav *et al*., 2004; Bidinger *et al*., 2005). However, this gain under stress was achieved at the cost of lower yield in the non drought environment. A major QLT mapped on Linkage Group 2 determining grain yield in post flowering drought stress environment (Yadav *et al*., 2002) has been transferred to drought sensitive pearl millet lines through marker assisted backcross breeding and its effects on yield performance under arid zone drought is being investigated.

Molecular marker assisted breeding has also been an important activity in breeding dowry mildew resistant hybrids of pearl millet. The first product from this approach has been released for drought prone environments in A_1 zone as HHB-67 improved (s. no. 17 of Table 1), which is downy mildew resistant replacement for popular extra early maturity hybrid HHB-67 (matures 65 days). The male parent of this hybrid was improved by backcrossing QTL determining downy mildew resistant hybrid HHB-67-improved in the first marker assisted field crop product in public domain in India that was released and marketed.

For salinity tolerance and grain Fe and Zn, lines with large contrasts for these traits have been identified and development of mapping populations (one each for salinity tolerance and micronutrients) is being pursued at ICRISAT.

11. FUTURE THRUSTS

- Genetic and cytoplasmic diversification of seed parents should be addressed more vigorously and on priority. More CMS lines should be developed on A_4 an A_5 cytoplasm.
- Dual-purpose cultivars to meet the need of crop-livestock production systems.
- More emphasis on genetic improvement targeting A_1 zone.
- More pro-active action to register both breeding material and commercial cultivars.
- In view of climate change and temperature-rise prediction, heat tolerance needs to be addressed in breeding programmes.
- Pearl millet is more nutritious than other cereals but general public is not aware of this. Awareness is needed to highlight pearl millet as nutritious food rather than branding it as poor man's crop.
- With increase in production of pearl millet, larger proportion of pearl millet grain is being diverted to feed for cattle and poultry and for beverage industry as well. However, there are no base-line surveys available on this.

REFERENCES

Baker T., Campos, H., Dolan,D. Edmeades, G., Habben, J., Schusser, J., Wright, D. and Zinselmeir, C. (2005). Improving drought tolerance in Maize. *Plant Breeding Reviews*, 25: 173–226

Bidinger F.R., Serraj, R., Rizvi, S.M.H., Howarth, C., Yadav, R.S. and Hash, C.T. (2005). Field evaluation of drought tolerance QTL effects on phenotype and adaptation in pearl millet. [*Pennisetum glaucum* (L.) R.Br.] top cross hybrids. *Field Crops Research*, 94: 14–32

Bidinger, F.R. and Yadav, O.P. (2009). Biomass heterosis as the basis for grain and stover yield heterosis in arid zone pearl millet (*Pennisetum glaucum* (L.) R.Br.) hybrids. *Crop Science*, 49: 107–112.

Bidinger F.R., Yadav, O.P., Sharma, M.M., Van Oosterom. E.J. and Yadav, Y.P. (2003). Exploitation of heterosis for simultaneous improvement in both grain and stover yields of arid zone pearl millet (*Pennisetum glaucum* (L.) R.Br). *Field Crops Research*, 83: 13–26.

Blum, A. (1988). Plant breeding for Stress Environments. CRC Press, Boca Ration

Khairwal, I.S. and Yadav, O.P. (2005). Pearl millet (*Pennisetum glaucum*) improvement in India retrospect and prospects. *Indian Journal of Agricultural Sciences,* 75: 183–191.

Rai, K.N., Ranjana Bhattacharjee, R.P. Thakur, Rajan Sharma, C.T. Hash, Vincent Vadez, BIG Haussmann, M.A. and Mgonja, H.D., Upadhyaya and michael blumm-genetic and cytoplasmic diversification of pearl millet hybrid parents http://ccer07.icrisat.org/Group2/GeneticandCytoplasmicDiversificationof PMHybrid Parents.pdf

Yadav, O.P., Weltzien-Rattunde, E., Bidinger, F.R. and Mahalakshmi, V. (2000). Heterosis in landrace-based topcross hybrids of pearl millet across arid environments. *Euphytica,* 112: 285–295.

Yadav O.P., Bidinger, F.R. and Singh, D.V. (2009). Utility of pearl millet landraces in breeding dual-purpose hybrids for arid zone environments of India. *Euphytica*, 166: 239–247.

Yadav, O.P., Rai, K.N., Khairwal, I.S., Rajpurohit, B.S. and R.S. Mahala (2011). Breeding pearl millet for arid zone of North-western India: Constraints, opportunities, and approaches. All India Coordinated Pearl Millet Improvement Project, Jodhpur. India, p. 28.

Yadav, R.S., Hash, C.T., Bidinger, F.R., Cavan, G.P. and Howarth, C.J. (2002). Quantitative trait loci associated with traits determining grain and stover yield in pearl millet under terminal drought stress conditions. *Theoritical and Applied Genetics* 104: 67–83.

Yadav, R.S., Hash, C.T., Bidinger, F.R., Cavan, G.P. and Howarth, C.J. (2002). Genomic regions associated with grain yield and aspects of post- flowering drought tolerance in pearl millet across stress environments and testers background. *Euphytica*. 136: 265–277.

Further Readings

Thakur R.P., H.S. Shetty and I.S. Khairwal (2006). Pearl Millet downy mildew research in India: Progress and prospective. SAT e journal/ejournal (icrisat.org 2 (1).

Gupta, S.C. (1999). Seed production procedures in sorghum and pearl millet. Information Bulletin no. 58. (*In*: En. Summaries in En, Fr.) Patancheru 502324, Andhra Pradesh, India: International Crops Research Institute for the Semi-Arid Tropics. p. 16. I SBN 92-9066-415-0. Ordercode IBE 058.

Rai, K.N., Kulkarni, V.N., Thakur, R.P., Haussmann, B.I.G. and Mgonja, M.A. (2006). Pearl millet hybrid parents research: Approaches and achievements. pp. 11–73. *In*: Hybrid Parents Research at ICRISAT (Gowda, C.L.L., Rai, K.N., Reddy, B.V.S. Reddy and Saxena, K.B. *Eds*.). Patancheru 502 324, Andhra Pradesh, India: International Crops Research Institute for the Semi-Arid Tropics. p. 216.

Serraj, R., Hash (*sic*), T.C., Buhariwalla, H.K., Bidinger, F.R., Folkertsma, R.T., Chandra, S., Gaur, P.M., Kashiwagi, J., Nigam, S.N., Rupakula, A. and Crouch, J.H. (2005). Marker-assisted breeding for crop drought tolerance at ICRISAT:

Achievements and prospects. pp. 217–238. *In:* Proceedings of the International Congress "In the Wake of the Double Helix: From the Green Revolution to the Gene Revolution", 27–31 May 2003, Bologna, Italy (Tuberosa, R., Phillips, R.L. and Gale, M.D. *Eds.*). Bologna, Italy: Avenue media.

Thakur, R.P. and Hash, C.T. (2004). Biotechnology in the management of pearl millet downy mildew. pp. 247–261. *In:* Biotechnological Approaches for the Integrated Management of Crop Diseases (Professor LV Gangawane Festschrift Volume) (Mayee, C.D., Manoharachary, C., Tilak, K.V.B.R., Mukudam, D.S. and Deshpande, J. *Eds.*). Daya Publishing House, New Delhi - 110035.

Thakur, R.P., Shetty, H.S. and Khairwal, I.S. (2006). Pearl millet downy mildew research in India: Progress and prospects. *International Sorghum and Millets Newsletter*, 47: 125–130.

Readings from web:

Floral biology of Pearl millet-
http://www.authorstream.com/Presentation/chhabra61-112093-pearl-millet-floral-biology-descriptors-education-ppt-powerpoint/

Pearl millet [*Pennisetum glaucum* (L.) R. Br.]diseases- A Compilation of Information of the Known Pathogens of Pearl Millet- http://www.tifton.uga.edu/fat/pearlmilletdiseases.htm

7

Sugarcane

Raman Kapur and M. Swapna

Division of Crop Improvement, Indian Institute of Sugarcane Research, Lucknow, India

Sugarcane breeding in India has come a long way since the release of the first inter–specific hybrid variety Co 205 in 1918. This variety marked a significant milestone in sugarcane improvement so much so that a plaque has been dedicated at Sugarcane Breeding Institute, Coimbatore to commemorate the achievement.

Internationally, sugarcane cultivars have been derived from a few clones of *Saccharum officinarum, S. barberi, S. sinense* and the wild species, *S. spontaneum* and *S. robustum*. Interestingly, despite an allegedly narrow genetic base, inter–varietal crossing has been the major avenue of evolving improved sugarcane varieties, owing possibly to the polyploid nature of sugarcane, species hybridity and its reservoir of genetic variability. In sporadic attempts at infusing variability, the basic

V.L. Chopra (Ed.): Breeding Field Crops II: Advances, Studium Press (LLC), USA.

philosophy and crossing techniques have remained unchanged but there is a much greater application of biotechnological tools. The last one and half decades have seen a tremendous input of molecular analyses for unraveling the sugarcane genome. These have helped in elucidating and genetic deciphering of hitherto unresolved issues regarding phylogeny, hybridity of the interspecific or intergeneric progeny, genomic composition and molecular characterization of present–day cultivars, developing molecular markers and genetic transformation, etc.

The sugarcane area, production and yield have steadily increased over the decades (Table 1) due primarily to improved sugarcane varieties and production technology.

Table 1: Sugarcane statistics in the early and the last decade

Statistic	***1930–39***	***2000–09***	***Fluctuations during the last decade***
Area (million hectares)	1.46	4.38	3.66 – 5.15
Cane prod (million tonnes)	55	291	237 – 356
Yield (t/ha)	35.0	68.0	59.4 – 70.0
Sugar recovery (%)	9.05	10.27	10.03 – 10.55
Sugar prod. (million tonnes)	0.58	19.01	12.69 – 28.36

Adapted from Indian Sugar, Sep. 2011; Nair, 2010

In the following we aim to briefly discuss the exciting developments under the broad subheads:

A. Conventional, and

B. Biotechnological approaches to sugarcane improvement.

A. CONVENTIONAL APPROACHES FOR IMPROVEMENT

I. Varietal Development

There is a huge network of 23 research stations engaged in varietal development with the entire crossing work carried out in the National Hybridization Garden with nearly 600 parental clones at the Sugarcane Breeding Institute, Coimbatore and the selection work at the respective breeding stations. This strategy has paid off for breeding locally adapted varieties for each zone with desired traits such as tolerance to water–logging, resistance to the prevalent races of red rot disease, sugar yield, maturity and ratooning ability, etc. (Table 2). It has also helped in containing the threat of pests and disease epidemics. The cane productivity in the tropical and subtropical regions is currently at 80 and 50 t/ha, respectively; the difference attributable to longer crop duration with more conducive growing conditions in tropics. Obviously, there is a greater scope for enhancing productivity in the subtropical region.

Breeding sugarcane for the stressful subtropical environment of relatively short growing period is more challenging. Use of species–specific molecular markers revealed that subtropical varieties had a greater proportion of the wild parental species, *S. spontaneum* as compared with tropical sugarcane varieties (Selvi *et al.,* 2005). Adaptability of sugarcane varieties is not a function of desirable gene complexes coming from adapted parents alone but possibly also a manifestation of the chromosomal complement from *S. spontaneum*. It was surmised that less number of nobilization is required for adaptability to stresses experienced by subtropical varieties (Roach, 1989; Srivastava *et al.,* 1994).

The greatest challenge faced by cane breeders is the reliable identification of superior genotypes in the unreplicated early selection stages (Zhou, 2005). In some countries, better performing families are identified first, before individuals are selected from these families. Family selection has also been shown to provide a superior way for estimating the breeding value of parental clones; being 2.5 times better than individual selection alone (Stringer *et al.,* 2010). There are as yet no ways of evaluating segregating progenies at the young seedling stage, since most of the economic traits manifest at the full grown stage. Even the red rot is a stalk disease, screening for which has to be carried out only after the cane formation. Thus, molecular markers can play a vital role in improving selection efficiency at an early stage.

Physiologically, the current set of varieties show better sugar accumulation because selection was done for higher sugar content (Lingle *et al.,* 2009) and higher harvest index than the species clones (Kapur *et al.,* 2011). In the USA, a highly effective recurrent selection for sugar content has resulted in varieties with higher sugar accumulation potential. An average annual gain of 0.80 kg sugar/t of cane, 0.10 t/ha of sugar and 0.31 t/ha of cane was realized by the Florida sugar industry in the last three decades, on account of a diverse parental gene pool (Edme *et al.,* 2005). In Australia, improvements in sugar yield showed declining contribution from sugar content (Jackson, 2005), possibly for two reasons. Firstly, the desirable gene combinations from

Table 2: Performance in AICRP (S) Zonal Varietal Trials (averaged over early and mid–late trials from the data pooled for years and locations)

Agro–climatic zone and the states represented	***Sugar yield (t/ha)***		***Cane yield (t/ha)***		***Sugar content (% cane)***		***Sugarcane varieties notified by central varietal release committee***
	Mean	***Range***	***Mean***	***Range***	***Mean***	***Range***	***in the last decade***
Peninsular Zone (Interior of Tamil Nadu and Andhra Pradesh, Gujarat, Maharashtra, Karnataka, Madhya Pradesh & Chhattisgarh)	11.57	9.58–13.00	92.4	76.8–105.4	12.41	11.61–13.34	Co 85004, Co 8371, Co 86032, Co 87025, Co 87044, CoM 88121, Co 91010, Co 94008, Co 99004, Co2001–13, Co 2001–15, Co 0218
East Coast Zone (Coastal Tamil Nadu, Andhra Pradesh & Orissa)	11.42	10.46–12.43	103.3	96.7–106.8	11.06	10.67–11.48	Co 86249, CoC 01061
North West Zone (Punjab, Haryana, Rajasthan, Western. & Central Uttar Pradesh, Uttarakhand)	9.21	7.82–10.79	76.4	61.8– 90.2	12.23	11.45–12.77	CoS 91230, CoH 92201, CoPant 90223, CoS 95255, CoS 94270, CoH 119, Co 98014, CoS 96268, CoS 96275, CoPant 97222, CoJ 20193 (CoJ 89), Co 0118, Co 0238, Co 0124, Co 0239
North Central and Eastern (Eastern Uttar Pradesh, Bihar, Jharkhand, West Bengal, Assam)	7.48	6.72–7.90	61.6	56.3– 65.6	11.92	11.58–12.49	Co 87263, Co 87268, Co 89029, CoSe 95422, CoSe 92423, BO 128, CoSe 96234, CoSe 96436, CoLk 94184, Co 0232 and Co 0233

Adapted from PI (Breeding)'s Report for AICRP on Sugarcane 2010–11 and sourced from Project Coordinator (Sugarcane), AICRP

the available variability might have been exhausted. Secondly, a negative correlation between yield and quality in the small number of oft–repeated progenitor clones may have resulted in increases in yield coming at the cost of quality. This is a valid concern warranting fresh inductions of new variability, since breeding for higher sugar content is 1.8 times better than achieving the same gain in sugar yield through cane yield (Inman–Bamber *et al.,* 2011). Yet, it has been surmised that improvements in cane yield have come at the expense of roots and capacity for water and nutrient uptake (Smith *et al.,* 2005). It is commendable that the steady genetic gains in sugar yield in India have mainly come through cane yield, without compromising on cane quality.

II. Broadening Genetic Base

Modern cultivars are derivatives of no more than 15–20 genotypes of nobilized cultivars that can be traced back to the initial nobilized genetic base developed in Java and India (Roach, 1989). The realization in 1970s that sugarcane varietal breeding is based on limited variability, besides *S. officinarum* being the only source of cytoplasm, prompted world–wide efforts to introgress desired variability from within the *Saccharum* complex. *Erianthus, Miscanthus, S. robustum, S barberi, S. sinense* and *S. spontaneum* accessions were crossed with *S. officinarum* or the commercial hybrids for providing diverse sources of genes for biotic and abiotic stress tolerance.

A 'Distant Hybridization Facility' was developed in 1999 at Agali in Kerala by ICAR under Sugarcane Breeding Institute, Coimbatore with a core collection of sugarcane. Apart from these, many improved species clones, advance interspecific and intergeneric hybrid clones along with elite hybrid varieties have been included for use of sugarcane breeders. A recent development has been the emergence of in–house crossing facilities by EID Parry group at Bangalore and by Vasantdada Sugar Institute at Amboli.

II.1. *Recent attempts at germplasm collection*

In the last decade, 10 explorations to Andaman–Nicobar islands, Mizoram, Arunachal Pradesh and the river basins or coasts of Tamil Nadu, Orissa, Andhra Pradesh, Kerala and Karnataka were launched by SBI, Coimbatore for *Erianthus arundinaceous, S. spontaneum, Miscanthus* and *S. officinarum* and over 400 clones have been collected (Nair, 2007). This continuing effort involves a lot of follow–up work of taxonomic verification, characterization, evaluation and maintenance before these accessions can be utilized for introgression.

II.2. *Intraspecific improvement*

The introgression attempts revealed the difficulties of hybridizing material with irregular flowering, poor pollen fertility and viability of hybrid seed. Moreover, the available accessions sometimes need genetic improvement before they can be involved in the crossing programme. A programme for intraspecific improvement was initiated in *S. spontaneum, S. officinarum* and *S. robustum* in 1988 at SBI with poly–crossing and selection. Substantial improvement in each species was achieved for cane yield in *S. officinarum*, and sugar content in *S. robustum*. These improved intraspecific hybrids have been used in interspecific hybridization with better results.

More recently, *species* germplasm was screened for resistance to red rot and smut (SBI Research highlights, 2008–09). Observations are summarized in Table 3. Hale *et al.* (2010) also reported high resistance to red rot in *S. spontaneum, S. barberi and Erianthus* accessions.

Table 3: Screening of *Saccharum* germplasm to major diseases

Species / Genus	***Percentage of accessions resistant to***	
	Red rot	***Smut***
S. officinarum	3.4	8.8
S. barberi	38.1	33.3
S. sinense	13.3	78.0
S. robustum	4.9	34.6
Erianthus arundinaceous	88.1	–

II.3. *Interspecific hybridization*

At the Sugarcane Breeding Institute, Coimbatore, base broadening programmes were taken up during 1970s and 1980s to improve productivity and impart better resistance to biotic and abiotic stresses. For this purpose, new cycles of matings and selection of hitherto unutilized accessions were carried out. The hybrids were intercrossed, backcrossed and crossed with commercial varieties. This resultant elite gene pool of more than 300 interspecific hybrids (designated as ISH clones) was evaluated extensively. The identified genetic stocks were then inducted into commercial breeding programme (Nair, 2005). The success of base broadening programmes is adequately reflected in the identification of 'Co' varieties like Co 95017, Co 97007, Co 97010, Co 97015, Co 97016 (INGR09052, registered with NBPGR and has *S.*

robustum), Co 97017, Co 98003, Co 99007, Co 99016 and Co 2000005 with ISH parentage (Nair, 2007). Similarly, some elite varietal clones developed at Indian Institute of Sugarcane Research, namely, CoLk 97147, CoLk 97169, CoLk 99270 and CoLk 05201 have ISH clones as one of the parents.

Bhat and Gill (1985) drew attention to the homozygosity accompanying formation of 2n gametes during interspecific hybridization involving *S. officinarum* and advocated using different noble clones or hybrid cultivars in subsequent matings to counter inbreeding depression. The evaluation of ISH clones of varied parental combinations (Bakshi Ram and Sahi, 2000; Sandhu *et al.,* 2011) has revealed that in most combinations where a commercial hybrid was involved, the progeny performance was better for economic traits, particularly for sugar content (Table 4).

II.4. *Intergeneric hybridization*

In the progeny of intergeneric crosses made between *S. spontaneum* and, *Sclerostachya, Narenga* and *Erianthus*, *S. officinarum* and *Sclerostachya* in 1960s, the juice quality was very poor. Serious efforts to use *Erianthus* crosses (because of its recognized resistance to biotic and abiotic stresses) with *S. officinarum* were made at SBI in 1990s. Some of the derived hybrids showed sprouting and growing capability in winter under subtropical conditions. Co *87008* is a hybrid between CoC 617 and *Erianthus* (Amalraj and Balasundaram, 2011). *S. spontaneum* has been used reciprocally as a bridging species between *Erianthus* and *S. officinarum*/hybrid varieties to overcome lack of flowering or sterility in the hybrids. *Saccharum* has also been successfully crossed with *Sorghum* and *Zea mays*. *Sorghum* was used to introduce earliness but the resulting spreading habit, bud sprouting and low productivity were the undesired effects (Nair, 2005).

II.5. *Cytoplasmic diversity*

Most of the present–day commercial varieties have *S. officinarum* cytoplasm from limited number of accessions. In an SBI initiative during early 2000s for diversifying cytoplasm sources, accessions of *S. spontaneum, S. barberi* and *S. sinense* were used as pistil parents with *S. officinarum* /commercial hybrids as male parents. These hybrids and their backcrossed derivatives have been tested for resistance to red rot and smut, yield and quality traits. These form a potential group of genetic stocks for base broadening programmes in sugarcane (Nair, 2007). For example, CYM 04–420, a derivative from [(*Erianthus* x *S. spontaneum*) x Co 775] x CoC 671, with *Erianthus* cytoplasm has been produced (SBI Research Highlights 2010–11),

Table 4: Influence of noble and commercial hybrid clones on sucrose % in various inter-specific hybrid mating groups

Group	*Mean*	*Range*	*Percent improvement*
First stage hybrids			
O-B	14.27	9.16–17.68	11.14
H-B	15.68	7.83–20.36	
O-R	12.76	8.14–16.97	10.19
H-R	14.06	10.06–18.45	
O-S	10.70	3.94–19.40	42.06
H-S	15.20	7.68–19.77	
Second stage hybrids			
(O-B)O	14.29	9.73–18.49	15.96
(O-B)H	16.57	11.10–19.94	
(H-B)O	13.58	8.59–19.25	12.89
(H-B)H	15.33	10.70–18.94	
(O-R)O	12.63	7.18–18.95	24.39
(O-R)H	15.71	11.47–19.67	
(H-R)O	15.98	9.05–11.97	4.82
(H-R)H	16.75	14.13–19.70	
(O-S)O	13.93	10.53–19.98	9.33
(O-S)H	15.23	8.57–22.10	
(H-S)O	14.68	5.33–19.36	15.40
(H-S)H	16.94	13.76–22.16	

O - *S. officinarum*, B - *S. barberi/S. sinense*, R - *S. robustum*, S - *S. spontaneum*, H - Commercial hybrids. Bakshi Ram and Sahi (2000), Bakshi Ram (2010).

III. Pre Breeding / Development of Breeding Stock for Specific Traits

Sugarcane economic breeding remains largely empirical. There is a dearth of new proven parents or crosses; old ones have been repeated to exhaustion. This has adversely affected breeding efficiency since very few clones breed true for majority of the desired traits. A study in Australia showed that the best set of new parents was not notably better as a group than the best old parents (Ratney *et al.,* 2004). The Indian Institute of Sugarcane Research, Lucknow started prebreeding work in 1990s and 2000s for specific economic traits, such as sugar accumulation potential, resistance to red rot and tolerance to top borer.

The effort to converge genes from diverse sources, including those from species accessions, for developing high sugar breeding stock has been the longest and the most comprehensive. Third cycle of matings among selections is now in progress. There is a perceptible impact on the population mean of the resulting progenies. Progeny tests have been used to identify breeding stock for transmitting the potential for early sugar accumulation to the progeny. LG 95053 (NBPGR registration No. INGR09054) is example of an ideal female parent with low pollen fertility (Table 5). However, wilt has been found to be a major deterrent in developing high sugar breeding stocks for sub–tropical conditions, as excellent selections do not survive beyond 2–3 clonal generations (Duttamajumder *et al.,* 2004). In similar efforts, selections with resistance to more than one race of red rot and clones showing tolerance to top borer have been developed at IISR, and are being studied for their progeny performance. Sporadic attempts in this direction have been made at many breeding centres, but this kind of sustained effort is needed for achieving genetic gains on long–term basis. A spin–off in sugarcane is that segregants with commercial merit are often obtained in the progeny, *e.g.*, CoLk 94184 is a selection from the progeny on selfing of a high sugar variety, CoLk 8001.

Table 5: Hand Refractometer Brix (TSS) in the last week of October /first week in November

Mating	*No. of seedlings studied*	*Family mean (% brix)*	*% Seedlings > CoJ 64* (>21%)*	*% close to CoJ 64(20–21%)*	*% high sugar progeny*
LG 94184 GC	45	18.35	2.2	15.6	17.8
LG 94184 Self	96	18.34	5.2	11.5	16.7
LG 95053 Self	22	20.00	18.2	40.9	59.1
LG 95053 GC	27	20.05	48.2	14.8	63.0
LG 95053 x HR 83–65	50	20.50	36.0	32.0	68.0
LG 95053 x LG 94184	61	20.19	32.8	32.8	65.6

*CoJ 64 is the standard for early high sugar (Kapur *et al.*, 2005)

IV. DUS Characterization

To meet the requirements of the new IPR regime a system for characterization of varieties for distinctness (D), uniformity (U) and stability (S) using morphological descriptors has been established at the Indian Institute of Sugarcane Research for subtropical varieties and the Sugarcane Breeding Institute for tropical varieties. Till date, 94 subtropical and 150 tropical sugarcane varieties have been characterized using 27 morphological descriptors and the passport data has been compiled.

B. MOLECULAR APPROACHES FOR IMPROVEMENT

Due to the genetic complexities of sugarcane, conventional techniques have not been helpful for deciphering the inheritance of important traits. With the advent of biotechnological tools, researchers are able to delve at the chromosomal as well as DNA/RNA level in order to unravel the sugarcane genome. A brief account of the recent developments is as follows:

Biotechnological interventions in sugarcane improvement

Intervention	*Tools used*	*Area of application*
Cytogenetical markers	GISH, FISH	Phylogeny studies, Basic chromosome number studies, Hybridity confirmation, Chromosomal transmission studies
DNA sequence based markers	RFLP, RAPD, SSRs, AFLP, SNP, DArT, etc.	Diversity studies, Phylogeny, Hybridity, DNA fingerprinting/Varietal characterization, Linkage mapping and QTL analysis, Marker Assisted Selection (MAS), Synteny, Disease diagnostics, Pathotype characterization
Gene identification and functional genomics	ESTs, Microarrays SAGE	Gene identification, RGAs, Marker development, Expression studies
In vitro culture based techniques	Micropropagation Somaclonal variation Genetic transformation	Production of true–to type healthy plantlets, *In vitro* conservation of germplasm, Production of somaclonal variants, Production of transgenics
Bioinformatics	Analytical software Databases	Storage, retrieval and utilization of the stored data for *in silico* analysis

I. Molecular Marker Applications

Molecular markers comprise of both molecular cytogenetic tools like Genomic *In Situ* Hybridization (GISH) and Fluorescence *In Situ* Hybridization (FISH) and DNA sequence based markers. These marker systems have been widely used for different applications in sugarcane:

I.1. *Molecular cytogenetics*

Information from molecular cytogenetic analyses on basic chromosome number, ploidy level (Octaploidy), pairing dynamics (predominantly bivalent formation as in an amphidiploid), etc. have played a significant role in structural genomic studies and for interpretation of genetic maps of modern cultivars. FISH of ribosomal gene clusters have helped prove the hypothesis that *S. officinarum* and *S. robustum* have a basic chromosome number of x=10, while *S. spontaneum* has x=8 (D'Hont *et al.*, 1994; reviewed in Ming *et al.*, 2006).

Differentiation of *S. officinarum* vs *S. spontaneum* chromosomes was first demonstrated using GISH in the cultivar R570 (D'Hont *et al.*, 1996). The origin of *S. sinense* Roxb. and *S. barberi* Jeswit was established through GISH as natural hybrids between *S. officinarum* and *S. spontaneum*. The demonstration of 2n+n transmission in *S. officinarum* x cultivar crosses (Piperidis *et al.*, 2010) revealed that such cytological peculiarities are not confined to interspecific crosses alone. Species–specific DNA markers and GISH were particularly useful for identifying true hybrids in *Saccharum* x *Erianthus* / *Miscanthus* progeny at seedling stage and monitoring subsequent generations of introgression (D'Hont *et al.*, 1995). FISH was employed to study the chromosomal composition of some modern cultivars (Cuadrado *et al.*, 2004). These studies have revealed that modern sugarcane cultivars have a somatic chromosome number ranging from 100–130 (Sreenivasan *et al.*, 1987) with unequal contribution from the two parental species, *S. officinarum* and *S. spontaneum*; 70–80% from the former and 10–23% from the later with 8–13% chromosomes showing interspecific chromosomal exchange on account of rather infrequent (1–2%) homeologous pairing (Piperidis *et al.*, 2010).

With the help of flow cytometry, D'Hont and Glaszmann (2001) could estimate the genome size of sugarcane revealing a total genome size of 7.7 pg (7440 Mbp) for *S. officinarum*, 6.2 pg (5990 Mbp) for *S. spontaneum* and 11 pg (10000 Mbp) for a modern sugarcane variety.

I.2. *Diversity studies and DNA fingerprinting*

The application of molecular markers to diversity studies and varietal characterization has confirmed the existing evolutionary relationship among the species clones. Many of the important sugarcane varieties have been DNA fingerprinted. The diversity analyses pointed to a comparatively narrow genetic base of Indian cultivated clones (Nair *et al.*, 1999, 2002, 2006; Selvi *et al.*, 2005). A preliminary assessment of functional genetic diversity with Unigene microsatellite markers (UGMS) indicated that the genetic base of the Indian sugarcane varieties was not narrow (Parida *et al.*, 2011). Variation with respect to specific gene sequences like those linked to sucrose metabolism, C_4 photosynthetic pathway etc., was also identified in this crop, which needs to be exploited for sugarcane breeding programmes.

I.3. *Identification of true hybrids*

The peculiarity of floral morphology and impracticality of emasculation in routine crossing activity result in some selfing even in bi–parental crosses. True hybrids need to be identified at a very early stage. This is particularly so for intergeneric and interspecific cross progenies since dependence on morphology alone could be misleading. Ribosomal DNA (rDNA) markers, satellite DNA markers, *Erianthus* specific Inter–*Alu* sequences, RAPD, SSR and AFLP markers were used to identify genuine hybrids in interspecific and intergeneric crosses (Cai *et al.,* 2005; Nair *et al.,* 2006; Selvi *et al.,* 2006). This opened up opportunities for utilizing drought and cold tolerant genes of *Erianthus* (Aitken *et al.,* 2007). The chromosome elimination encountered in hybrids and backcross progenies may be a concern in the effective application of markers for this purpose.

I.4. *Linkage mapping and QTL analysis for marker assisted selection (MAS)*

Linkage mapping and QTL analysis to locate trait–linked loci resulted in the development of a large number of linkage maps and QTLs (Reviewed in Ming *et al.,* 2006). But as evident from different reports, there is still scope with respect to the level of saturation and coverage of the maps.

Gene for resistance to brown rust in R570 was the first one mapped (Daugrois *et al.,* 1996; Asnaghi *et al*., 2004). Other traits taken up for mapping were sugar yield, sugar content, fibre content, plant height and flowering, yellow spot disease, etc. Even though several maps have been constructed and loci mapped to specific locations on these maps, a complete linkage map which corresponds to the known basic chromosome number of *Saccharum* has not become a reality till now. The recombinational heterogeneity in the genome and the larger coverage of *S. spontaneum* chromosomal component revealed through these studies has significance in the context of linkage mapping and mapping of economically important traits. It was revealed that when more than one locus or allele are present, multiple doses of loci with favourable alleles do not always lead to increased phenotypic expression. This suggests epistasis among the loci (Ming *et al*., 2001). Another hypothesis regarding these non–additive multi–dose QTLs is their probable role in imparting stability across environments. In sugarcane, the molecular markers are yet to be put to use in the actual field situation. Breeding for such economic traits as ratooning ability and tolerance to water–logging would gain the most from molecular markers.

I.5 *Comparative genomics*

In the light of the concept "grasses as a single family", comparative studies in monocot crops have revealed that sorghum, also a C_4 plant, is the closest diploid relative of sugarcane. Collinearity is perfect among the genomes of sugarcane, maize and sorghum at several chromosomal locations. This has helped in identifying corresponding QTLs controlling plant height and flowering in sorghum and sugarcane (Ming *et al.*, 2001) and also in cloning of rust resistance gene in sugarcane, referred to earlier. Conservation among these genomes with respect to the functional components was also confirmed in sugar gene related studies (Singh *et al.*, 2011), particularly with respect to a sugar transporter gene (Wang *et al.*, 2010). Thus the scope of comparative genomics has widened from mere comparison of genome organization or study of synteny and collinearity, to comparison of trait–specific genes in terms of their location and their role in phenotypic variation etc.

I.6. *Disease diagnostics*

Sugarcane is host to a large number of diseases caused by bacteria, fungi, phytoplasma, etc., necessitating sensitive and reliable diagnostic methods for their early and efficient detection. This is important for healthy seed production. DNA based diagnostic methods involving pathogen–specific primers have been developed for diseases like ratoon stunting, grassy shoot, smut, wilt, etc., using intergenic spacer regions, AFLP, etc.. PCR based diagnostic kits for detection of incipient infection of smut and red rot was also developed (Annual Report, Department of Biotechnology 2006–07). Identification and use of Resistant Gene Analogues (RGAs) is another powerful strategy for map–based cloning and marker identification. These biotechnological tools will help in precision breeding for disease resistance and in proper understanding of plant-pathogen interactions.

II. Gene Identification/Functional Genomics

For discovering agronomically important sugarcane genes, an International Consortium of Sugarcane Biotechnology with Australia, Brazil, South Africa and the USA as members has been constituted. The Brazilian programme, SUCEST, has been the most ambitious, accounting for >90 % of the currently listed 300,000 sugarcane EST sequences, with 33,620 unigenes in the GenBank (http://sucest.lad.ic.unicamp. br/en/) (Genetics and Molecular Biology 2001 Vol. 24 No. 1–4). The EST approach has helped identify a vast array of genes, including 'Resistance Gene Analogues' (RGAs). In India, a collaborative DBT initiative by the Indian Institute of Sugarcane Research and the University of Delhi resulted in 26453 ESTs, some of these from subtractive cDNA libraries generated against red rot disease, drought and water–logging (Gupta *et al.*, 2010). Development of markers and mapping studies for important economic traits has made use of this valuable EST resource on a large scale. The Sugarcane Genome Sequencing Initiative (SUGESI) Consortium is another collaborative attempt for unravelling the complexities of sugarcane genome/transcriptome, mainly through BAC sequencing strategy in R570, a cultivar with the best well–characterized genome. The library developed was used in building a physical map around rust resistance gene (Le Cunff *et al.,* 2008) and others.

III. *In Vitro* Culture Based Technologies

III.1. *Micropropagation*

Micropropagation is particularly important in sugarcane, where plants are rapidly multiplied *in vitro* resulting in true–to–type plantlets. The rate of multiplication through micropropagation can be 1:1000s as compared to 1:10 in conventional propagation, thus contributing to the rapid increase of seed material, especially for newly released varieties. It can also be used to rejuvenate pathogen–afflicted old run–down sugarcane varieties. Micropropagation employs either shoot tip (for quick multiplication) or the apical meristem (for the production of disease–free planting material). Technique for cost–effective micropropagation, however, has been a challenge even when successful micropropagation has been demonstrated with several varieties (review by Jalaja *et al.*, 2008). At Vasantdada Sugar Institute, Pune, an excellent micropropagation facility has been developed to produce up to 1 lakh plantlets/ month (also at EID Parry, India) and good experience has been gained on handling sugarcane plantlets, crop geometry and husbandry (Pant, 2007). In Australia and South Africa, micropropagation technology has been patented as Smartset and NovaCane^{R}, respectively.

Another area of great utility in sugarcane is the *in vitro* conservation of germplasm instead of maintaining live field collections. At SBI in India, CIRAD in France and BSES in Australia, sugarcane shoot tips/ meristems are grown on minimal culture media or at low temperature with transfer to fresh media every few months with no change on revival (Sreenivasan and Sreenivasan, 1985; Lakshmanan *et al.*, 2005; Chandran, 2010).

III.2. *Somaclonal variation*

Starting 1990s, a lot of effort was invested for more than a decade to produce somaclonal variants at many sugarcane research centres. It made sense in sugarcane because sexual rectification of a defect is not feasible without losing the genetic integrity of a variety. The polyploid nature of the crop and the inability of directing a desired change *in vitro* are limitations. In many cases, the enhanced vigour was only a manifestation of the cleaning–up process and not a genetic change. Yet, reports of salt tolerance, increased sugar content, disease resistance, preferred morphological changes, etc. of somaclonal variants abound from most sugarcane tissue culture laboratories. Development of varieties has also been reported but none has made it through AICRP varietal trials.

III.3. *Transformation studies*

On account of its vegetative nature and complex genetic inheritance, sugarcane is highly suitable and amenable candidate for defect elimination/value addition through genetic transformation. Transformation and plant regeneration protocols are now well established in sugarcane. After the first report of protoplast transformation with gene *nptII*, many genes have been transferred to calli or shoot meristem tissues. The first report of a sugarcane transformant of agronomic importance was in NCo310 using *bar* gene for herbicide resistance through microprojectile bombardment (Gallo–Meagher and Irvine, 1996). Genetic engineering for various other traits like disease resistance, pest resistance, herbicide resistance, sucrose pathway engineering, etc. have been attempted (reviewed in Lakshmanan *et al.*, 2005; Ming *et al.*, 2006, Subramonian and Premachandran, 2007; Watt *et al.*, 2010; Manners and Casu, 2011). Transformation and pyramiding in sugarcane plants containing bovine pancreatic trypsin inhibitor (Aprotinin) with *Cry1 Ab* gene, driven by maize ubiquitin (*Ubi*–1) promoter was also successfully attempted against top and early shoot borer (Christy *et al.*, 2009; Arvinth *et al.*, 2010). At Sugarcane Breeding Institute, confined field trials of Co 86032 transgenics carrying *Cry 1Ab* or pyramided with Aprotinin gene are underway (SBI Highlights 2010–11).

The most thrilling development was of transgenic plants exhibiting doubling of total sugar content in mature internodes of an elite high sugar variety overcoming the constraints of osmotic limits (Wu and Birch, 2007). High yields of isomaltulose (IM) and trehalulose (TH), sucrose isomers of microbial origin and resistant to degradation by plant invertases, were obtained by vacuole–targeting of efficient isomer–specific synthase without disrupting the growth and development (Hamerli and Birch, 2011). The field trials involving these transgenics indicate a strong potential to develop sugarcane for commercial scale production of IM and TH (Basanayake *et al.*, 2012).

There are still different areas in transgenic research that need to be mastered. Somaclonal variation in the transformants still causes concern since most transformation protocols still find callus cultures the most amenable to regeneration. Transgene silencing and expression, promoter availability for temporal and spatial regulation of transgene expression, transgene containment and the coupled socio–legal environmental issues still need to be reconciled.

IV. Bioinformatics

With the advent of genome sequencing projects, voluminous data that needs to be handled has increased many–fold. This calls for a very efficient bioinformatics support. An example is the SUCEST–FUN database which assembles different databases like SUCEST, SGI, SUCAST, SUCAMET, GRASSIUS, etc. With the publicly available databases and specific software and programmes, sugarcane genomics has also picked up pace in the recent years. The development of new markers systems and their utilization, new approaches for mapping leading to increased coverage and saturation of linkage maps, QTL and association mapping, developing sequence databases and their application in functional genomics, gene manipulation strategies, etc. would not have been possible without a very fast evolving field of bioinformatics in sugarcane research.

Future research thrusts

1. Precision breeding through development of breeding stocks for specific traits. This would require the convergence of desired variability through introgression.
2. The availability of such relatively true breeding stocks shall lead to better understanding of the inheritance and genetic control of important traits.
3. Better molecular maps and functional genomic studies could be carried out with such stocks for gene identification and trait–specific molecular markers for MAS.
4. More physiological genetic studies focusing on rooting pattern for better nutrient use efficiency, harvest index and assimilate allocation towards sugar accumulation and, ratooning ability.

5. Enhancing breeding efficiency by greater reliance on breeding value of parents and family evaluation and selection; also by utilizing molecular markers for better selection efficiency.
6. Transgenics for defect elimination and value addition, more importantly for resistance/tolerance to biotic and abiotic stresses, and possibly sugar metabolism.

ACKNOWLEDGEMENTS

Sincere thanks are expressed for late Dr. B. L. Srivastava, Head, Division of Crop Improvement for leading the project on development of breeding stock for high sugar and Dr. S.K. Duttamajumder, Head, Division of Crop Protection for being an active Co–Investigator. Mr. Ram Kumar is thanked for rendering technical assistance. Respective Directors of the Indian Institute of Sugarcane Research and the Sugarcane Breeding Institute are thanked for kindly providing facilities of work and scientific interaction.

REFERENCES

Aitken, K., Li, J., Wang, L., Qing, C., Fan, Y.H. and Jackson, P. (2007). Characterization of intergeneric hybrids of *Erianthus rockii* and *Saccharum* using molecular markers. *Genet. Resour. Crop. Evol.,* 54: 1395–1405.

Amalraj, V.A. and Balasundram, N. (2011). Status of sugar–cane genetic resources in India. FAO 2011. *PGR Newsletter Issue No.* 148: 26–31.

Arvinth S., Arun, S., Selvakesavan, R.K., Srikanth, J., Mukunthan, N., Ananda Kumar, P., Premachandran, M.N. and Subramonian, N. (2010). Genetic transformation and pyramiding of aprotinin–expressing sugarcane with *cry1Ab* for shoot borer (*Chilo infuscatellus*) resistance. *Plant Cell Rep.,* 29: 383–395.

Asnaghi, C., Roques, D., Ruffel, S., Kaye, C., Hoarau, J.Y., Telismart, H., Girard, J.C., Raboin, L.M., Risterucci, A.M., Grivet, L. and D'Hont, A. (2004). Targeted mapping of a sugarcane rust resistance gene (Bru1) using bulked segregant analysis and AFLP markers. *Theor. Appl. Genet.,* 108: 759–764.

Bakshi Ram (2010). Breeding for higher sucrose content in sugarcane –achievements and prospects. Proc. of the 9th Joint Conv. of STAI and SISSTA: pp. 74–84.

Bhat, S.R. and Gill, S.S. (1985). The implication of 2n egg gametes in nobilization and breeding of sugarcane. *Euphytica*, 34: 377–384.

Cai, Q., Aitken, K., Deng, H.H., Chen, X.W., Fu, C., Jackson, P.A. and McIntyre, C.L. (2005). Verification of introgression of *Erianthus arundinaceus* germplasm into sugarcane using molecular markers. *Plant Breed*, 124: 322–328.

Chandran, K. (2010). *In vitro* multiplication and conservation of *Saccharum* germplasm. *Indian J. Plant Genetic Res.,* 23: 65–68.

Christy, L.A., Arvinth, S., Saravanakumar, M., Kanchana, M., Mukunthan, N., Srikanth, J., George Thomas and Subramonian, N. (2009). Engineering sugarcane cultivars with bovine parncreatic trypsin inhibitor (aprotinin) gene for protection against top borer (*Scirpophaga excerptalis* Walker). *Plant Cell Rep.,* 28: 175–184.

Cuadrado, A., Acevedo, R., Moreno, Dias de la Espina, S., Jouve, N. and de la Torre, C. (2004). Genome remodelling in three modern *S. officinarum x S. spontaneum* sugarcane cultivars. *J. Exp. Bot.,* 55: 847–854.

Daugrois, J.H., Grivet, L., Roques, D., Hoarau, J.Y., Lombard, H., Glaszmann, J.C and D'Hont, A. (1996). A putative major gene for rust resistance linked with an RFLP marker in sugarcane cultivar R 570. *Theor. Appl. Genet.,* 92: 1059–1064.

D'Hont A., Lu, Y.H., Le'on, D.G.D., Grivet, L., Feldmann, P., Panaud, E. and Glaszmann, J.C. (1994). A molecular approach to unravelling the genetics of sugarcane, a complex polyploid of Andropogoneae tribe. *Genome,* 37: 222–230.

D'Hont, A., Grivet, L., Feldmann, P., Rao, P.S., Berding, N. and Glaszmann, J.C. (1996). Characterization of the double genome structure of modern sugarcane cultivars (*Saccharum* spp.) by molecular cytogenetics. *Mol. Gen. Genet.,* 250: 405–413.

D'Hont, A. and Glaszmann, J.C. (2001). Sugarcane genome analysis with molecular markers, a first decade of research. *Proc. Int. Soc. Sugar Cane Technol.,* 24: 556–559.

Duttamajumder, S.K., Srivastava, B.L. and Kapur, R. (2004). Wilt disease a major deterrent in breeding for early sugarcane genotypes in subtropical India. *Indian J. Agricultural Sci.*, 74: 80–83.

Edme,x S.J., Miller, J.D., Glaz, B., Tai, P.Y.P. and Comstock, J.C. (2005). Genetic contribution to yield gains in the Florida sugarcane industry across 33 years. *Crop Sci.,* 45: 92–97.

Gupta V., Raghuvanshi, S., Gupta, A., Saini, N., Gaur, A., Khan, M.S., Gupta, R.S., Singh, J., Duttamajumder, S.K.D., Srivastava, S., Suman, A., Khurana, J.P., Kapur, R. and Tyagi, A.K. (2010). The water–deficit stress– and red–rot–related genes in Sugarcane. *Functional and Integrative Genomics*, 10(2): 207–214.

Hamerli, D. and Birch, R.G. (2011). Transgenic expression of trehalulose synthase results in high concentrations of the sucrose isomer trehalulose in mature stems of field–grown sugarcane. *Plant Biotech. J.,* 9: 32–37.

Jalaja N.C., Neelamathi, D. and Sreenivasan, T.V. (2008). Micropropagation for quality seed production in sugarcane in Asia and the Pacific. Food and Agriculture Organization of the United Nations, Rome; Asia–Pacific Consortium on Agricultural Biotechnology, New Delhi; Asia–Pacific Association of Agricultural Research Institutions, Bangkok, p. i–x + 46.

Kapur, R., Srivastava, B.L. Duttamajumder, S.K., Madhok, H.L., Ram Hit and Kumar, R. (2011). A plant–type concept in sugarcane for the subtropics. Proc. Int. Sugar Conf. pp. 514–516.

Lakshmanan P., Geijskes R.J., Aitken K.S.,Grof, C.L.P., Bonnett, G.D. and Smith G.R. (2005). Sugarcane biotechnology: The challenges and opportunities. *In vitro Cell. Dev. Biol. Plant,* 41: 345–363.

Lingle, S.E., Viotar, R.P., Johnson, R.M., Tew, T.L. and Boykin, D.L. (2009). Recurrent selection for sucrose content has altered growth and sugar accumulation in sugarcane. *Field Crops Res.,* 113: 306–311.

Ming, R., Moore, P.H., Wu, K.K., D'Hont, A., Glaszmann, J.C. and Tew, T.L. (2006). Sugarcane improvement through breeding and biotechnology. *Plant Breed. Rev.,* 27: 15–118.

Nair, N.V., Selvi, A., Sreenivasan, T.V., Pushpalatha, K.N. and Mary, S. (2006). Characterization of inter–generic hybrids of *Saccharum* using molecular markers. *Genet. Resour. Crop. Evol.,* 53: 163–169.

Nair, N.V. (2007). Sugarcane genetic resources: An indian perspective. *In*: Singh S.B., Rao G.P., Solomon, S. and Gopalasundaram, P. (*Eds.*) Sugarcane Crop Production and Improvement. Studium Press LLC, Texas, USA, pp. 1–20.

Pant, N.M. (2007). Sugarcane micropropagation and production of good quality planting material. *In*: Singh, S.B., Rao, G.P., Solomon, S. and Gopalasundaram, P. (*Eds.*) Sugarcane Crop Production and Improvement. Studium Press LLC, Texas, USA, 2007, pp. 247–266.

Piperidis, G., Piperidis, N. and D'Hont, A. (2010). Molecular cytogenetic investigation of chromosome composition and transmission in sugarcane. *Mol. Genet. Genomics*, 284: 65–73.

Sandhu, S.K., Uppal, S.K., Sharma, B. and Jain, J. (2011). Genetic enhancement in sugarcane –potential of interspecific introgression lines for cane yield, quality and red rot resistance. Proc. Int. Sugar Conf., pp. 539–545.

Selvi A., Nair, N.V., Noyer, J.L., Singh, N.K., Balasundaram, N., Bansal, K.C., Koundal, K.R. and Mohapatra, T. (2005). Genomic constitution and genetic relationship among the tropical and subtropical Indian sugarcane cultivars revealed by AFLP. *Crop Sci.,* 45: 1750–1757.

Selvi A., Nair, N.V., Noyer, J.L., Singh, N.K., Balasundaram, N., Bansal, K.C., Koundal, K.R. and Mohapatra, T. (2006). AFLP analysis of the phenetic organization and genetic diversity in the sugarcane complex, *Saccharum* and *Erainthus.*Genetic Resour. *Crop Evol.,* 68: 831–842.

Singh, R.K., Singh, S.B., Singh, S.P. and Sharma, M.L. (2011). Identification of sugarcane microsatellites associated to sugar content in sugarcane and transferability to other cereal genomes. *Euphytica*, 182: 335–354.

Smith, D.M., Inman–Bamber, M.G. and Thorburn, P.A. (2005). Growth and function of the sugarcane root system. *Field Crops Res.,* 92: 169–183.

Srivastava B.L., Cooper, M. and Mullins R.T. (1994). Quantitative analysis of the effect of selection history on sugar yield adaptation of sugarcane clones. *Theor. Appl. Genet.,* 87: 627–640.

Stringer, J.K., Cox, M.C., Atkin, F.C., Wei, X. and Hogarth, D.M. (2010). Family selection improves the efficiency and effectiveness of selecting original seedlings and parents. *Proc. ISSCT*, 27: 1–8.

Subramonian, N. and Premachandran, M.N. (2007). Biotechnological and molecular approaches for genetic improvement in sugarcane *In*: Singh, S.B., Rao, G.P., Solomon, S. and Gopalasundaram, P. (*Eds.*) Sugarcane Crop Production and Improvement. Studium Press LLC, Texas, USA, pp. 125–146.

Watt, D.A., Sweby, D.L., Potier, B.A.M. and Snyman, S.J. (2010). Sugarcane genetic engineering research in South Africa: From gene discovery to transgene expression. *Proc. Int. Soc. Sugar Cane Technol.,* 27: 1–8.

Wu, I. and Birch, R.G. (2007). Doubled sugar content in sugarcane plants modified to produce a sucrose isomer. *Plant Biotechnol. J.,* 5: 109–117.

Zhou, M. (2005). Potential of using physiological parameters to enhance sugarcane selection. *Proc. S. Afr. Sugar Technol. Assoc.,* 79: 521–529.

8

Pigeonpea

N. Nadarajan

Indian Institute of Pulses Research, Kanpur, India

The area, production and productivity of pigeonpea during the first decade of this century have shown some increase (Table 1). The increased production is mainly attributable to increase of area. The production and productivity trends in major production. States are summarized in Table 2. Pigeonpea varieties released since 2000 are given in Table 3.

HYBRID PIGEONPEA TECHNOLOGY

ICAR has allocated considerable effort and resources to develop productive hybrids of pigeon pea and this has resulted in the release of a number of hybrids. These include ICPH 8 in 1991 (by ICRISAT) and some genetic male sterility-based (GSM) hybrids from the national programme. The first among these was PPH 4 in 1993 from Punjab Agricultural University, Ludhiana. The Tamil Nadu Agricultural University released a short duration hybrid IPH 732 (CoPH 1) in 1994 and hybrid CoPH 2 in 1997. Subsequently, two more pigeonpea hybrids AKPH 4101 and AKPH 2022 were released by PDKV, Akola in 1997 and 1998, respectively.

In spite of release of six pigeonpea hybrids at the national level, the hybrid seed cost being within affordable levels and the yield advantage of hybrids being stable, the hybrid technology could not be extended by the seed sector. This is because the technology suffers major bottlenecks

V.L. Chopra (Ed.): Breeding Field Crops II: Advances, Studium Press (LLC), USA.

Table 1: All India area, production and productivity of pigeonpea

Year	*Area (m ha)*	*Production (m tonnes)*	*Productivity (kg/ha)*
2000–01	3.63	2.25	618
2001–02	3.33	2.26	679
2002–03	3.36	2.19	651
2003–04	3.52	2.35	670
2004–05	3.52	2.35	667
2005–06	3.58	2.74	765
2006–07	3.56	2.31	650
2007–08	3.73	3.08	826
2008–09	3.38	2.27	671
2009–10	3.53	2.46	697
2010–11	4.42	3.19	721

Table 2: Trends in area (K ha), production (t) and yield (kg/ha) of pigeonpea in different states of India

State		*1996-2000*	*2001-05*	*2006-10*
Andhra Pradesh	Area	376.25	473.20	450.25
	Production	130.00	198.80	241.50
	Productivity	338.50	419.20	529.50
Bihar	Area	67.15	39.48	32.90
	Production	85.20	48.38	36.57
	Productivity	1274.50	1223.20	1117.50
Gujarat	Area	380.55	302.86	268.50
	Production	329.60	197.06	263.50
	Productivity	866.25	665.60	987.75
Karnataka	Area	462.70	534.50	618.75
	Production	209.68	228.24	378.25
	Productivity	446.00	423.60	609.25
M.P	Area	363.45	312.88	322.87
	Production	298.70	232.32	233.62
	Productivity	821.50	741.60	724.00
Maharashtra	Area	1025.17	1062.02	1101.25
	Production	686.20	712.66	823.75
	Productivity	668.00	672.40	742.00
Orissa	Area	136.30	131.30	135.45
	Production	78.87	82.94	109.32
	Productivity	578.75	635.00	806.50
Rajasthan	Area	32.20	20.42	18.97
	Production	27.30	11.14	13.20
	Productivity	867.75	550.20	700.75
Tamil Nadu	Area	78.62	51.28	30.80
	Production	50.65	32.54	20.15
	Productivity	653.00	627.00	662.50
Uttar Pradesh	Area	450.88	382.92	367.88
	Production	521.75	418.78	324.38
	Productivity	1161.75	1089.80	885.25
West Bengal	Area	3.18	4.16	2.20
	Production	2.23	3.38	1.48
	Productivity	700.25	29.00	722.25

when it comes to large-scale seed production (Niranjan *et al.*, 1998). These production constraints can be overcome if an efficient cytoplasmic nuclear male-sterility (CMS) based hybrid seed production technology can be developed.

Development of CMS Systems in Pigeonpea - A Major Breakthrough

Considering the limitations of large-scale hybrid seed production in GMS hybrids, the development of cytoplasmic nuclear male-sterility (CMS) became imperative. So far, five such systems have been reported in pigeonpea of varying degrees of success. Out of these, A_2 cytoplasm has shown promise because of its stability in different agro-climatic zones and availability of good maintainers and restorers. A brief description of A_1 to A_5 CMS systems of pigeonpea is presented below.

Cajanus sericeus (A_1) cytoplasm

The CMS lines derived from this species (Saxena *et al.*, 1997) are not stable at low temperatures. At or below 10°C the male-sterile plants revert back to male fertility (Saxena, 2005). This tendency is more pronounced in the early maturing male-sterile lines. In the long duration types this instability is not so prominent. The CMS derived from *C. sericeus* produces good heterosis (K.B. Saxena, unpublished). However, the occurrence some pollen shedders in the female line, and absence of good maintainers makes it commercially non-viable for hybrid breeding.

C. scarabaeoides (A_2) cytoplasm

The CMS system derived from this species is reported to be very stable (Tikka *et al.*, 1997; Saxena and Kumar, 2003). This system has shown promise in terms of yield (IIPR, 2007).

C. volubilis (A_3) cytoplasm

Wanjari *et al.* (2001) isolated CMS genotypes with maternal inheritance from derivatives of a cross between *C. volubilis* and *C. cajan* (var. ICPL 83024). However, the CMS lines developed from this species could not become popular due to their fertility restoration problems.

C. cajanifolius (A_4) *cytoplasm*

C. cajanifolius is the most closely related wild species of pigeonpea and a single gene differentiated progenitor of cultivated type (De, 1974). The CMS system derived from this species is the best among the CMS systems developed so far. This CMS system has good number of maintainers and restorers. The male-sterile lines are stable across locations and years. The F_1 hybrid plants produce excellent pollen load and pod set.

C. cajan (A_5) *cytoplasm*

Mallikarjuna and Saxena (2005) found that when *C. cajan* is used as female parent and *C. acutifolius* as pollen parent the resulting hybrid

Table 3: Pigeonpea varieties released after 2000

Zone	*Variety/Hybrid**	*Year of identification/ Release*	*Area of adoption/origin*
North West Plain Zone	Pusa 992 (E)	2002	IARI, New Delhi
	PAU 881 (E)	2007	PAU, Ludhiana
	PA291 (E)	2008	G.B.P.U.A.&T., Pantnagar
North East Plain Zone	Laxmi (Kanke3) (ML)	2000	BAU, Ranchi
	MA 6 (L)	2002	BHU, Varanasi
	NDA 981 (NDA 2) (L)	2004	NDUA&T, Faizabad
	MAL 13 (L)	2005	BHU, Varanasi
	NDA 996 (NDA 3) (L)	2006	NDUA&T, Faizabad
Central Zone	AKT 811 (E)	2000	PDKV, Akola
	BSMR 853 (M)	2001	ARS, Badnapur
	GT 101 (E)	2002	SDAU, SK Nagar
	GAUT 001E (E)	2003	SDAU, SK Nagar
	BDN 708 (M)	2004	ARS, Badnapur
	GTH 1 (CGMS Hybrid)	2004	SDAU, SK Nagar
	JKM 189 (M)	2006	Khargone
	Vipula(M)	2006	Rahuri
	TT 401(M)	2007	BARC, Trombay
	TJT501 (M)	2008	BAARC & Khargone
South Zone	Vamban 2 (ML)	2000	NPRC (TNAU),Vamban
	TS 3 (ML)	2001	UAS, Gulbarga
	WRP 1 (ML)	2002	ARS, Gulbarga
	CoRG 9701(E)	2004	TNAU, Coimbatore
	GC 11 -39 (E)	-	ARS, Gulbarga
	BRG 1 (M)	2005	UAS, Bangalore
	BRG 2 (M)	2009	UAS, Bangalore

*E = early; ML = medium late, L= late

is sterile because of nuclear-cytoplasmic interaction. The system shows perfect fertility restoration by cultivated accessions. However, the male-sterility is maintained only by its wild relative parent. Attempts are underway to breed for its maintainers among the cultivated types (K.B. Saxena, personal communication).

DEVELOPMENT OF CGMS BASED HYBRIDS

Following two hybrids have been developed:

1. GTH-1

World's first CGMS based pigeonpea hybrid, 'GTH-1' developed at S D A U, S K Nagar, has been identified and released for cultivation in central zone. Parents of this hybrid are GT 288 A (CMS line/female having cytoplasm of *C. scarabaeoides*) and GTR-11 (restorer/male). Based on yield trials (2000-2003), GTH 1 (1760 kg/ha) gave 42% yield superiority over the best check AKPH 4101 (1240 kg/ha) a latest released GMS based hybrid and 32% yield superiority over the best local variety, GT 101 (1330 kg/ha). This hybrid is early in maturity (140 days) and possesses indeterminate plant type and large white seeds. This hybrid was used as check in multilocation trials conducted for evaluation of pigeonpea hybrids in IHT and AHT. In these trials this hybrid gave the highest yields in Central Zone as compared to the experimental hybrids. Therefore, this hybrid was released by CVRC in 2007 for its cultivation in whole Central Zone. This hybrid has got notified in 2007.

2. ICPH 2671

Hybrid ICPH 2671 (Pushkal) was developed by ICRISAT, and released by Madhya Pradesh state for its cultivation in Madhya Pradesh. But this hybrid is still not notified. This hybrid possesses a high yield potential with the added advantages of wilt and sterility-mosaic virus resistance, a trait absent in many high yielding varieties.

Genome Sequencing

In a recent development, the pigeonpea genome has been sequenced by a group of Indian scientists (Singh *et al.,* 2011) and ICRISAT scientists (Varshney *et al.,* 2011). A total 47,004 protein coding genes have been identified in the genome, of which 1,213 genes are for disease resistance and 152 genes for tolerance to drought, heat and salinity.

Both these reference genome sequences will, in the long run, facilitate identification of the genetic basis of agronomically important traits, and accelerate the development of improved pigeonpea varieties.

REFERENCES

De, D.N. (1974). Pigeonpea. *In*: Hutchinson, J. (*Ed.*) Evolutionary Studies in World Crops. Cambridge University Press, London. pp. 79–87.

IIPR. (2007). Enhancing yield and stability of pigeonpea through heterosis breeding. Progress Report ISOPOM Project. Indian Institute of Pulses Research, Kanpur, India p. 61.

Mallikarjuna, N. and Saxena, K.B. (2005). A new cytoplasmic nuclear male-sterility system derived from cultivated pigeonpea cytoplasm. *Euphytica*, 142: 143–148.

Niranjan, S.K.D.F.F., Bantilan, M.C.S., Joshi, P.K. and Saxena, K.B. (1998). Evaluating hybrid technology for pigeonpea. pp. 231–240. *In*: Proceedings of the International Workshop on Joint Impact Assessment of NARS/ICRISAT Technologies for the Semi-Arid Tropics, held 2-4 December 1996 at ICRISAT Center, A P, India.

Palmer, R.G., Gai, J., Dalvi, V.A. and Suso, M.J. (2011). Male sterility and hybrid production technology. *In*: Pratap, A. and Kumar, J. (*Eds.*) Biology and Breeding of Food Legumes. CABI, Oxfordshire, U.K. pp. 193–207.

Saxena, K.B. (2005). Pigeonpea (*Cajanus cajan* (L.) Millsp.) pp. 85–115. *In*: Genetic Resources, Chromosome Engineering, and Crop Improvement, Vol 1: Grain Legumes (Singh, R.J. and Jauhar, P.P. *Eds.*) CRC Press, USA.

Saxena, K.B., Kumar, R.V., Singh, L. and Raina, R. (1997). Development of a cytoplasmic-male-sterility system in pigeonpea. Progress report forth consultative group meeting on cytoplasmic male-sterility in Pigeonpea March 3-4, 1997. Bhabha Atomic Research Centre, Trombay, India. p. 19.

Saxena, K.B. and Kumar, R.V. (2003). Development of a cytoplasmic-nuclear male-sterility system in pigeonpea using *C. scarabaeoides* (L.) Thouars. *Indian Journal of Genetics and Plant Breeding*, 63(3): 225–229.

Singh, N. K. Gupta, D.K., Jayaswal, P.K., Mahato, A.K., Dutta, S., Singh, S., Bhutani, S., Dogra, V., Singh, B.P. and Kumawat, G. (2011). The first draft of the pigeonpea genome sequence. *J. Plant Biochem.,* Biot. doi: 10.1007/s13562-011-0088-8.

Tikka, S.B.S., Parmar, L.D.Q and Chauhan, R.M. (1997). First record of cytoplasmic-genic male-sterility system in pigeonpea (*Cajanus cajan* (L.) Millsp.) through wide hybridization. *Guj. Agri. Uni. Res. J.,* 22: 160–162.

Varshney, R.K., Chen, W., Li, Y., Bharti, A.K., Saxena, R.K., Schlueter, J.A., Donoghue, M.T.A., Azam, S., Fan, G., Whaley, A.M., Farmer, A.D., Sheridan, J., Iwata, A., Tuteja, R., Penmetsa, R.V., Wu, W., Upadhyaya, H.D., Yang, S.P., Shah, T., Saxena, K.B., Michael, T., McCombie, W.R., Yang, B., Zhang, G., Yang, H., Wang, J., Spillane, C., Cook, D.R., May, G.D., Xu, X. and Jackson, S.A. (2011). Draft genome sequence of pigeonpea (*Cajanus cajan*), an orphan legume crop of resource-poor farmers. *Nat. Biotechnol.,* doi: 10.1038/nbt.2022.

Wanjari, K.B., Patil, AN., Manapure, P., Manjaya, J. G. and Manish, P. (2001). Cytoplasmic male-sterility in pigeonpea with cytoplasm from *Cajanus volubilis*. *Ann. Pl. Physiol.,* 13: 170–174.

9

Chickpea

R.M. Singh

Department of Genetics and Plant Breeding, Institute of Agricultural Sciences, Banaras Hindu University, Varanasi, India

Chickpea (*Cicer arietinum*) is an annual self-pollinated crop which occupies third position in legume production globally. It is a diploid (2n=2x=16) with moderate genome size of approximately 740 Mb. The global chickpea area, production and productivity during 2010 were 11.97 m ha, 10.89 m t and 913 kg/ha, respectively (FAOSTAT, 2010). India is the largest chickpea producing country, with 68% of the world chickpea production. Estimated area, production and productivity during 2011-12 are 9.01 m ha, 7.58 mt and 841 kg/ha, respectively. Seven states (Madhya Pradesh, Uttar Pradesh, Rajasthan, Maharashtra, Gujarat, Andhra Pradesh and Karnataka) share 95% of the total area under chickpea cultivation in the country. During the past decade, the productivity of chickpea has increased by 1.74% but the gross chickpea production has gone up by 6.32% along with growth in area by 4.43%.

Out of 350 chickpea varieties developed globally, more than 180 varieties have been released in India under AICRP for varying agro-climatic conditions having resistance/tolerance to different biotic and abiotic stresses. Some of the important varieties tolerant to various stresses are listed below (Table 1).

V.L. Chopra (Ed.): Breeding Field Crops II: Advances, Studium Press (LLC), USA.

Table 1: Recently released chickpea varieties with specific trait of interest

	Traits	***Varieties***
1a	Wilt	Avrodhi, KWR 108, JG 315, JG 74, ICCV 10, Pusa 256, ICCC 37, Vishwas, Vijay, Vishal, and Haryana Channa 1
1b	*Ascochyta* blight	Gaurav,Vardan, Samrat, PBG 1 and BG 261
2	Short duration varieties	BG 372, KPG 59, Pant G 186, ICCV 2, ICCV 10, BGM 547, Vijay Rajas and JG 11
3	Kabuli varieties	KAK 2, BG 1053, BG 1003, JGK 1, Phule G 95311, Virat
4	Salt tolerance	CSG 8962
5	Excessive moisture	DCP 92-3, HC 3 and HC 5

GENETIC RESOURCES

Out of total germplasm of 57,037 held by different institutions, the combined share of ICRISAT, NBPGR and ICARDA is around 85%. Development of multi-parent advanced generation inter-cross (MAGIC) population is in progress at ICRISAT for accumulation of large number of recombination events for increasing novel alleles to bring in greater genetic diversity. These populations can be used for gene discovery, characterization and their deployment for studying complex traits (Glaszmann *et al.*, 2010). In addition to trait mapping, the highly recombined MAGIC population may be used directly as source material for the development of better varieties.

Garg *et al.* (2011) have generated about 2 million high-quality sequences of average length of 372 bp using pyro-sequencing technology and have identified more than 4000 simple sequence repeats, which can be developed as functional molecular markers in chickpea. Further, a comprehensive genetic map spanning 845.56 cM was developed using RILs from *Cicer arietinum* (ICC 4958) × *C. reticulatum* (PI 489777) which included 1291 markers on eight linkage groups, with the highest number of markers being on LG3 (218) and the lowest on LG8 (68) with an average inter-marker distance of 0.65 cM. Using flow cytometry and PCR-based primers, Zatloukalova *et al.* (2011) assigned LG8 to chromosome H, LG5 to chromosome A, LG4 to chromosome E and LG3 to chromosome B. The chromosomes C and D which could not be sorted out were jointly assigned to LG6 and LG7. Likewise, LG1 and LG2 were assigned to chromosomes F and G. This procedure could be used to discover genes and determine order of low- copy genic regions on the chromosomes.

Plant genetic resources (PGR) are vital for future genetic upgradation and safe-guarding against unforeseen threat to production. Under 'Generation Challenge Programme'of the Consultative Group of

International Agricultural Research (CGIAR), Upadhyaya *et al.* (2008) extracted a reference set of 300 accessions using SSR markers and reported its genetic structure, diversity and allele richness. *Desi* and *kabuli* shared 436 alleles, whereas, wild *Cicer* shared 17 and 16 alleles with *desi* and *kabuli,* respectively. *Desi* chickpea contained higher proportion of rare alleles (53%) than *kabuli* (46%), where wild *Cicer* accessions were devoid of rare alleles. This reference set captured 1315 alleles (78%) of the 1683 alleles representing diversity from the entire spectrum of composite collection. A number of accessions with desirable traits have been identified (ICRISAT unpublished data) from this set which is therefore a valuable genetic resource for locating lines that can be used for functional and comparative genomics, for mapping and cloning genes and for enhancing genetic potential of chickpea. Some important accessions with resistance to disease are listed in Table 2.

Table 2: Sources of stable resistance/tolerance to disease and other important traits

Disease	***Varieties***
Wilt	MPJG 2001-12, Phule G 9425-9, IPC 97-29, ICP 2000-32, PBG 1, H 82-2 GNG 1381, JG 2000-7, JG 2004-03, JG 2003-14-16
Dry root rot	ICC 2644, 10384, 10630, MPJGK 3
Ascochyta blight	GNG 1581, H 01-80, E 100Y, GLK 1232, GLK 83145
BGM (MR)	BG 276, GL 90159, 91040, 91071, 92162

Variability for Traits of Interest in Wild *Cicer* Species

Most of the annual wild *Cicer* species are repositories of resistance against *Ascochyta* blight, *Fusarium* wilt, *Helicoverpa* (pod borer), leaf minor, *Botrytis* grey mould, Bruchids, and cold. Resistance against *Phytophthora* root rot, Botrytis grey mould and drought are observed only in *C. echinospermum, C. bijugum* and *C. microphyllum* respectively. Among wild species, *C. reticulatum* showed the largest morphological variability followed by *C. pinnatifidum, C. echinospermum* and *C., bijugum* which included variability for wide leaflets in *C. chorassanicum*, number of branches in *C. bijugum* and *C. reticulatum* and early flowering in *C. judaicum.* Members of the primary gene pool of the chickpea consisting of 38 accessions of *C. arietinum*, six of *C. reticulatum* and four of *C. echinospermum* grown in India were investigated using 100 SSR markers to analyse their genetic structure, diversity and relationships. Most of the diversity was confined to wild species, which had higher value for polymorphic information content (PIC), gene diversity and heterozygosity than the cultivated species, implying a narrow genetic base of cultivated chickpea (Choudhary *et al.*, 2012).

Nguyen *et al.* (2004) observed that world chickpea germplasm has a narrow genetic base and lacks useful traits for varietal improvement and therefore use of wild species (which have wider genetic base) is desirable as potential source of resistance against biotic and abiotic stresses (Singh *et al.*, 2008a). It is equally important to enlarge the number of accessions in the primary gene pool to enhance available genetic diversity for incorporation into *C. arietinum*, the cultivated species.

Species Relationship

Though successful crosses of *C. bijugum, C. pinnatifidum* and *C. judaicum* with *C. arietinum* have been reported by various workers, using ovule and embryo culture techniques, their true hybrid nature has not been established. Therefore, these three wild species along with *C. chorassanicum, C. yamashitae* and *C. cuneatum* have been placed in tertiary gene pool. Thus, primary gene pool comprises only of *C. arietinum* and *C. reticulatum* (Singh *et al.*, 2008, Fig. 1). Therefore, for utilizing wild species in chickpea improvement, greater efforts should be deployed to create crosses using *C. reticulatum* with other wild species to develop pre-breeding lines for successful introgression of desirable genes into cultivated species.

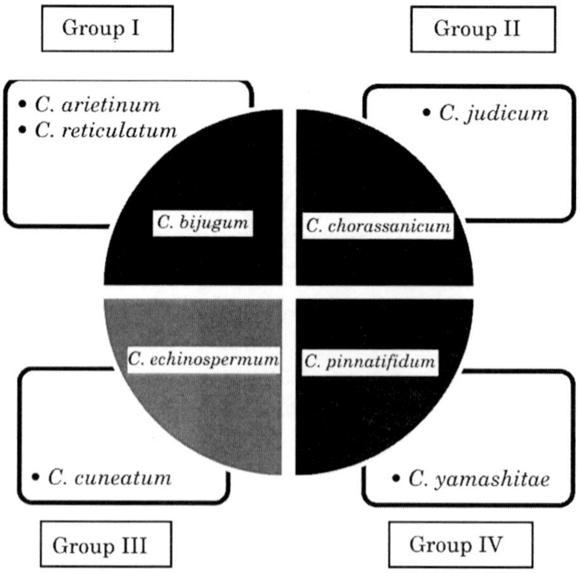

Fig.1: Schematic diagram depicting relationship between nine annual *Cicer* species based on different methods. Controversial species are placed in the centre and the rectangles indicate the groups in which they can be placed (*Source* Singh *et al.*, 2008).

The major problem in transferring desirable genes from wild species to cultivated ones is the occurrence of post-zygotic abortion of immature embryos following interspecific hybridization. Microscopic studies suggest that suitable time to rescue chickpea × *C. bijugum* hybrids is at the early globular stage of embryogenesis (2-7 days-old), which requires complex tissue culture medium. In contrast, hybrids between chickpea × *C. pinnatifidum* abort later (up to 15-20 days old) at the heart-shaped or torpedo stages, and are easier to rescue *in vitro* (Wilson *et al.*, 2006).

In addition, it is essential to determine physiological, nutritional and hormonal requirements of developing chickpea embryos (Croser, 2002; Croser *et al.*, 2003a) for efficient interspecific hybridization among cultivated and wild species (Ahmad *et al.*, 2005).

Cytogenetics and Tissue Culture

The karyotypes of primary gene pool species resemble closely to those of tertiary gene pool species. The longest chromosome pair is associated with nucleolar organizing region (Ahmad, 2000) and thus is considered satellited chromosome in *C. arietinum, C. reticulatum*, and *C. echinospermum.*

Using flow cytometry, the individual chickpea chromosomes have been used for mapping specific DNA sequences and genes on them (Vlacilova *et al.*, 2002). Spatial resolution of neighboring loci has been demonstrated by using FISH analysis on super-stretched flow-sorted chickpea chromosomes (Valarik *et al.*, 2004).

Tissue culture of shoot tip explants of chickpea on MS medium supplemented with thidiazuron (3.0 mg/L) led to multiple shoot proliferation (2-10 per explant) within three weeks of culture. These were cultured on the same medium for elongation. The elongated shoots were transferred to Indolebutyric acid (IBA, 1.0 mg/L-5.0 mg/L) containing medium for rooting which was achieved within two weeks of culture. Rooted plantlets were successfully hardened under culture conditions and later established under field conditions with 86% survival. *In vitro* regenerated plants yielded viable seeds (Parveen *et al.*, 2012).

Genetic Linkage Mapping

A genetic map constructed from intraspecific crosses is preferable over maps from interspecific crosses because these are free from uneven recombination of homeologous chromosomes and distorted segregation ratios (both occurring in interspecific crosses). Using STMS markers and including 47 defense response genes, Winter *et al.* (2002) and Pfaff and Kahl (2003) have extended the linkage map to cover a distance of 2500 cM in 12 linkage groups. This makes the map one of the most elaborate ones.

Breeding Methods

For any crop improvement programme, the knowledge of nature of gene action controlling different characters is most desirable. Both additive and non-additive types of gene actions for most traits have been reported by various workers in chickpea. Flowering, primary and secondary branching, seeds/pod, 100 seed weight and grain yield per plant showed low heritability. Plant height and pods/plant had high heritability. Therefore, breeding procedures for self-pollinated crops would be most suitable for chickpea improvement. Non-additive gene action can only be utilized if CMS becomes available along with a flower mutant form that permits cross-pollination.

Preference should be given to selection of crosses rather than plants within crosses because of intermediate growth habit. Bulk method, in different modified forms, is now the most common selection method used after hybridization along with back-cross for transferring genes of interest in chickpea. Single-seed descent (SSD), a modified form of bulk method, procedure has been commonly used for developing recombinant inbred lines (RILs) for genome mapping studies.

Marker-Assisted Backcrossing (MABC) approach is being used at ICRISAT to transfer desirable major genes/QTLs into elite cultivars from donor parents without linkage drag. A genomic region spanning several QTL for root related traits, harvest index, yield and yield contributing traits on LG4 from two intra-specific RIL population was introgressed to one *desi* (JG 11) and two *kabuli* cultivars (KAK 2 and Chefe) and the BC_3F_4 lines are being evaluated at multiple locations in India, Ethiopia and Kenya. Also this approach is adopted for combining resistance to *Fusarium* wilt and *Ascochyta* blight. The advantages of MABC include: (i) the Foreground Selection (FS) can be used for identifying heterozygous plants at the seedling stage itself, (ii) recessive alleles can be identified, (iii) it can be used to develop near isogenic lines (NILs) which are commonly used for genomic analysis, and (iv) linkage drag can be minimized (Hospital, 2005). Molecular markers associated with double podding, resistance to *Fusarium* wilt and *Ascochyta* blight have been identified (Cho *et al*., 2002; Rajesh *et al*., 2002a) which will prove valuable in marker assisted selection (MAS). Among molecular markers, STMS markers are quite amenable to multiplex PCR and electrophoresis. But in chickpea, their number is too small for such use. The use of genomic approaches will immensely improve the efficiency of breeding programs and reduce the time for cultivar development.

Globally, 40–50% decline in yield occurs due to drought in chickpea. Screening for high root biomass, smaller leaf area, osmoticum adjustment and early maturity have provided some valuable germplasm for use in breeding programmes to develop drought tolerant cultivars. Large root traits of ICC 4958 and fewer pennules trait of ICC 5680 have been combined together to develop drought tolerance in chickpea (Saxena, 2003). Varshney *et al.* (2012), based on genotypic data from mapping populations of drought tolerant into sensitive crosses (ICC 4958 × ICC 1882 and ICC 8261 × ICC 283), have identified QTLs for root related traits involved in drought tolerance. This genomic region also has QTLs for other drought related traits like C-isotope discrimination, yield and harvest index under rain-fed conditions. Thus, this genomic region has been targeted for incorporation in elite chickpea lines for enhancing drought tolerance using marker-assisted backcrossing programme. Because of low heritability (h^2=0.20 to 0.33) for osmotic adjustment, the gains from selection for this trait are likely to be low (Abbo *et al.*, 2002b). The development of short duration varieties to escape terminal heat stress has received greater emphasis in the recent past at ICRISAT, leading to release of ICCV 2 which is a extra short duration *kabuli* variety. This was achieved from a multiple cross involving five parents (2 *kabuli* and 3 *desi*). This variety has helped expansion of chickpea to tropical conditions. Two super-early *desi* chickpea lines (ICCV 96029, 96030) developed at ICRISAT are being widely used as donors in crossing programmes in India and Canada to breed short duration chickpea varieties.

Croser *et al.* (2003b) have reviewed the effect of low temperature and its implications in chickpea improvement and observed that most cultivars which are susceptible to chilling temperature at flowering are adversely affected in productivity. Stable sources of cold tolerance were found in germplasm line ILC 8262, the mutant ILC 8617 and the breeding line FLIP87-82C which have been used to develop cold tolerance combined with *Ascochyta* blight resistance in several countries.

Late-embryogenesis abundant (LEA) proteins are reported to be closely associated with desiccation tolerance during seed development and response of plant to drought, salinity, and freezing, etc. Gu *et al.* (2011) isolated a LEA gene, CarLEA4, from chickpea based on a c-DNA library constructed with chickpea seedling leaves treated with polyethylene glycol (PEG). CarLEA4 contained two exons and one intron within the genomic DNA and encoded a putative polypeptide of 152 amino acids which may be a nuclear protein. The transcripts of CarLEA4 were detected in many chickpea organs, such as seedling roots, stems, leaves, flowers, and young pods. The transcript was highly expressed in developing seeds at 15 days-after-flowering and was induced by many abiotic stresses including exogenous hormonal treatments. It is suggested that CarLEA4 is a protein of LEA group 4 and may be involved in various plant developmental processes and abiotic stress responses.

In another study, microarray data showed that 4,815 differentially expressed unigenes were $\geq$ 2-fold up or $\leq$ 0.5-fold down-regulated in at least one of the five time points during drought stress. Also, 2,623 and 3,969 unigenes were time-dependent and differentially expressed only in drought- stressed root and leaf, respectively, and 110 pathways in two tissues were found to respond to drought stress. Compared to control, 88 and 52 unigenes were expressed only in drought stressed root and leaf respectively, while nine unigenes were expressed in both the tissues (Wang *et al*., 2012). On the basis of sequence similarity search (BLASTN and BLASTX), abiotic stress responsive genes have been identified in chickpea which can be used for developing varieties with improved yields under stress conditions using molecular plant breeding procedures (Roorkiwal and Sharma, 2012).

Pod borer *(Helicoverpa armigera)* and cowpea aphid (*Aphis craccivora*) cause major losses in chickpea. Researchers at Assam Agricultural University in collaboration with Commonwealth Scientific and Industrial Research Organization (CSIRO), Australia have developed marker-free transgenic chickpea plants expressing *Bt* gene (Cry2Aa) which provides resistance to pod borer infestation. Some transformed plants showed a near complete protection (over 98% larvae mortality) against pod borer. The field trials of chickpea transgenic plants are in progress. It has been found that *Allium sativum* (garlic) Leaf Lectin (ASAL), a manose binding homodimeric protein, is antagonistic to various sucking pests in the homopteran group including cowpea aphid. Researchers at Bose Institute, Kolkata have developed transgenic chickpea plants containing ASAL gene which showed improved resistance to the sucking pests. For the above technologies, ISCB has established a detailed licensing agreement with private partner, Mahyco and a public sector partner (the University of Agricultural Sciences Dharwad), the latter to be funded by ISCB for further development of the two technologies. Mehrotra and Singh (2011) used modified cry1Ab and cry1Ac genes under two different constitutive promoters to transform chickpea plants through *Agrobacrerium-*

mediated transformation of pre-conditioned cotyledonary nodes. Insect bioassay performed with transgenic plants showed relatively higher toxicity for plants expressing Cry1Ac protein as compared to Cry1Ab to *Helicoverpa armigera* larvae. Pyramided transgenic plants with moderate expression levels (15-20 ng per mg of total soluble protein) showed high level of resistance and protection against pod borer *H. armigera* larvae as compared to high level expression of a single toxin. Chickpea lines have also been transformed with Cry1X (*in planta* transformation) and plants harbouring cry1X were resistant to *H. armigera* (Asharani *et al.*, 2011).

Two new hybrid chickpea varieties (Amber and Neelam) have been developed by Khan and Siddique (2012) at the University of Western Australia that have high yield potential and resistance to diseases. Amber is an early flowering variety that gives high yield due to its bushy growth habit and profuse branching along with high degree of *Ascochyta* blight resistance. Neelam is a mid- flowering variety with wide adaptation, average seed size and attractive seed appearance combined with very high yield. These varieties will help the farmers to raise the production of chickpea in India.

Even though genetic sources of resistance to *Fusarium* wilt are available, due to lack of effective screening techniques, progress of developing resistant chickpea varieties has been slow. Selection of genotypes for *Fusarium* wilt resistance based on pollen bioassay is as reliable as marker assisted selection and/or sporophytic selection. The genotypes which were classified as wilt resistant, using molecular markers and *in vivo* seedling reaction to fusaric acid, were also resistant under *in vitro* pollen bioassay and required higher concentration of toxin for pollen growth inhibition (Babu, 2007).

Tekeoglu *et al.* (2000) reported that genes imparting resistance to races 1, 4, and 5 of *Fusarium* wilt are clustered on the same chromosome, unlike race 0 which is found in different region of the genome. They proposed the gene symbols foc-0, foc-4 and foc-5 for corresponding races of the pathogen, *Fusarium oxysporum*. Search and further evaluation of new gene clusters for *Fusarium* wilt resistance in chickpea will help in transferring resistance genes to new varieties.

Srinivasan and Gaur (2012) identified a spontaneous open flower mutant (OFM 3) in chickpea in which reproductive organs were exposed facilitating natural outcrossing. This trait was found to be recessive and controlled by a single gene. Large number of pods (73%) remained unfilled in OFM 3. So far, three independent genes (*ofl 1, ofl 2 and ofl3*) for open flower have been identified in chickpea. Breeding lines with open flower trait and higher percentage of filled pods have been developed from the progenies of the crosses of OFM 3 with normal-flowered lines. Thus open flower trait offers bright prospects for exploiting hybrid technology in chickpea.

Mutation breeding is another option for developing cultivars for traits whose genetic variability is narrow in chickpea. Two cultivars of chickpea, namely Pb-91 and C-44, were irradiated with ^{60}Co gamma rays at 300 and 400 Gys. Mutants derived from Pb-91 ranged in grain yield from 1838 to 2011 kg/ha as against 1538 kg/ha of the standard check –NIFA-2005. Likewise, mutants of C-44 were also significantly higher yielding than check (1598 kg/ha) ranging from 1803-2018 kg/ha. Some of these mutants had higher 100 seed weight than the two parents (Khattak *et al.*, 2012). Some mutant varieties released are RS 11, RSG 2 (Kiran), and WCG 2 (Surya).

In another study, combination of resistance for Fe-deficiency chlorosis and leaf minor has been achieved through gamma rays mutagenesis in chickpea (Toker *et al.*, 2012). Considerable yield losses occur due to iron (Fe) deficiency chlorosis in chickpea when grown in calcareous soil with high pH. Variety ICC 6119 (*desi*), which is leaf minor (*Liriomyza cicerina* Rond) resistant but susceptible to Fe-deficiency chlorosis, was irradiated with 200, 300, and 400 Gy gamma rays. In M_3 generation one mutant was identified for resistance to Fe-deficiency chlorosis and leaf minor using SSD method. Active Fe and chlorophyll contents in Fe efficient mutant were higher than parent ICC 6119. This mutant can be used as donor in chickpea improvement programme to develop high yielding cultivars which are also resistant to Fe-deficient chlorosis and leaf minor. In addition, mutant ACC 3204 (*kabuli* type) with high level of active Fe and chlorophyll content is an example of genetic biofortification which would prove valuable to overcome iron deficiency in human nutrition.

Selections made from interspecific crosses (*C. arietinum* x *C. echinospermum*) having resistance to root lesion nematode, *Phytophthora* and *Ascochyta* blight have been released as successful cultivars in Australia (Knights *et al.*, 2002).

It has been observed that seed filling duration in wild species *C. judaicum* was 15 days shorter than in cultivated species due to rapid loss of moisture content and hence resulted in earlier maturity and reduced seed biomass. This was associated with high acid invertase and low sucrose synthase activities in wild species. While transferring desirable genes from wild species into cultivated species such linkage drags must be eliminated to maintain the high productivity of chickpea cultivars (Kaur *et al.*, 2012). Some wild species have vernalization requirement, which may be beneficial in certain environments to avoid early flowering and consequent frost damage, but undesirable in normal situations (Abbo *et al.*, 2002). This calls for caution in using wild species in hybridization programme.

REFERENCES

Abbo, S., Lev-Yadun, S. and Galway, N. (2002a). Vernalization response of wild chickpea. *New Phytologist*, 154: 695–701.

Abbo, S., Turner, N.C., French, R.J. and Berger, J.D. (2002b). Breeding for osmotic adjustment in chickpea (*Cicer arietinum* L.). *In*: McComb, J.A. (*Ed.*) Proceedings of the 12th Australian Plant Breeding Conference, Australian Plant Breeding Association, Perth, Australia, pp. 463–467.

Ahmad, F., Gaur, P.M. and Croser, J.S. (2005). Chickpea (*Cicer arietinum* L). *In*: Singh, R.J. and Jauhar, P.P. (*Eds.*) Genetic Resources, Chromosome Engineering and Crop Improvement - Grain Legumes, Vol. 1, CRC Press, p. 376.

Asharani, B.M., Ganeshaiah, K.N., Raja, Amrita, Kumar, Vinod, and Makarla, Udayakumar (2011). Transformation of chickpea lines with Cry1X using in planta transformation and characterization of putative transformants T1 lines for molecular and biochemical characters. *J. Plant Breeding and Crop Science*, 3: 413–423.

Babu, D.R. (2007). Molecular evidence to *in vitro* pollen bioassay for wilt resistance in chickpea., Ph.D. Thesis, University of Agricultural Sciences, Dharwad, Karnataka, India.

Cho, S., Kumar, J., Shultz, J.L., Anupama, K., Tefera, F. and Muehlbauer, F.J. (2002). Mapping gene for double podding and other morphological traits in chickpea. *Euphytica*, 128: 285–292.

Choudhary, P., Khanna, S.M., Jain, P.K., Bharadwaj, C., Kumar, J., Lakhera, P.C. and Srinivasan, R. (2012). Genetic structure and diversity analysis of the primary gene pool of chickpea using SSR markers. *Genetics and Molecular Research*, 11: 891–905.

Clarke, H.J., Wilson, J.G., Kuo, I., Lülsdorf, M.M., Mallikarjuna, N., Kuo, J. and Siddique, K.H.M. (2006). Embryo rescue and plant regeneration *in vitro* of selfed chickpea (*C. arietinum* L.) and its wild annual relatives. *Plant Cell, Tissue and Organ Culture*, 85: 197–204.

Croser, J.S. (2002). Haploid and zygotic embryo culture of *Cicer*. Ph. D. Thesis, Univerity of Melbourne, Australia.

Croser, J.S., Ahmad, F., Clarke, H.J. and Siddique, K.H.M. (2003a). Utilization of wild *Cicer* in chickpea improvement-progress, constraints, and prospects. *Australian J. Agricultural Res.*, 54: 429–444.

Croser, J.S., Clarke, H.J., Siddique, K.H.M. and Khan, T.N. (2003b). Low- temperature stress: Implications for chickpea (*Cicer arietinum* L.) improvement. *Critical Review in Plant Sciences*, 22: 185–219.

Garg R., Patel, R.K., Jhanwar, S., Priya, P., Bhattacharjee, A., Yadav, G., Bhatia, S., Chattopadhyay, D., Tyagi, A.K. and Jain, M. (2011). Gene discovery and tissue-specific transcriptome analysis in chickpea with massively parallel pyro-sequencing and web resource development. *Plant Physiology*, 156: 1661–1678.

Gaur, P.M., Jukanti, A.K. and Varshney, R.K. (2012). Impact of genomic technologies on chickpea breeding strategies. *Agronomy*, 2: 199–221.

Glaszmann, J.C., Kilian, B., Upadhyaya, H.D. and Varshney, R.K. (2010). Accessing genetic diversity for crop improvement. *Curr. Opin. Plant Biol.,* 13: 1–7.

Gu, H., Jia, Y., Wang, X., Chen, Q., Shi, S., Ma, L., Zhang, J., Zhang, H. and Ma, H. (2012). Identification and characterization of a LEA family gene CarLEA4 from chickpea (*Cicer arietinum* L.). *Molecular Biology Report*, 39: 3565–3572.

Hospital, F. (2005). Selection in backcross programmes. *Philos. Trans. R. Soc. B*, 360: 1503–1511.

Kaur, H., Gupta, A.K., Kaur, N. and Sandhu, J.S. (2012). High acid invertase activity for a prolonged period in developing seeds/pod wall of wild chickpea is detrimental to seed filling. *Ind. J. Exptl. Biology*, 50: 735–743.

Khattak, G.S. S., Saeed, I. and Rehman, S. (2012). Breeding high yielding desi chickpea (*Cicer arietinum* L.) mutants. *Pak. J. Bot.,* 44: 255–258.

Knights, E.J., Southwell, R.J. and Schwinghamer, M.W. (2003). Evaluation of wild *Cicer* species for resistance to phytophthora root rot. *In*: Sharma, R.N. *et al.* (*Eds.*) Chickpea Research for the Millennium. Proc. Intl. Chickpea Conference. Indira Gandhi Agric. University, Raipur, India, p. 54.

Nguyen, T.T., Taylor, P.W.J., Redden, R.J. and Ford, R. (2004). Genetic diversity estimates in *Cicer* using AFLP analysis. *Plant Breed.,* 123: 173–179.

Parveen, S., Venkateshwarlu, M., Srinivas, D., Reddy, K.J.M. and Ugandhar, T. (2012). Direct *in vitro* shoot proliferation of chickpea (*Cicer arietinum* L.) from shoot tip explants induced by thidiazuron. *Bioscience Discovery*, 3: 1–5.

Pfaff, T. and Kahl, G. (2003). Mapping of gene specific markers on the genetic map of chickpea (*Cicer arietinum* L.). *Mol. Genet. Genomics*, 269: 243.

Rajesh, N., Tullu, A., Gil, J., Gupta, S., Ranjekar, K. and Muehlbauer, J. (2002). Identification of an STMS marker for double podding gene in chickpea. *Theor. Appl. Genet.,* 105: 604–607.

Roorkiwal, M. and Sharma, P.C. (2012). Sequence similarity based identification of abiotic stress responsive genes in chickpea. *Bioinformation*, 8: 92–97.

Saxena, N.P. (2003). Management of drought in chickpea – A holistic approach. *In*: Saxena, N.P. (*Ed.*) Management of agricultural drought – Agronomic and Genetic options, Oxford and IBH Publishing Co. Pvt. Ltd., New Delhi, p. 103.

Singh, R., Sharma, P., Varshney, R.K. and Sharma, S.K. (2008). Chickpea improvement: Role of wild species and genetic markers. *Biotechnol. Genet. Eng.,* 25: 267–314.

Srinivasan, S. and Gaur, P.M. (2012). Genetics and characterization of an open flower mutant in chickpea. *J. Heredity*, 103: 297–302.

Tekeoglu, M., Tullu, A., Kaiser, W.J. and Muehlbauer, F.J. (2000). Inheritance and linkage of two genes that confer resistance to *Fusarium* wilt in chickpea. *Crop Science*, 40: 1247–1251.

Toker, C., Canci, H., Inci, N.E., Ceylan, F.O., Uzun, B., Sonmez, S., Citak, S. and Ikten, C. (2012). Pyramiding of resistance to Fe-deficiency chlorosis and leaf minor (*Liriomyza cicerina* Rond.) in chickpea (*Cicer arietinum* L.) by mutation breeding. *Turkish J. Field Crops.,* 17: 41–45.

Upadhyaya, H.D., Dwivedi, S.L., Baum, M., Varshney, R.K., Udupa, S.M., Gowda, C.L., Hoisington, D. and Singh, S. (2008). Genetic structure, diversity, and allele richness in composite collection and reference set in chickpea (*Cicer arietinum* L.). *BMC Plant Biology*, 8: 106.

Válárik, M., Bartos, J., Kovárová, P., Kubaláková, M., de Jong, J.H. and Dolezel, J. (2004). High resolution FISH on super-stretched flow-sorted plant chromosomes. *The Plant J.,* 37: 940–950.

Varshney, R., Kudapa, H., Roorkiwal, M., Thudi, M., Pandey, M.K., Saxena, R.K., Chamarthi, S.K., Mohan, S.M., Mallikarjuna, N., Upadhyaya, H., Gaur, P.M., Krishnamurthy, L., Saxena, K.B., Nigam, S.N. and Pande, S. (2012). Advances in genetic and molecular breeding of three legume crops of semi-arid tropics using next-generation sequencing and high- throughput genotyping technologies. *J. Biosci.,* 37: 811–820.

Vlácilová, K., Ohri, D., Vrána, J., Cíhalíková, J., Kubaláková, M., Kahl, G. and Dolezel, J. (2002). Development of flow cytogenetics and physical genome mapping in chickpea (*Cicer arietinum* L.). *Chromosome Res.,* 10: 695–706.

Wang, X., Liu, Y., Jia, Y., Gu, H., Ma, H., Yu, T., Zhang, H., Chen, Q., Ma, L., Gu, A., Zhang, J., Shi, S. and Ma, H. (2012). Transcriptional responses to drought stress in root and leaf of chickpea seedling. *Molecular Biology Reports*, 39: 8147–8158.

Winter, P., Benko-Iseppon, A.M., Hüttel, B., Ratnaparkhe, M., Tullu, A., Sonnante, G., Pfaff, T., Tekeoglu, M., Santra, D., Sant, V.J., Rajesh, P.N., Kahl, G. and Muehlbauer, F.J. (2000). A linkage map of chickpea (*Cicer arietinum* L.) genome based on recombinant inbred lines from *C. arietinum* x *C. reticulatum* crosses. Location of resistance genes for Fusarium wilt races 4 and 5. *Theor. Appl. Genet.,* 101: 1155–1163.

Zatloukalová, P., Høibová, E., Kubaláková, M., Suchánková, P., Simková, H., Adoración, C., Kahl, G., Millán, T. and Doležel, J. (2011). Integration of genetic and physical maps of the chickpea (*Cicer arietinum* L.) genome using flow-sorted chromosomes. *Chromosome Res.,* 19: 729–739.

10

Mungbean

M.N. Singh and A.K. Singh

Department of Genetics and Plant Breeding, Institute of Agricultural Sciences, Banaras Hindu University, Varanasi, India

MUNGBEAN AREA, PRODUCTION AND PRODUCTIVITY

Mungbean or greengram (*Vigna radiata* L. Wilczek) is an important self-pollinated food legume crop ($2n=2x=22$) of South and Southeast Asia. It has high levels of folate and iron (Keatinge *et al.*, 2011). A mungbean crop can increase the yield of a succeeding rice crop by up to 8% through the nitrogen it fixes in the soil and by reducing pest and disease problems (Weinberger, 2003).

India is ranked first in mungbean production followed by China and Myanmar (Nair *et al.*, 2013). Out of ~6 m ha area at global level, India has a share of 3.55 m ha with total production of 1.82 m tonnes and productivity of 512 kg/ha, during 2010-11 (Table 1, Fig. 1). Being a short duration crop, it offers the best option for enhancing pulse production by fitting into varied cropping sequences. For example, mungbean area in rice fallows as well as spring summer crop has increased from 5.56 lakh ha in 2001-02 to 6.27 lakh ha in 2005-06.

Drastic reduction in area has occurred in Madhya Pradesh and Orissa where as considerable increase has been recorded in Rajasthan and Karnataka. Productivity of mungbean is the highest in Punjab (820 kg/ha) followed by Bihar (602 kg/ha). However, productivity is low in

V.L. Chopra (Ed.): Breeding Field Crops II: Advances, Studium Press (LLC), USA.

Karnataka (233 kg/ha), Orissa (247 kg./ha) and Madhya Pradesh (303 kg/ha). In terms of area and production, the top ranking states are Rajasthan, Maharashtra and Andhra Pradesh. There is ample scope for increase in area to the tune of 2-3 m ha under rice fallow, rice-wheat and rice-rice cropping systems which will require considerable improvement in production technology and seed supply of quality seed of improved varieties (Singh *et al.,* 2010).

Table 1: Area, production and productivity of mungbean in India (1964-65 to 2010-11)

Year	*Area (m ha)*	*Production (m tonnes)*	*Yield (kg/ha)*
1964-65	1.99	0.60	302
1970-71	2.07	0.70	339
1980-81	2.83	0.98	344
1990-91	3.36	1.38	413
2000-01	3.01	1.02	340
2010-11	3.55	1.82	512

Important diseases affecting mungbean include yellow mosaic virus, powdery mildew (*Erysiphe polygoni* D.C.), *Cercospora* leaf spot (*Cercospora canescens* Illis & Martin) and common bacterial blight (*Xanthomonas* spp.). The common pests are bruchids (*Callosobruchus* spp.), pod borer (*Maruca vitrata*), bean flies (*Ophiomyia* spp. and *Melanagromyza* spp.) and thrips (*Megalurothrips* spp.). Post harvest damage to mungbean seed by bean weevils or bruchids (*Callosobruchus chinensis* L. and *C. maculatus* F.) is common and

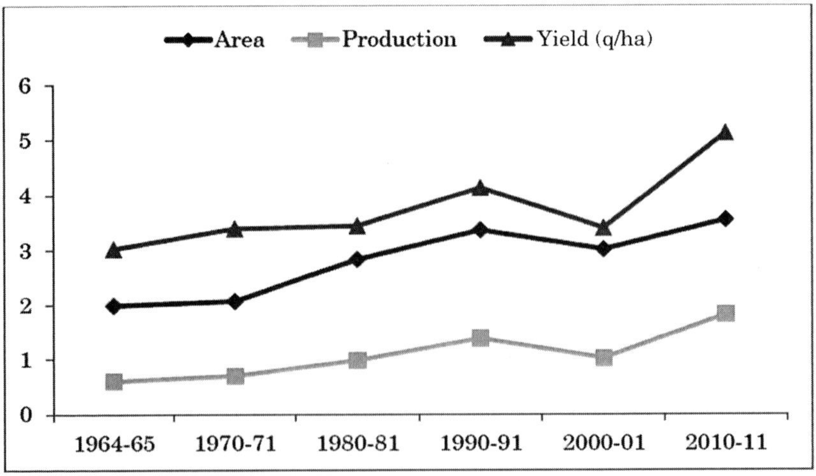

Fig. 1: Area, production and productivity of mungbean in India (1964-65 to 2010-2011)

can lead up to 100% loss (Zhang *et al.*, 2002). The genotypes resistant to bruchids have significantly higher trypsin inhibitor. Seed globulin content adversely affects bruchid egg deposition (Landerito *et al.*, 1993).

Crop improvement

Over the last decade (2000 to 2011), a total of 35 mungbean varieties, including four *viz.*, HUM 2, HUM 6, HUM 12 and HUM 16 from BHU Centre, have been released. In respect of maturity period, HUM 16 is the earliest maturing (55 days) followed by SML 668, Pusa 9531, Pusa Vishal, HUM 12, OUM11-5, COGG 912, IPM 02-3, IPM 02-14, Pant Mung 5 and PKV AKM 4 (60-62 days) [Table 2] (Project Coordinator's report MULLaRP, 2011-2012).

Germplasm Identification for Desired Traits

The following donors for desirable traits were identified in mungbean during last decade (2000-2010).

Donor	***Desirable traits***
MH 2-15	Resistant to MYMV, CLS, Anthracnose, Leaf Crinkle Virus, Leaf curl, Stem necrosis & *Macrophomina* blight
IPM 02-3	Resistant to MYMV, Root rot, Leaf Crinkle Virus & Stem necrosis
Sattya	Resistant to MYMV, CLS and Web blight
LGG 450, Shining moong 1	Resistant to Anthracnose and CLS
MH 96-1, ML 1108, ML 1194	Resistant to Anthracnose and Web blight
EC 304793, BDYR 2, VL 112	Resistant to *Meloidogyne javanica*
V2 802 and V2 709	Resistant to Bruchids (*C. chinensis* and *C. maculatus)*
HUM 16, Pusa Vishal, Pusa 9531, SML 668, Pant Moong 5, HUM 12	Large seeded (~6.0/100 seeds) with early maturity (55-60 days)
IPM 409-4 and IPM 205-7	Extra early maturity (46-48 days)

The following recommendations were made to raise a good mungbean crop in the recent past

Crop production

(i) Pre-emergence application of pendimethalin 30 EC @1.0 kg/ha + Imazethapyr 2 EC (Vellore 32) @ 0.75 kg/ha PE were most effective for controlling weeds.

Table 2: Mungbean varieties released during the decade (2000-2010) in India.

Sl. no.	*Name of variety*	*Developed by*	*Pedigree*	*Year of release*	*Maturity (days)*	*Average yield (q/ha)*	*Area of adoption*	*Characteristics*
1.	Pusa 9531	IARI, New Delhi	Sel. from NM 9473	2000	60	9.0	CZ for Summer	Res. to MYMV
2.	Pusa Vishal (Pusa bold 1)	-do-	Sel. from NM 2	2000	62	11.0	NWPZ for Summer	Large seed (6g/100 seeds)
3.	HUM 2 (Malaviya Jagriti)	BHU, Varanasi	Sel. from TVCM 3	2000	60	10.5	UP & U. Khand for Zaid	Mod. res. to MYMV
4.	HUM 6 (Malaviya Janpriya)	-do-	Sel. from BHUM 54	2001	68	10.0	UP for Zaid	High protein content (25%) & MR to MYMV
5.	Gaga 8	ARS, Sriganganagar	K851 × Pusa 105	2001	72	9.2	NWPZ for Kharif	Mod.res. to MYMV, CLS etc.
6.	SML 668	PAU, Ludhiana	Sel. from NM 94	2002	60	11.3	Punjab for Spring/Summer	Large seed & tol. to MYMV
7.	OUM 11-5	OUAT, Berhampur(0)	Mutant of Dhauli	2002	62	7.3	SZ for Kharif	MR to PM, MYMV & CLS-
8.	OBGG 52	OUAT, Berhampur(0)	Mutant of K 851	2002	65	8.0	Orissa, all seasons	
9.	HUM12 (Malaviya Janchetna)	BHU, Varanasi	HUM5 × DPM90-1	2003	60	11.2	NEPZ for Summer	MR to MYMV
10.	Meha(IPM 99-125)	IIPR, Kanpur	Pant Mung 3 × APM 36	2004	66	9.8	NEPZ for Spring	Res. to MYMV
11.	TMB 37	BARC, Mumbai	Kopergaon × TARM 2	2005	65	11.0	NEPZ for Summer	MR to MYMV
12.	COGG 912	TNAU,	MGG 336 ×	2005	62	8.0	SZ for Kharif	Res. to MYMV &

Table 2: (*Contd....*)

Table 2: (*Contd....*)

Sl. no.	*Name of variety*	*Developed by*	*Pedigree*	*Year of release*	*Maturity (days)*	*Average yield (q/ha)*	*Area of adoption*	*Characteristics*
		Coimbatore	COGG 902					CLS
13.	HUM16 (Malaviya Jankalyani)	BHU, Varanasi	Pusa bold1 × HUM 8	2006	55	10.9	NEPZ for Summer	Res. to MYMV
14.	MH 2-15 (Sattya)	HAU, Hisar	PDM 116 × Gujarat 1	2007	67	10.5	NWPZ for Kharif	Res. to MYMV & CLS
15.	Pant Mung 5	GBPUA&T, Pantnagar	Sel. from VC 6368	2007	60	12.0	U.P. & U. Khand for Zaid	Res. to MYMV
16.	Pant Mung 6	GBPUA&T, Pantnagar	Pant M 2 × AMP 36	2007	96	10.5	NHZ for Kharif	Res. to YMV, CLS, BLS & L crinkle
17.	KM 2241	CSAU, Kanpur	Samrat × PDM 54	2008	65	10.0	NHZ Kharif	Res. to MYMV
18.	IPM 02-3	IIPR, Kanpur	Meha × Pusa bold 2	2009	62	11.0	NWPZ for Kharif & Spring	Large seeded and Res. to MYMV
19.	PKV AKM 4	PKV, Akola	BM4 × PS 16	2009	62	10.0	CZ for Kharif	Res. MYMV
20.	Pusa 0672	IARI, New Delhi	11/395 × ML 267	2009	64	10.0	NHZ for Kharif	Res. to MYMV
21.	IPM 02-14	IIPR, Kanpur	Meha × Pusa bold 2	2010	60	11.0	SZ for Spring /Summer	Res. to MYMV

MR =Moderately resistant, Res.= Resistant, MYMV= Mungbean yellow mosaic virus, CLS= Cercospora leaf spot, PM= Powdery mildew
Source: Project Coordinator's Report MULLaRP, 2011-2012

(ii) Sole application of Imazethapyr, a post-emergence weedicide, @ 55 g ai/ha after 20 days of sowing was equally effective for controlling almost all kinds of weeds.
(iii) Two sprays of 2% urea at the interval of 10-15 days at flower initiation have given 20% yield increase in mungbean.
(iv) Inoculation of Rhizobium (AUBR 10) with PGPR (CRB 2 of *Pseudomonas* spp.) registered 40% increase in yield.

Crop protection

(i) Seed treatment with carbendazim @ 2 g/kg along with foliar spray of fungicides, propiconazole and difenoconazole were effective in reducing the severity of foliar diseases.
(ii) Seed treatment with imidacloprid (17.8 SL) @ 5 ml/kg was effective for controlling major insect-pest.

BREEDING STRATEGIES

Conventional Breeding

Pure line selection, recombination breeding, and mutation breeding have been successfully employed to develop new varieties of mungbean (Fernandez and Shanmugasundaram, 1988; Tickoo *et al.*, 2006). For broadening the genetic base and for obtaing rare recombinants, crosses with related species especially with blackgram have been employed and four high yielding and disease resistant varieties namely, HUM 1, Meha, Pant Mung 4 and Pant Mung 6 have been developed and notified during the last decade. The deployment of related species in breeding programs will become increasingly important for incorporating traits of interest into *Vigna* species

Improving Protein Quality

Protein malnutrition remains a major nutrition problem in Asia and affects children most severely (WHO 2000; UNSCN 2010). About 150 million children worldwide are underweight and 182 million are stunted of which about 70% are concentrated in Asia. Mungbean, with an average protein content of about 24% can play an important role in balancing vegetarian diets. However, the nutritional value of mungbean protein is limited by its low content of sulfur-amino acids that include methionine and cysteine. Black gram (accession no.VM 2164), is richer in methionine and qualifies as important donor for upgrading methionine content of mungbean through inter-specific hybridization.

Gamma-glutamyl-methionine (γ-Glu-Met) is the major dipeptide in blackgram seeds, whereas gamma-glutamyl-S-methyl-cysteine (γ-Glu-S-metCys) is the dominant dipeptide in mungbean. Both peptides were found in F_2 hybrids between mungbean and black gram. AVRDC has made an interspecific cross between VM 2164 and NM 94 for developing a recombinant inbred line (RIL) population for mapping of genes involved in high methionine content, molecular marker development for improving locally adapted lines for methionine content through backcrossing. Inter-specific hybridisation thus has latent potential for mungbean improvement (Nair *et al.*, 2013).

Germplasm Resources

The genus *Vigna* includes about 150 species distributed in warm and tropical regions of the world; 22 species are native to India and 16 to Southeast Asia, but the largest number of species are found in Africa (Polhill and van der Maesen, 1985). The genus *Vigna* includes 13 cultigens, including important legumes *viz.*, mungbean, cowpea and adzuki bean. To enable rational and cost effective evaluation of the large germplasm holdings, a core collection approach has been widely applied. Marker systems are now being used to a core set. A core set aims to represent about 90% of the allelic diversity in 10% of the accessions, while a mini-core collection aims to represent about 80% of the allelic diversity in 1% of the accessions (Upadhyaya and Ortiz, 2001).

For efficient utilization of the different mungbean gene pools for developing promising lines or superior varieties, an exhaustive characterization of the available germplasm collection in different gene pools is necessary. Core collections have been developed for the major *Vigna* crops *viz.*, mungbean, cowpea and adzuki bean. Molecular diversity in the large collection of mungbean germplasm has not been adequately studied. However, Sangiri *et al.* (2007) analysed a collection of mungbean germplasm encompassing 415 cultivated, 189 wild (*V. radiata* var. *sublobata*) and 11 intermediate accessions using 19 adzuki bean SSR markers. Their results revealed that mungbean has the highest diversity in South Asia, supporting the view of its domestication in the Indian subcontinent and showing that Australia and Papua New Guinea is a centre of diversity for wild mungbean. Germplasm collection reports focussing on Asian *Vigna* collection can be accessed online (http:/ /www.gene.affrc.go.jp/ publications.php?type=report&section=plant).

Asian Vegetable Research and Development Centre (AVRDC) in Taiwan was designated as international base collection centre for

Table 3: Details of gene(s)/QTL(s) associated with different biotic stresses in mungbean.

Disease/ Insect-pest	***Causal organism***	***Population used***	***Type of marker used***	***Gene/ QTL***	***Name of markers***	***Distance (cM)/ Phenotypic variation (%)***	***Reference(s)***
Yellow Mosaic disease	Mungbean Yellow Mosaic Virus (MYMV)	F_8 RIL derived from the cross *V. mungo* (cv. TU 94-2) with *V. mungo* var. *silvestris*	ISSR	Gene	*ISSR811*$_{1357}$	6.8 cM	Souframanien and Gopalakrishna (2006)
-do-	-do-	167 $F_{2:8}$ RILs from the cross TM-99-37 (R) x Mulmarada (S)	RAPD	*mymvr*-gene	*OPB-07*$_{600}$ *OPC-06*$_{1750}$ *OPB-12*$_{820}$	6.8 cM 22.8 cM 25.2 cM	Dhole and Reddy (2012)
Powdery mildew	*Erysiphe polygoni*	147 RILs from Berken (Res.) x ATF 3640 (Sus.) cross	RFLP	QTL	-	86.0 %	Humphry *et al.* (2003)
-do-	-do-	170 F_7 RILs derived from the cross Kamphaeng Saen 1 (Sus.) x VC6468-11-1A (Res.)	SSR	*qPMR-1* *qPMR-2*	(CEDG282 and CEDG191) (MB-SSR238 and CEDG166)	20.10% 57.81%	Kasettranan *et al.* (2010)
Cercospora leaf spot (CLS)	*Cercospora-canescens*	F_2 (KPS1 x V4718) and BC_1F_1 [(KPS1 x 4718) x KPS1] populations	SSR	*qCLS*	*CEDG117* and *VR393*	65.5-80.53 %	Chankaew *et al.* (2011)
Bruchid insect-pest	*Callosobruchus* spp.	200 F_{12} RILs from the cross NM 92 x TC1966 *V. sublobata* (Roxb.) Bairig. accession	7 CAPs marker derived from RAPD marker *OPW02a4*	*Br* gene	*W02s2*, *W02s3*, *W02s4*, *W02s6*, *W02s9*, *W02s10* and *W02s11*	1.3-1.4 cM	Chen *et al.* (2007)

mungbean by the International Board for Plant Genetic Resources (IBPGR). Worldwide, a total of 43,027 mungbean accessions are held here *ex situ*. The major Institutes responsible for collections are: (1) The Institute of Plant Breeding, University of the Philippines, Los Baños, Philippines (6889 accessions); the University of Georgia, Griffin, USDA-ARS, USA (3900 accessions); the National Bureau of Plant Genetic Resources (NBPGR), New Delhi, India (3147 accessions); the Regional Station (NBPGR) Jodhpur, Rajasthan, India (2466 accessions); the Field Crops Research Institute of the Department of Agriculture, Bangkhen, Thailand (2250 accessions); and the National Institute of Agrobiological Sciences, Tsukuba-shi, Japan (1579 accessions) (WIEWS 2012).

AVRDC has recently updated its Vegetable Genetic Resources System (AVGRIS) for mungbean and has made characterization data for a total of 9198 accessions available online, covering mungbean, blackgram, and ricebean (AVGRIS, 2012b). To date, the AVRDC has 10,733 accessions of *Vigna* and related species, including 5900 of *V. radiata var. radiata* obtained from different sources (Shanmugasundaram *et al.*, 2009). In India, seven AICRP centres *viz.*, Dharwad, Akola, Coimbatore, Badnapur, CSA Kanpur, IIPR Kanpur and Durgapur maintain 1694 accessions of mungbean during Kharif, 2011.

Molecular Marker-Assisted Breeding

Molecular markers are indispensable for designing molecular breeding strategies for crop improvement, but unfortunately limited molecular markers were developed specifically for mungbean. Gene tagging and QTL mapping efforts have been made in mungbean using different molecular marker systems *viz.*, RAPD, AFLP, RFLP, SSR, SCAR, CAPS etc. (Table 3). The use of markers based on single nucleotide polymorphisms, copy number variation, and insertions/deletions, as well as genotyping by sequencing, is becoming popular for genetic mapping and analyses of quantitative trait loci associated with various agronomic traits, biotic and abiotic stresses.

Genetic Linkage Maps

Earlier genetic linkage maps of mungbean were constructed using RFLP, RAPD and SSR markers. These maps have been used to map genes for resistance to adzuki bean weevil, seed colour and QTLs for seed weight, hard-seededness, powdery mildew and *Cercospora* leaf spot resistance.

Recently, a genetic linkage map of mungbean was constructed by Isemura *et al.* (2012) using 430 SSR and EST-SSR markers from mungbean and its related species. This is the first mungbean linkage map containing 11 linkage groups spanning a total of 727.6 cM map length and identified a total 105 QTLs and genes for 38 domestication related traits. QTLs associated with some important traits *e.g.*, seed size, pod dehiscence and pod maturity have also been identified in mungbean, and these QTLs may play important role as new gene resources for other Asian *Vigna* species.

Genetic Transformation

During the last three decades significant progress has been made in devising transformation protocols for grain legumes including *Vigna* species. In mungbean, genetic transformation *via* microprojectile gun, electroporation and *Agrobacterium*-mediated gene transfer using either cotyledonary node or axillary bud region of node has been reported but the efficiency was still low. *A. tumefaciens* mediated transformation is the preferred method.

Chloroplast Genome Sequencing

Tangphatsornruang *et al.* (2010) obtained the complete chloroplast (cp) genome sequence of mungbean using 454 pyro-sequencing technology. They found that the gene content and structural organization of mungbean cp genome is similar to that of *Phaseolus vulgaris*, its close relative in the family *Phaseoleae*. The mungbean cp genome is 1,51,271 bp in length and includes a pair of inverted repeats (IRs) of 26,474 bp separated by a small single-copy region of 17,427 bp and a large single-copy region of 80,896 bp. The genome contains 108 unique genes and 19 of these genes are duplicated in the IR. Out of the 108 genes, 75 are predicted to be protein-coding genes, 4 ribosomal RNA genes and 29 tRNA genes. The cp genome sequencing of *V. radiata* will explore the use of polymorphic microsatellites at the intra-and inter-specific levels among *Vigna* species.

Mungbean Genomic Resources

A wide range of genomic resources are now available that can be used to accelerate legumes improvement. These include expressed sequence tag (EST) database, genome sequences (whole or partial), physical maps, molecular maps, DNA chips and bacterial artificial chromosome (BAC) or similar genomic libraries. Mungbean has small genome size estimated to be 0.59 pg/1C (574 Mbp) and thus offers a great opportunity for doing functional genomics. High-throughput genome analysis is now possible by the introduction of a massive parallel pyro-sequencing technology developed by 454 Life Sciences Technology. Tangphatsornruang *et al.* (2009) generated and characterized a total of 4,70,024 genome shotgun sequences covering 100.5 Mb of the mungbean genome and identified 1,493 SSR motifs. Moe *et al.* (2011), similarly used 454 sequencing technology for transcriptome sequencing and generated a total of 1,50,159 and 1,42,993 reads producing 5,254 and 6,374 large contigs (≥500 bp) with an average length of 833 and 853bp for two mungbean varieties Sunhwa and Jangan, respectively. A total of 2,964 SSR motifs and 8,249 single nucleotide polymorphisms (SNPs) and indels were identified in these two mungbean varieties. Large resource sequence data generated by such studies could be used to develop potential molecular markers for mungbean improvement. Diversity Array Technology (DArT), a technique for quickly generating large numbers of molecular markers, was established for mungbean by Hang Vu *et al.* (2012). The mungbean library comprising of approximately 1600-1800 polymorphic DArT clones may be applied to construct linkage maps, QTLs mapping studies and comparative genome studies of mungbean. Some plant genome and gene family tools are available at legume information system (LIS; http://www.comparative-legumes.org/pages/resources). Similarly, SSR marker information of *Vigna* species is available at http://www.gene. affrc.go.jp/databases-marker information en.php website.

TILLING (Targeting Induced Local Lesions in Genomes) is an innovative, molecular technique that allows rapid identification of induced mutations in a population. This method is a reverse genetics approach that allows scientists to rapidly screen for mutations in a gene of interest without the creation of transgenic material. EcoTILLING is a similar molecular technique, except that it targets natural genetic variation as opposed to induced mutations. Barkley *et al.* (2008) employed EcoTILLING to identify small SNPs and insertions/deletions (INDELS) ranging from 4-24 and 1-6, respectively in a collection of *V. radiata* var. *radiata* accessions and found it very useful for crop plants having narrow genetic base such as *V. radiata* or looking for variation in highly conserved genes.

Developing Hybrids

Nair *et al.* (2013), in their review on mungbean, discussed the feasibility of hybrid production in mungbean. Heterotic effects up to 200% have been reported in mungbean for grain yield (Tickoo *et al.*, 2006). Cleistogamy is one of the major barriers to hybrid seed production in mungbean. Sorajjapinun and Srinives (2011) reported a chasmogamous flower mutant developed through gamma irradiation, which showed an increase of 9.6% in cross pollination compared with the wild type. However, stable male sterile systems will be required to test the feasibility of developing commercial hybrid mungbean.

REFERENCES

AVGRIS (2012).AVRDC vegetable genetic resources information system. Available at: http://203.64.245.173/result_vigna.asp?ACCNO=&TEMPNO=&SPECIE=RADIATA&SUBTAX=& PEDCUL=&COUNTR= (accessed on 25 May 2012).

Barkley, N.A., Wang, M.L., Gillaspie, A.G., Dean, R.E., Pederson, G.A. and Jenkins, T.M. (2008). Discovering and verifying DNA polymorphisms in mungbean [*V. radiata*(L.). Wilczek] collection by EcoTILLING and sequencing. BMC Research Notes.

Chankaew, S., Somta, P., Sorajjapinun, W. and Srinives, P. (2011). Quantitative trait loci mapping of *Cercospora* leaf spot resistance in mungbean, *Vigna radiata* (L.) Wilczek. *Mol. Breed.,* 28: 255–264.

Chen, H.M., Chien, A.L., Kuo, C.G., Chien, C.M., Sun, H.C., Huang, C.C., Lin, Y.C. and Ku, H.M. (2007). Development of a molecular marker for a bruchids (*Callosobruchus chinensis* L.) resistance gene in mungbean. *Euphytica*, 157: 113–122.

Dhole, V.J. and Reddy, K.Y. (2012). Development of a SCAR marker linked with a MYMV resistance gene in mungbean (*Vigna radiata* L. Wilczek) doi:10.1111/pbr.12006 *Plant Breed.,*

Fernandez, G.C.J. and Shanmugasundaram, S. (1988). The AVRDC mungbean improvement program: The past, present, and future. *In*: Shanmugasundaram, S. and McLean, B.T. *Eds*., Proc. Second International mungbean symposium, Shanhua, Taiwan: The World Vegetable Center (AVRDC).

Hang Vu, T.T., Lawn, R.J., Bielig, L.M., Molnar, S.J., Xia, L. and Kilian, A. (2012). Development and initial evaluation of diversity array technology for soybean and mungbean. *Euphytica*, 186: 741–754.

Humphry, M.E., Magner, T., McIntyre, C.L., Aitken, E.A.B. and Liu, C.J. (2003). Identification of a major locus conferring resistance to powdery mildew (*Erysiphe polygoni* DC) in mungbean (*Vigna radiata* L. Wilczek) by QTL analysis. *Genome,* 46(5): 738–744.

Isemura, T., Kaga, A., Tabata, S., Somta, P. and Srinives, P. (2012). Construction of a genetic linkage map and genetic analysis of domestication related traits in mungbean (*Vigna radiata*). *PLoS ONE*, 7(8): e41304. doi:10.1371/journal.pone.0041304.

Kasettranan, W., Somta, P. and Srinives, P. (2010). Mapping of quantitative trait loci controlling powdery mildew resistance in mungbean (*Vigna radiata* (L.) Wilczek). *J. Crop Sci. Biotech.,* 13(3): 155–161.

Keatinge, J., Easdown, W., Yang, R., Chadha, M. and Shanmugasundaram, S. (2011). Overcoming chronic malnutrition in a future warming world: The key importance of mungbean and vegetable soybean. *Euphytica,* 180: 129–141.

Landerito, E.O., Mendoza, E.M.T., Laurena, A.C. and Garcia, R.N. (1993). Physicochemical and biochemical factors in mungbean (*Vigna radiata*, (L.) Wilczek) and blackgram (*Vigna mungo*) associated with bruchid (*Callosobruchus chinensis* L.) resistance. *Philippine Journal of Crop Science,* 18: 153–163.

Moe, K.T., Chung, J.W., Cho, Y.I., Moon, J.K., Ku, J.H., Jung, J.K., Lee, J. and Park, Y.J. (2011). Sequence information on simple sequence repeats and single nucleotide polymorphisms through transcriptome analysis of mungbean. *J. Integr. Plant Biol.,* 53(1): 63–73.

Nair, R.M., Schafleitner, R., Kenyon, L., Srinivasan, R., Easdown, W., Ebert, A.W. and Hanson, P. (2013). Genetic improvement of mungbean, SABRAO Jl. *Breed. and Genet.*, 44(2) 177–190.

Polhill RM and van der Maesen LJG (1985). Taxonomy of grain legumes. *In*: Summerfield, R.J. and Roberts, E.H. (*Eds*.). Grain Legume Crops. Collins, London, England, pp. 3–36.

Sangiri, C., Kaga, A., Tomooka, N., Vaughan, D. and Srinives, P. (2007). Genetic diversity of the mungbean (*Vigna radiate*) Leguminosae gene pool on the basis of microsatellite analysis. *Aust. J. Bot.,* 55: 837–847.

Shanmugasundaram, S., Keatinge, J.D.H. and Hughes, J.A. (2009). The mungbean transformation, diversifying crops, defeating malnutrition. IFPRI Discussion Paper 00922 on project, 2020 Vision Initiative, supported by the CGIAR.

Singh, B.B., Dixit, G.P. and Katiyar, P.K. (2010). Vigna research in india (25 years of Research Achievements), All India Coordinated Research Project on MULLaRP, IIPR, Kanpur, p. 2.

Sorajjapinun, W. and Srinives, P. (2011). Chasmogamous mutant, a novel character enabling commercial hybrid seed production in mungbean. *Euphytica,* 181: 217–222.

Souframanien, J. and Gopalakrishna, T. (2006). ISSR and SCAR markers linked to the mungbean yellow mosaic virus (MYMV) resistance gene in blackgram [*Vigna mungo* (L.) Hepper]. *Plant Breed.,* 125: 619–622.

Tangphatsornruang, S., Sangsrakru, D., Chanprasert, J., Uthaipaisanwong, P., Yoocha, T., Jomchai, N. and Tragoonrung, S. (2010). The chloroplast genome sequence of mungbean (*Vigna radiata*) determined by high–throughput pyrosequencing: Structural organization and phylogenetic relationships. *DNA Res.,* 17: 11–22.

Tangphatsornruang, S., Somta, P., Uthaipaisanwong, P., Chanprasert, J., Sangsrakru, D., Seehalak, W., Sommanas, W., Tragoonrung, S. and Srinives, P. (2009). Characterization of microsatellites and gene contents from genome shotgun sequences of mungbean (*Vigna radiata* (L.) Wilczek). BMC Plant Biol

Tickoo, J.L., Lal, S.K., Chandra, N. and Dikshit, H.K. (2006). Mungbean breeding, *In*: Ali, M. and Kumar, S. *Eds*., Advances in mungbean and urdbean, IIPR, Kanpur, pp. 110–148.

UNSCN (2010). 6th Report on the World Nutrition Situation. Geneva: United Nations Standing Committee on Nutrition.

Upadhyaya, H.D. and Ortiz, R. (2001). A mini core subset for capturing diversity and promoting utilization of chickpea genetic resources in crop improvement. *Theor. Appl. Genet.,* 102: 1292–1298.

Weinberger, K. (2003). Impact analysis of mungbean research in South and Southeast Asia. AVRDC: Shanhua, p. 19.

WHO (2000). Turning the tide of malnutrition – responding to the challenge of the 21th century. World Health Organization /NHD/00.7.

WIEWS (2012). World Information and Early Warning System on PGRFA. WIEWS Germplasm Report. http://apps3.fao.org/wiews/germplasm_report.

Zhang, Y.W., Zhao, X.Y. and Cheng, X.Z. (2002). Screening and identification of bruchid–resistant materials of mungbean. *In*: Proc. on Technology and Utilization of Mungbean in China. China Agriculture Press, Beijing. pp. 116–118.

11

Soybean

S.K. LAL AND R.L. SAPRA

Division of Genetics, Indian Agricultural Research Institute,
New Delhi, India

INTRODUCTION

Soybean continues to be the most important oilseed crop in the world. It contributes more than 50 percent of the total oilseed production worldwide. One of the major achievements has been the development of transgenic soybeans that are tolerant to herbicide glyphosate. The GM soybean had a major impact on the area, production and productivity of soybean worldwide. The world's leading producers of soybean are USA (33%), Brazil (27%), Argentina (21%) and China (7%). In 2010, 102 m ha area was under soybean, out of 102 m ha, 75 m ha was occupied by GM soybean. GM soybean is grown in USA, Brazil, Argentina, China, Paraguay, Uruguay, Canada, Bolivia, South Africa, Mexico, Chile and Costa Rica. GM soybean has helped in adoption of soybeans in many

Table 1: Global coverage of soybean (2010)

Country	*Area (m ha)*	*Yield (t/ha)*	*Production (m t)*
World	102.30	2.55	261.57
United States	31.00	2.92	90.60
Brazil	23.29	2.94	68.52
Argentina	18.13	2.9	52.68
China	8.51	1.77	15.08
India	9.21	1.06	9.81
Paraguay	2.67	2.79	7.46
European Union	2.74	1.75	4.79

Source: www.fao.org/statistics

V.L. Chopra (Ed.): Breeding Field Crops II: Advances, Studium Press (LLC), USA.

new areas where it was never grown before. Argentina has emerged as the third largest producer of soybean beating China to 4th position.

Soybean has grown from the position of third most important oilseed crop in India to number one oilseed crop. India remains at the 5th position in terms of area and production of soybean among all soybean producing countries (Table 1).

SOYBEAN CULTIVATION IN INDIA

Soybean is grown in approximately 10 million hectares in India and soybean ranks first among the oilseed crops in India. Soybean provides 10% of total oil requirements and earns Rs. 7000 crores worth of foreign exchange through export of oil meal. Because of its importance as the number one oilseed crop in India and the importance of soybeans and soybean products in earning foreign exchange, major problems related with the production and productivity of soybean along with the end-use quality characteristics of soybeans can have a significant impact upon the economy of the nation. Continued breeding efforts are needed for meeting future demands for changes in soybean seed yield, seed composition and pest resistance.

Area under soybean cultivation in India continues to grow. However, there is marginal improvement in productivity but production has increased due to growth in area (Table 2). The growth in area to cover over 10 million ha is as predicted by Holt *et al.* (1997). But the prediction of reaching productivity of 1.5 t/ha by 2010 has not been met. Hence there is short fall in meeting the predicted production goal of 15 million t by over 3 million t.

Table 2: Area, production and productivity of soybean during *kharif* 2011 (Estimated)

Country/State	*Area (m ha)*	*Yield (t/ha)*	*Production (m t)*
India	10.3300	1.155	11.9400
Madhya Pradesh	5.7300	1.077	6.1700
Maharashtra	3.0600	1.256	3.8400
Rajasthan	0.8960	1.393	1.2500
Andhra Pradesh	0.1550	1.050	0.1628
Karnataka	0.2170	1.070	0.2320
Chhattisgarh	0.1500	1.025	0.1538
Other States	0.1241	1.010	0.1253

Source: www.sopa.org

Yellow seeded soybean was introduced in India in late sixties. In 1970 total area under soybean cultivation was 30,000 ha which has increased to over 10 million ha with a production of over 10 million metric tons. The productivity has increased from 400 kg/ha in 1970 to 1.2 tons/ha in 2010. More than 90 varieties of soybean have been developed and released for cultivation in India (Table 3).

Table 3: List of released varieties of soybean in India

State	***Recommended varieties***
Andhra Pradesh	LSb-1, Pratikar (MAUS 61), Pooja (MAUS 2), MACS 450, Pant Soybean 1029, MACS 124, Monetta and Bragg
Arunachal Pradesh & Assam	Ahilya 1 (NRC 2), JS 80-21, Samrudhi (MAUS 71), Pratap Soya (RAUS 5), Bragg, Indira Soya 9
Bihar	PK 416, Pusa 16, Pusa, 24, Pant Soybean 564, Pant Soybean 1024, Pant Soybean 1042, Bragg and SL 525, Pusa 9814, Pusa-12
Chhattisgarh	Ahilya 1 (NRC 2), Ahilya 3 (NRC 7), Ahilya 2 (NRC 12), Ahilya 4 (NRC 37), JS 71-05, JS 335, JS 80-21, JS 75-46, MACS 58, JS 9041, Indira Soya 9, JS 93-05, Kalitur, Parbhani Sona (MACS 47), Pratishta (MAUS 61-2), Monetta, Punjab-1, PK 472, Shakti (MAUS 81), Samrudhi (MAUS 71) and Pratap Soya (RAUS 5) , JS-93-05, JS-95-60, JS-97-52
Delhi	PK 416, Pusa 9712, Pant Soybean 564, Pant Soybean 1024, Pant Soybean 1042, Bragg and SL 525, Pusa 9814, Pusa 12, Pusa 14.
Gujarat	Gujarat Soybean 1, Gujarat Soybean 2, JS 93-05, JS 335, JS 80-21, JS 75-46, MACS 58, Monetta, Parbhani Sona (MAUS 47), Pratishta (MAUS 61-2), Shakti (MAUS 81) and PK 472
Haryana	Punjab-1, PK 416, Pusa 16, Pant Soybean 564, Pant Soybean 1024, Pant Soybean 1042 and SL 525
Himachal Pradesh	VL Soya 2, VL Soya 47, Shilajeet, Shivalik, Bragg, Pusa 16, Hara Soya and Palam Soya
Jharkhand	Birsa Soybean 1, Ahilya 1 (NRC 2), JS 80-21, Samrudhi (MAUS 71), Pratap Soya (RAUS 5) and Bragg
Karnataka	Hardee, Bragg, Sneh (KB 79), KHSb-2, Pratikar (MAUS 61), Pant Soybean 1029, MACS 124, MACS 450 and Pooja (MAUS 2)
Madhya Pradesh	Ahilya 1 (NRC 2), Ahilya 3 (NRC 7), Ahilya 2 (NRC 12), Ahilya 4 (NRC 37), JS 71-05, JS 335, JS 80-21, JS 75-46, MACS 58, JS 90-41, Indira Soy 9, JS 93-05, Kalitur, Parbhani Sona (MAUS 47), Pratishta (MAUS 61-2),Monetta, Punjab-1, PK 472 and Shakti (MAUS 81) , JS-93-05, JS-95-60, JS-97-52
Vidarbha and Marathwada region of Maharashtra	Ahilya 1 (NRC 2), JS 335, JS 93-05, JS 80-21, MACS 58, Parbhani Sona (MAUS 47), Pratishta (MAUS 61-2), Shakti (MAUS 81), MACS 13, Monetta, Prasad (MAUS 32) PK 472, Shakti (MAUS 81), TAMS-38 and Phule Kalyani (DS-228), JS-93-05, JS-95-60, JS-97-52
Southern Maharashtra	MACS 124, MACS 450, Pant Soybean 1029, Pooja (MAUS 2), Pratikar (MAUS 61), Prasad (MAUS 32), MACS 13, Monetta and Phule Kalyani (DS-228)
Manipur, Meghalaya and Nagaland	Ahilya 1 (NRC 2), JS 80-21, Samrudhi (MAUS 71), Pratap Soya (RAUS 5), Bragg, Indira Soya 9
Orissa	JS 80-21, Pusa 24, Indira Soya 9, Ahilya 1 (NRC 2), Ahilya 2 (NRC 12), Ahilya 3 (NRC 7), Ahilya 4 (NRC 37) and Pusa 16

Table 3: (*Contd...*)

Table 3: (*Contd....*)

State	*Recommended varieties*
Punjab	PK 416, Pusa 16, Pant Soybean 564, Pant Soybean 1024, Pant Soybean 1042, SL 295, Bragg and SL 525
Rajasthan	Pratap Soya (RAUS), Bragg, Punjab-1, PK 472, MACS 58, JS 80-21, JS 335, Ahilya 4 (NRC 37), Parbhani Sona (MAUS 47), JS 93-05, Pratishta (MAUS 61-2) and Shakti (MAUS 81) , JS-93-05, JS-95-60, JS-97-52
Sikkim	NRC 2, JS 80-21, MAUS 124, MAUS 71, RAUS 5, Pusa 16, Bragg and Indira Soya 9
Tamil Nadu	Co 1, Co Soya 2, ADT-1, MACS 124, MACS, 450, Pooja (MAUS 2), Pratikar (MAUS 61) Hardee, Pant Soybean 1029, and Bragg
Uttar Pradesh	Pusa 16, Pant Soybean 1092, Pant Soybean 1042, Pant Soybean 1024, Pant soybean 564, PK 472, PK 472, PK 416, Pratishta (MAUS 61-2), JS 93-05, Ahilya 4 (NRC 37), JS 335, SL 525, PS 1241, PK 262 and PK 327, Pusa 9814, Pusa-12
Uttarakhand	Hara Soya, Palam Soya, Punjab-1, Pusa 16, PS 1241, VL Soya 1, VL Soya 2, VL Soya 21, VL Soya 47, Shilajeet and Pant Soybean 1092

Source: www.DSR.org

ADVANCES IN SOYBEAN BREEDING

One of the major achievements has been the development of transgenic soybeans that are tolerant to herbicide glyphosate. The GM soybean had a major impact on the area, production and productivity of soybean worldwide. The world's leading producers of soybean are USA (33%), Brazil (27%), Argentina (21%) and China (7%). In 2010, 102 m ha area was under soybean. Out of 102 m ha, 75 m ha was occupied by GM soybean. It is grown by USA, Brazil, Argentina, China, Paraguay, Uruguay, Canada, Bolivia, South Africa, Mexico, Chile and Costa Rica. GM soybean has helped in adoption of soybeans in many new areas where it was never grown before.

The field of plant breeding in general has witnessed a major change with the development of molecular biology. Molecular biology tools have emerged as a major technological breakthrough in providing solutions to the limitations of classical breeding methods. Soybean has all the requisite information available in public domain for application of molecular techniques to supplement the classical breeding methods (http//soyBase). Soybean (cv.Williams 82) is one of the few pulses for which the whole genome sequencing has been completed. From the genome sequencing it is apparent that soybean is a palaeopolyploid with highly duplicated genome with nearly 75% of the genes present in multiple copies (Schmutz *et al.*, 2010). A well saturated genetic linkage map is the first prerequisite for application of molecular tools for breeding. It helps in identification of useful traits/markers, genetic diversity analysis and genome structure of genetic resources. The latest highly saturated molecular linkage map contains over 5500 markers spanning the entire 2,300 cM soybean genome. The first soybean linkage map was constructed by using RFLP marker (Keim *et al.*, 1990). This map consisted of 26 genetic linkage groups containing 150 RFLP markers. Subsequently, linkage maps based on RFLP markers were developed by Lark *et al.* (1993) and Shoemaker and Specht (1995). Akkaya *et al.* (1995) for the first time used simple sequence repeat markers (SSRs) to develop an integrated genetic linkage map of soybean. A major breakthrough was reported by Song *et al.* (2004) who for the first time established the 20 linkage groups. Song *et al.* (2004) used JoinMap software to join five maps into an integrated genetic map. The map contained 1,849 markers including 1,015 SSRs. Choi *et al.* (2007) for the first time reported an integrated genetic linkage map using SNP markers.

Large number of markers are required for QTL mapping and map-based cloning for identifying genes. SNP markers are ideal for such type of work. Till recently approximately over 5000 SNPs were available in the public domain (www.soybase.org). Wu *et al.* (2010) reported 39,022 putative SNPs. They used Illumina/Solexa sequencing system for identifying the SNPs. The validation rates of these putative SNPs predicted with low and high stringency were 72% and 85%, respectively. Apart from this, 50,000 SNPs are under development which will be used to create haplotype maps of more than 19,000 accessions of the USDA soybean collection (Boerma *et al.*, 2010).

Genetic diversity

Low genetic diversity among the released cultivars of soybean is a major concern through out the world. This relates to loss of diversity during domestication and loss of diversity due to breeding. The genetic base of North American cultivar is narrow as 80% of the released cultivars trace their ancestry to 12 genotypes (Gizlice *et al.*, 1996). Brazilian cultivars (Priolli *et al.*, 2010) and Indian cultivars (Lal and Rana, 2000) are also said to be of narrow genetic base. The impact of genetic bottlenecks on soybean genome diversity was investigated, by sequencing 111 fragments from 102 genes in four soybean populations representing the populations before and after "genetic bottlenecks". The results indicated that there was less loss of diversity due to breeding than due to domestication. The low sequence diversity present in the wild species was halved, 81% of the rare alleles were lost and 60% of the genes exhibited evidence of significant allele frequency changes due to domestication (Hyten *et al.*, 2006). Homologous chromosome meiotic recombination in the creation of soybean cultivars is important. Cross over events were identified using 143 simple sequence repeat markers and the conclusion was that the soybean varieties have been developed using new within chromosome allele combinations (Stefaniak *et al.*, 2006). Studies on heritability and correlation among traits in four-way soybean crosses for development of population for recurrent selection have indicated that the four-way crosses could be an efficient method to increase the genetic base of populations for recurrent selection or cultivar development (Alliprandini and Vello, 2004).

Early generation selection

Gains from selection in early generation is a highly debated issue. The studies on the role of early generation testing in increasing the gain from selection revealed that there was no gain in selection form early selection before F_4 generation (St. Martin and Xie, 2000). A comparison between three procedures for early generation testing of soybean indicated that all the three selection procedures had no advantage over the lines derived without early generation selection (St. Martin and Geraldi, 2002). Studies on heritability in early generations of two way crosses in soybean indicated that, additive genetic variance accounted for the larger part of the variance for most of the characters. This means that the simple selection methods can be used (Costa *et al.*, 2008).

The use of early generation selection for yield contributing traits in interspecific crosses of soybean was studied. The data from two crosses indicated that the selection for dry matter, pods/plant, plant height, harvest index and yield *per se* is an effective criterion for yield improvement in early generation in interspecific cross of soybean (Karnwal *et al.*, 2009).

Breaking negative correlation

There is a negative correlation between high seed protein content and seed yield in soybean. Cober and Voldeng (2000) compared the advantage of backcross over single cross in breaking the negative correlation between high protein content and high seed yield in soybean. To break negative correlation between seed yield and protein content, seed density was used as selection criteria, which are positively correlated with protein concentration. Male-sterile line was used for recurrent selection and selections were made for lines with the highest as well as lowest seed density. But the results suggested that the single plant selection for seed density was not effective in increasing seed density or seed protein concentration (Li and Burton, 2002).

Breeding methods

A comparison between single pod descent method and single pod descent with selection (SPDS) for advancement of segregating generations in soybean was made (Destro *et al.*, 2003). SPDS yielded superior mean on the basis of all traits/generation combinations. SPDS has an advantage in handling because of smaller population size. A comparison between "Bulk selection method", "Pedigree selection method" and "Modified single seed descent method", for selection in segregating generations indicated that the modified single seed descent method gave the best results (Miladinovic *et al.*, 2011).

Family selection vs. Line selection

A comparison of family and line selection for developing elevated palmitate (Bravo *et al.*, 1999); reduced palmitate, saturates and linolenate of soybean (Streit *et al.*, 2001a), or seed yield of soybean (Streit *et al.*, 2001b) revealed that family selection had no advantage over the line selection method.

Development of high protein lines

Relationship between assimilate supply per seed during seed filling and soybean seed with high protein content was studied. The results suggest that high protein lines maintain assimilate supply per seed at or near saturating levels during seed filling. Improving seed protein levels in high yielding varieties will require increasing assimilate supply per seed without sacrificing seed numbers (Rotundo *et al.*, 2009).

Biotic stress

Gravina *et al.* (2004) studied the inheritance of resistance to *Cercospora sojina* Hara using a diallel mating design. The additive, dominant and epistatic genetic effects were more important for the expression of the resistance, although the additive genetic effect was the most important component.

Abiotic stress

Selection for higher water use efficiency is one of the most potent strategies for breeding drought tolerant varieties. However, high water use efficiency is because of low stomatal conductance under drought conditions. Such genotypes may be associated with low total water use, slow growth and poor yield. Therefore, increasing photosynthetic rate is one alternate strategy. Studies on variation in photosynthetic capacity and stomatal conductance revealed that the variation in photosynthetic capacity and stomatal conductance was independent of each other. The higher photosynthetic capacity at mild drought levels was correlated with maximal rate of Rubisco carboxylation. It was concluded that it is possible to employ a selection strategy of breeding water-saving soybeans with high photosynthetic capacities to compensate for otherwise reduced photosynthesis in genotypes with lower stomatal conductance (Gilbert *et al.*, 2011).

G x E interaction

Weikan and Istavan (2002), used biplot analysis of test sites and trait relationship to find a selection strategy over environments. They found that selection for seed yield alone was the simplest and the most efficient method at the early stages of breeding programme. Pacheco *et al.* (2005) reported use of supplementary genotypes in AMMI analysis. The paper suggests a new equivalence method developed from singular value decomposition used in AMMI and the spectral decomposition used in principal components analysis.

Screening techniques

Seed protein

A novel method for selection for protein in soybean from single F_2 seed by near infrared reflectance spectroscopy was reported by Lee *et al* (2010).

Insect resistance

A rapid screening method for estimating defoliation in soybean breeding programme has been developed (Bowers *et al.*, 1999).

Physiological traits

It has been shown that there are crop specific traits which help different crop to adopt under moisture stress condition (Richards, 2002). In soybean low sensitivity to biological nitrogen fixation under moisture stress conditions is the trait which imparts drought tolerance. A three stage screening method for identification of soybean genotypes with N_2 fixation tolerance to water deficit is now available (Sinclair *et al.*, 2000).

Selection for physiological traits has been suggested as a new strategy to break the stagnation in yield among major crop plants. Subrahmanyam (2002), found a positive correlation between leaf gas-exchange, area leaf mass and yield in soybean. They suggested leaf gas exchange and leaf area mass can be used as a surrogate for higher photosynthetic efficiency. Kumudini (2002) reviewed the role of physiological traits in increasing the yield in soybean. She concluded that selection for improving source-sink ratio will be more effective than improving assimilate supply alone as soybean has a complex feedback system which maintains a cultivar-specific source-sink ratio. Morrison *et al.* (1999), reviewed the physiological changes in short season soybean over 58 years of genetics improvement in soybean cultivars in Canada. They concluded that the increase in yield was due to increase in harvest index, photosynthesis and stomatal conductance and decrease in leaf area index.

Heat tolerance

Screening for pollen viability under higher temperature regime was found to be an effective method in common bean (Warrag and Hall, 1983). Salem *et al.* (2007), evaluated pollen based screening of soybean genotypes for high temperatures. They found that heat tolerant genotype were less sensitive to high temperature than the heat susceptible genotypes. They concluded that the temperature-dependent pollen response functions will be useful in screening heat tolerant genotypes.

Linolenic acid content

Biometrical analysis of linolenic acid content of soybean seeds revealed that additive genetic variance contributed more and there was no role of dominance deviation. Therefore, the trait can be transferred through backcross breeding (Gesteria *et al.*, 2003).

ADVANCES IN MOLECULAR BREEDING

Marker Assisted Selection in Soybean

Marker assisted selection is widely being used in soybean breeding programme by private seed companies since last ten years and some Universities in USA. It has been widely used for biotic stress (nematode, rust, sudden death syndrome.), seed quality traits (linolenic acid, lioleic acid, palmitic acid, stearic acid, oleic acid, kunitz trypsin inhibitor, phytic acid and raffinose). But there are no examples where it has been used successfully for transferring resistance to abiotic stress or yield and yield related traits. Chapman *et al.* (2003) used SSR markers to select QTLs for seed protein concentration, seed oil concentration, seed size, plant height lodging and maturity in F_2 and $F_{4:6}$. The results suggested that selection based solely on QTLs produced limited gain.

Oil Content and Oil Quality

Soybean is the leading oil seed crop in the world. It accounts for more than 50% of the total edible oil produced world wide. Soybean oil, on an average basis contains five major fatty acids, palmitic acid (10%), stearic acid (4%), oleic acid (18%), linoleic acid (55%) and lenolenic acid (13%). Although soybean oil is widely consumed throughout the world yet there are many issues related with its quality which needs to be addressed such as oxidative stability (increasing stearic and oleic acid content and reducing lenoleic acid and lenolenic acid content; increasing vitamin E content) and increasing total oil content. The advances in understanding of lipid biosynthesis pathway in general (Cases *et al.*, 1998) and specific information pertaining to inheritance of specific fatty acids in soybean (Fehr *et al.*, 1992; Bubeck *et al.*, 1989; Bilyeu *et al.*, 2003; Spencer *et al.*, 2003; Byfield *et al.*, 2006; Bachlava *et al.*, 2008; Zang *et al.,* 2008; Reinprecht *et al.*, 2009) and health related issues have led to rapid developments in developing soybean with modified fatty acid content. The major issues are oxidative stability, trans-fats and functionality of the soybean oil. The molecular tools which have been developed recently to address these issues are summarized in Table 4.

Protein Content and Quality

Soybean contains around 40% protein and soy protein is used for human consumption as well as for animal feed. To a soybean oil processor in India 1/3rd of the total revenue comes from the sale of the oil where as the 2/3rd of the revenue comes from the export of the oil meal. Thus the value of the protein is much higher than the oil. The value of soybean protein is lowered due to the presence of anti-nutritional factors such as protease inhibitors, like kunitz trypsin inhibitor (KTI), undesirable beany flavour, presence of allergens and phytic acid, poor digestibility and deficiency in sulphur containing amino acids. Both classical breeding methods and molecular biological approaches have been used to solve these problems. Now many molecular tools are available to address the issues related to protein content and protein quality (Table 4).

Breeding for Drought Tolerance

Soybean is cultivated in more than 10 million hectares in India, but 90% of the soybean is cultivated as a rainfed crop. As a result, though the genetic yield potential of currently available varieties is 2.5 to 3 tons/ha, the average national productivity is only 1 ton/ha (Lal and Rana, 2000). Drought accompanied by heat has been attributed as the major reason for low yield (Holt *et al.*, 1997). Soybean growing period spreads over four months of the year *viz.*, June, July August and September. Moisture Availability Index (MAI) (Hargreaves, 1977) is a good indicator of drought. Lal *et al.* (2008) studied standard deviation, coefficient of variation and range of MAI values of Indore and Jabalpur regions of Madhya Pradesh. The analysis indicated that the crop undergoes a severe to acute moisture stress in both the regions at the initial stages when the germination or seedling establishment takes place. The crop undergoes an acute to moderate stress in the Indore region again at the terminal stage when the seed development takes place. The MAI data of the two stations exhibit a definite pattern of seasonally recurrent excess moisture during the month of July and August ($1.2 \leq MAI \leq 4.45$) with almost 97% occurrence due to heavy rains during the intermediate stage of crop growth. Excess moisture in the month of July and August, leads to water logging resulting in reduced root growth, reduced nodulation, lower nitrogen fixation, decreased photosynthesis, biomass accumulation and decreased stomatal conductance. For drought tolerance role of water use efficiency in soybean was explored by several authors (Mian *et al.*, 1996, 1998; Specht *et al.*, 2001). But a viable product could not be developed using WUE. Currently, tolerance to biological nitrogen fixation under moisture stress and slow wilting (Sincliar, 2007) has been identified as the two most important traits for drought tolerance in soybean. Tolerance to salinity (Lee *et al.*, 2004) and alkalinity (Tuyen *et al.*, 2010) have been reported in soybean. Apart from drought, water logging, salt is the major problem in soybean growing areas in India. To develop dependable selection

Table 4: Molecular tools used in soybean for improvement of different traits

Trait	***QTL/marker***	***Remarks***	***Reference***
Meta analysis of oil content	An integrated map of QTLs for oil content in soybean was made using data from 13 different published reports with the help of Software BioMercator 2.1 for Meta-analysis.	Soybean is the leading oil seed crop in the world. Oil content in the seed is an important trait, which is highly influenced by the environment. A number studies have been carried out to map the QTL for oil content, with varying results.	Qi *et al.*, 2011
Palmitic acid (16:0)	Allele specific primers for GmFATB1a	Imparts oxidative stability but has negative effect on cardio-vascular health	Cardinal *et al.*, 2007
Stearic acid (18:0)	Point mutation in SAD-C enzyme	Imparts oxidative stability and is neutral on cardio-vascular health. Target to increase from 4% to 20%.	Zhang *et al.*, 2008
Oleic acid (18:1)	Six QTLs were mapped in a F_2 population. Satt211 (LGA1), Satt389 (LGD2), Satt394 (LG-G), Satt191 (Lg-G), Satt418 (LG-L) and Satt561 (LG-L)	Oleic acid content is important from the nutrition point as well as the oxidative stability of the oil.	Monteros *et al.*, 2008
Oleic acid content	SNP markers for FAD2 -1A, FAD2-1B and FAD2-2B	Delta12 FAD enzyme catalyses the conversion of oleic acid to linolenic acid is coded by FAD 2 gene	Bachlava *et al.*, 2008
Oleic acid seed content	SSR Sat_309		Bachlava *et al.*, 2009
Linolenic acid (18;3)		Linolenic acid constitutes up to 13%. To increase oxidative stability and to stop off flavour reduction of linolenic acid is important. It is an essential fatty acid.	
Linolenic acid	Satt534 and Satt560, LG-B2	FAD3 gene which codes for the enzyme delta15 fatty acid desaturase which catalyses the conversion of linoleic acid into linolenic acid by adding the third double bond is the target.	Spencer *et al.*, 2004
Linolenic acid	Molecular markers specific to GmFAD3A and Gm FAD3C	Three genes FAD3A, 3B and 3C have been identified. But role of GmFAD3A is most important in reducing the lenolenic acid content	Beuselinck *et al.*, 2006
Lenolenic acid content	Marker aided selection for development of low lenolenic acid lines	First report of MAS for Linolenic acid in soybean	Sauer *et al.*, 2008
Vitamin E content	Satt376 (LG-C2), Satt286(LG-C2), Satt266 (LG-D1b)	Vitamin E is important for nutronal value as well as oxidative stability of the oil. Besides its application in medicines, cosmetics and animal feed	Li *et al.*, 2010

Trait	*QTL/marker*	*Remarks*	*Reference(s)*
A Meta analysis of seed protein concentration QTL	An integrated map of QTLs related to seed protein concentration in soybean has been constructed, based on the public genetic map, soymap2 as a reference map, along with a set of 107 QTLs reported in the literature over the past 20 yr.	Increasing the protein content, which is influenced by the environment, is the target.	Qi *et al.*, 2011.
Sulphur containing amino acids	Three QTL for glycinin (Satt461, Satt292, and Satt156) on LG D2, I, and L. Two QTL for conglycinin (Satt461 and Satt249) on LG D2 and J.	A major limitation of soybean seed storage proteins is their low levels of the sulfur-containing amino acids, methionine and cysteine. Glycinin (11S) and β-conglycinin (7S) are important seed storage proteins in soybean. Glycinin contains significantly more S-containing amino acids than does β-conglycinin.	Panthee *et al.*, 2004
Sulphur containing amino acids	Identified a four base pair insertion in the first exon of *cg-2*, which introduced a premature stop codon. The insertion co-segregated with the lack of A-subunit, indicating that this mutation is the cause of the A-subunit deficiency.		Ishikawa *et al.*, 2006
β-conglycinin	Co-dominant DNA marker for the β-conglycinin a' subunit (Cgy-1) gene for use in marker-assisted selection (MAS).	The development of soybean varieties that lack the β-conglycinin a' subunit is an attractive goal because the β-conglycinin a' subunit exerts a negative influence on nutrition and to gelation, and is also a major allergen.	Kim *et al.*, 2011
Soybean kunitz Trypsin inhibitor	Satt 409, Satt 228, and Satt 429.	Elimination or development of null-Kunitz trypsin inhibitor genotype is the target.	Kim *et al.*, 2006
Kunitz trypsin inhibitor	Validated the SSR markers Satt 228 and Satt 409		Anita *et al.*, 2011
Seed phytate	SSR markers Satt 237 and Satt 561	Low seed phytate is desirable	Scaboo *et al.*, 2009.
Phosphorus and Sugar Composition	Satt 453 marker,	Development of soybeans [*Glycine max* (L.) Merr.] with modified phosphorus (P) and sugar composition adds value to soybean meal.	Maupin *et al.*, 2011

(Contd....)

(Contd....)

Trait	***QTL/marker***	***Remarks***	***Reference(s)***
Lipoxygenase	Satt522 and Sat074 and three SNPs.	Lipoxygenase-2 (Lx_2) in soybean seed is mainly responsible for generation of grassy-beany and bitter flavours.	Kim *et al.*, 2004
Lipoxygenase	Seed lipoxygenases L-1 and L-2 mapped as single major genes to the same location on LG G13-F. L-3 mapped to LG G11-E.	Development of null-lipoxygenase lines is required to reduce the 'beany flavour" of soybean food products.	Reinprecht *et al.*, 2006
Seed cadmium accumulation	Seven SSR markers, SatK138, SatK139, SatK140 (0.5 cM), SatK147, SacK149, SaatK150 and SattK152 (0.3 cM), were linked to Cda1 in soybean seed.	Daily consumption of cadmium (Cd) contaminated foods poses a risk to human health. Cultivar selection is an important method to limit Cd uptake and accumulation; however, analyzing grain Cd concentration is costly and time-consuming. Developing markers for low Cd accumulation will facilitate marker assisted selection (MAS).	Jegadeesan *et al.*, 2010
Ca content in soybean seed	Four QTL designated as Ca1, Ca2, Ca3, and Ca4 on linkage groups (LG) A2, I, and M were identified by both single-marker analysis and composite-interval mapping		Zhang *et al.*, 2009
Soybean sprout yield & quality	RFLP; L154,AO89,A668N,B046		Lee *et al.*, 2001
Isoflavonoids	Satt144 on LG F that was associated with Daidzein; LG M detected by Satt540 was associated with TI	Soybean isoflavones are valued in certain medicines, cosmetics, foods and feeds. Selection for high-isoflavone content in seeds along with agronomic traits is a goal of many soybean breeders	Zeng *et al.*. 2009
Isoflavonoids	Validated a major locus which alone accounted for up to 10% of the phenotypic variance for glycitein, and 35-37% for genistein, daidzein and the sum of all three soybean isoflavones.	Seeds of soybean [*Glycine max* (L.) Merr.] accumulate more isoflavones than any tissue of any plant species. In other plant parts, isoflavones are usually released to counteract the effects of various biotic and abiotic stresses. Because of the benefits to the plant and positive implications that consumption may have on human health, increasing isoflavones is a goal of many soybean breeding programs.	Gonzalez *et al.*, 2011

criteria which are free from environmental effects many authors have explored the molecular tools for breeding resistance against abiotic stresses (Table 5).

Biotic Stresses

Soybean is attacked by several pests, which include bacteria, fungi, viruses and insects. The zone wise biotic stresses are summarized below.

(a) *North Hill Zone*: Bacterial pustule, Soybean yellow mosaic virus, Frog eye leaf spot, Alternaria leaf spot, Target leaf spot etc.,

(b) *North Plain Zone*: Soybean yellow mosaic virus, Soybean mosaic virus, Bud blight, Sudden death syndrome, Bacterial blight, Bacterial pustule, Alternaria leaf spot, Frog eye leaf spot, charcoal rot, Rhizoctonia aerial blight etc,.

(c) *North Eastern Zone*: Rust, bacterial pustule, soybean yellow mosaic virus.

(d) *Central Zone*: Rust, Charcoal rot, Soybean yellow mosaic virus, Soybean mosaic virus, Bud blight, Sudden death syndrome, Bacterial blight, Bacterial pustule, Alternaria leaf spot, Frog eye leaf spot, Rhizoctonia aerial blight etc,.

(e) *Southern Zone*: Rust, charcoal rot, wilt, collar rot etc,.

Insect pests

Cultivation of yellow seeded soybean started in late sixties in India. There were merely two-dozen insect pests known to infect soybean in 1970s. But their number increased with the rapid increase in area under soybean cultivation, continuous cropping of soybean without any crop rotation and other factors like imbalance in use of insecticides and fertilizers. More than 270 insect pests, one mite, two millipedes, 10 vertebrates and one snail are reported to prey on soybean (Singh and Singh, 1990). But out of these only ten are important. These insects are

(a) *Stem borers*: Stem fly (*Melanagromyza sojae* (Zehnt.) and Girdle Beetle (*Oberopsis brevis*)

(b) *Pod borers*: *Helicoverpa armigera* and *Cydia ptychora*

(c) *Defoliators*: Green semilooper (*Chrysodeixis acuta* Walker), Grey semilooper (*Gesonia gemma* (Swinehoe), Tobacco caterpillar (*Spodoptera litura* (Fabr.) and Leaf miner (*Bilobata subseciella* (Zell).

(d) *Sap feeders*: Jassid (*Apheliona maculosa* Dist.), Aphis (*Aphis gossypii* Glov.) and Whitefly (*Bemisia tabaci* (Gen).

Breeding for resistance against insect pests is difficult due to the quantitative inheritance of the resistance and the linkage drag associated with the wild species being the resistance sources. However, one major and two minor QTLs have been identified for resistance to

Table 5: Summary of molecular studies on breeding for abiotic stress tolerance

Abiotic stresses	*QTL/marker*	*Remarks*	*Reference(s)*
Drought	Review paper		Manavalan *et al.*, 2009.
Root traits	Five QTLs Satt383, Satt339, Sct_191, Satt429 and Sat_299 were identified which together explained 51% of the variation in root score.	Plant introduction PI416937 was shown to possess fibrous roots and has been used as a parent in breeding programs to improve soybean productivity.	Hussein *et al.*, 2011.
Water logging	These STS primer sets were designated GmWT01-GmWT06 and GmWS01-GmWS04		Kang *et al.*, 2010
Salt tolerance			
Salinity	SSR markers Sat_091 and Satt237		Lee *et al.*, 2004.
Salt tolerance	SSR markers Sat_164 and Sat_358		Chen *et al.*, 2008.
Alkaline salt tolerance	On LG D2 a region between satt669 and sat_300 was identified which accounted for 50.2 and 13.0% of the total variation for alkaline Salt Tolerance Rating in the F_6 and the F_2 populations, respectively.	Salt-affected soils are generally classified into two main categories, sodic (alkaline) and saline	Tuyen *et al.*, 2010.
Iron deficiency chlorosis	Satt211, Satt481, and Sat_104.		Charlson *et al.*, 2005.

H. zea. The major QTL was linked to the RFLP marker A584 on LGM (Rector *et al.*, 1998). The minor QTLs were linked to the RFLP markers R249 (LG'H) and Bng047 (LG D1). Using RFLP markers, Narvel *et al.* (2001a) identified QTLs associated with insect resistance from PI 229358 and PI 171451. Soybean aphid resistance genes in the soybean cultivars, Dowling and Jackson map to linkage group M (Li *et al.*, 2007). Resistance to defoliating insects in soybean is expressed as a combination of antibiosis and antixenosis and both of these modes of resistance are inherited quantitatively (Rector *et al.*, 2000). The recent advances in the application of molecular tools to solve these problems are summarized in Table 6.

Table 6: Summary of molecular markers associated with major biotic stress tolerance

Pest and disease resistance	***QTL/marker***	***Remarks***	***Reference(s)***
Resistance against bacterial pustule	Rxp locus was mapped 3.9 cM from Satt372 and 12.4 cM from Satt014 on linkage group D2	Soybean bacterial leaf pustule (BLP) is a prevalent disease caused by *Xanthomonas axonopodis* pv. *glycines*	Kim *et al.*, 2010
Partial resistance to *Phytophthora sojae* Kauffmann and Gerdemann	The QTL on MLG D1b+W		Burnham *et al.*, 2003
Cyst nematode	Markers Satt309 and GMENOD2B	Resistance to race 14	Silva *et al.*, 2007
Cyst nematode	BACR-Satt309,		Li *et al.*, 2009
Cyst nematode	Satt309, Satt275, Satt574.	The rhg1 associated QTL found in earlier studies were confirmed	Kazi *et al.*, 2010
Meloidogyne incognita	Markers Satt201, Satt358, Satt487 and Satt590,	Race 2	Hendrika *et al.*, 2008
Root knot nematode	We discovered three SNPs in Satt358 source-sequences located near a major Mi-resistant QTL on LG-O and three SNPs in Satt199 source -sequences located near a minor Mi-resistant QTL on LG-G.		Ha *et al.*, 2007
Reniform nematode			
Frogeye leaf spot	The *Rcs3* gene was positioned near Satt244, SNPs AZ573TA150 and AZ573CA393.	Allele specific oligonucleotide probes specific for MAS were developed for a direct hybridization assay on a Luminex 100 flow cytometry platform.	Mian *et al.*, 1999; Missaoui *et al.*, 2007
Phytophthora Resistance gene *Rps1*	Sat_186 and Satt152,	*Rps1*-d Gene.	Uimoto *et al.*, 2008

Table 6: (*Contd....*)

Table 6: *(Contd....)*

Pest and disease resistance	***QTL/marker***	***Remarks***	***Reference(s)***
Phytophthora root rot	8 QTL underlying tolerance to PRR were identified,		Li *et al.*, 2010.
Soybean Rust			
Rust	The RB resistance gene mapped between Satt460 and Satt307 on linkage group (LG)-C2.		Monteros *et al.*, 2007
Rust	Placed *Rpp3* between SSR markers BARC_ Satt460 and BARC_Sat_263 on linkage group C2	For SBR resistance conditioned by *Rpp3*	Hyten *et al.*, 2009
Rust	SSR markers Sat_275+Sat_280	For SBR resistance conditioned by gene *Rpp5*	Morceli *et al.*, 2008
Purple seed strain	The candidate resistance gene was mapped between Sat_308 (6.6 cM) and Satt594 (11.6 cM) on molecular linkage group G		Jackson *et al.*, 2008
Rhizoctonia root and hypocotyl rot,	Satt281, Satt177, and Satt245		Zhao *et al.*, 2005
Powdery mildew	SSR marker Sat_224 and SNP marker BARC-021875-04228 at distances of 9.6 and 1.3 cM from the markers, respectively.	Powdery mildew (caused by *Microsphaera diffusa* Cooke & Peck). The previously reported *Rmd* locus of soybean for resistance to powdery mildew was mapped on soybean molecular linkage group J (chromosome 16).	Kang and Mian, 2010
Pyramiding of soybean mosaic virus resistance genes by Marker-Assisted Selection			Maroof *et al.*, 2008
SMV	Gene:Rsv1_ Satt114, Satt510, and Sat_154 (LG-F)Gene: Rsv3- Satt726 and Sat_424 (LG-B2)	Pathogen race G-1	Shi *et al.*, 2008

Table 6: *(Contd....)*

Table 6: (*Contd....*)

Pest and disease resistance	***QTL/marker***	***Remarks***	***Reference(s)***
SMV	Gene: Rsv4- Satt296 and Satt542, Race:G-1 Gene:Rsv1_ Satt114, Satt510, and Sat_154 (LG-F) Gene: Rsv3- Satt560 and Satt063Gene: Rsv4- AI856415-g		Shi *et al.*, 2009
SMV	Simple sequence repeat (SSR) markers with bulked segregation analysis (BSA) were used to conduct genetic mapping of the resistance to SMV1 in the segregating populations. The results indicated that resistance to SMV1 in adult- plants and the resistance to seed coat mottling in Dongnong 8143 and Tie 6915 was separately controlled by one single dominant gene.	Soybean mosaic virus (SMV) could lead to adult-plant system diseases, and cause mottling of soybean seeds.	Wang *et al.*, 2010
SMV	SNPs for the three candidate genes Rsv1,3 and 4	Three independent loci for SMV resistance have been identified in soybean germplasm.	Shi *et al.*, 2011
Brown stem rot	SSR markers Satt215 and Satt431 were linked to Rbs_1 and that Satt244 and Satt431 were linked to Rbs_2.		Bachman *et al.*, 2001
Lepidopteron insects	Synthetic Bt Cry1Ac transgene & SSR linked to antibiosis/anti xenosis QTL on linkage group M	Gene pyramiding	Walker *et al.*, 2002
Aphid		The discovery of biotype diversity of soybean aphid (SA: *Aphis glycines* Matsumura) in North America emphasizes the necessity to identify new aphid-resistance genes.	
Aphid	Mapped to Soybean Linkage Group M between the Markers Satt435 and Satt463. *rag1* Mapped		Li *et al.*, 2007

Table 6: (*Contd....*)

Table 6: (*Contd....*)

Pest and disease resistance	***QTL/marker***	***Remarks***	***Reference(s)***
	4. 2 cM from Satt435 and 7. 9 cM from Satt463. the Jackson Gene Mapped 2. 1 cM from Satt435 and 8. 2 cM from Satt463.		
Aphid	SSR markers Satt334 and Sct_033 on LG F.		Mian *et al.*, 2008
Aphid resistance	fine map Rag2.		Kim *et al.*, 2010
Aphid resistance	Two quantitative trait loci (QTLs) controlling the aphid resistance were found using the composite interval mapping method. These two QTLs were localized on linkage groups (LGs) F and M.		Zhang *et al.*, 2009
Pod borer	Three QTL directly associated with the resistance to SPB judged by PDS at harvest were identified. qRspb-1 (Satt541–Satt253) and qRspb-2 (Satt253–Satt314) were both on linkage group (LG) H and qRspb-3 (Satt288–Satt199) on LG GB.		Zhao *et al.*, 2008
Resistance to corn earworm (Lepidoptera: Noctuidae) and Soybean looper (Lepidoptera:	Simple-sequence repeat (SSR) markers were used previously to map soybean insect resistance (SIR) quantitative trait loci	The development of superior soybean, *Glycine max* (L.) Merr. cultivars exhibiting resistance to insects has been hindered due to linkage drag, a common phenomenon when introgressing alleles from exotic germplasm.	Warrington *et al.*, 2008

Genetically Modified Soybean

The herbicide resistant transgenic soybean was developed in 1996. The comparison between area, production and productivity of soybean in 1997 and 2010 will indicate the impact of herbicide tolerant soybean. In 1997 worldwide total area under soybean cultivation was 67 million hectares, which increased to 102 million hectares in 2010, an increase of 52% area. Of the 35.44 m hectares increase in area word wide, South America (Brazil, Argentina, Paraguay and Bolivia) accounted for 25.76 m hectares or 73% of the increase in area. The world wide production registered an increase of 90%. The soybean production increased from 144.535 million metric tons (m m tons) to 261.57 m tons. Of the total 130.27 m tons increase in production, South America accounted for 89 m m tons or 765 of the increase in production. Brazil and Argentina contributed to this change by adopting herbicide tolerant soybean. In 1997 GM soybean accounted for 10% of the total soybean grown in Brazil and 22.6% in Argentina in 2010 the share of GM soybean has increased to 76% in Brazil and 99% in Argentina. At the same time the area has more than doubled in Brazil and grown three times in Argentina. The productivity has increased from 22.97 q/ha to 29.41 q/ha in Brazil and 17.24 q/ha to 29.05 q/ha in Argentina. As a result of increase in productivity and area, production has increased 160% in Brazil and 379% in Argentina.

Table 7: Changes in area and productivity of soybean in major countries following adoption of glyphosate-tolerant GM varieties

	Area (million ha)		***Production (million metric tons)***		***Productivity (q/ha)***	
	1997	***2010***	***1997***	***2010***	***1997***	***2010***
World	66.94	102.38	144.535	261.57	21.56	25.54
USA	27.96	31.00	27.960	90.60	21.60	29.22
Brazil	11.48	23.29	11.480	68.52	22.97	29.41
Argentina	6.39	18.13	6.390	52.68	17.24	29.05
Paraguay	0.93	2.60	0.930	7.46	28.47	27.90

Source: FAO Statistics

Regeneration and transformation

Somatic embryogenesis in soybean was first reported by Christianson *et al.* (1983). Since then the protocol has been refined several times. Now immature cotyledon is routinely used for induction of somatic embryogenesis in soybean using high dose of cytokinin (2,4,-D, 40 µg/l). However, the induction and proliferation of somatic embryos from immature cotyledons are genotype dependent. Somatic embryogenesis in soybean is genetically controlled was shown by Kita *et al.* (2007). They backcrossed a genotype with poor somatic embryogenesis potential with Jack, a genotype with somatic embryogenesis potential. The backcross derived line showed a significant improvement in the potential for somatic embryogenesis. However, Bailey *et al.* (1993) reported genotypic difference among varieties to different steps of somatic embryogenesis.

Transformation in Soybean

Agrobacterium mediated transformation (Hinchee *et al.*, 1988) and particle bombardment method of transformation (Christou *et al.*, 1988) was first reported in soybean in 1988. Both the methods have been used by several workers and numerous refinements have been reported in the protocols. But transformation frequency is low in soybean as compared to other crops. Secondary somatic embryos proliferating directly from the terminal portion of the primary embryos are preferred for transformation for their unicellular origin. Herbicide tolerant transgenic soybean is the most successful commercial product. But attempts to develop transgenic soybean with modified seed protein, oil, amino acid, secondary seed compounds, insect resistance, nematode resistance, virus resistance, fungus resistance, drought stress tolerance, iron deficiency stress tolerance and several other traits have been reported.

Future Thrust

The three major constraints in soybean production in India are drought; weed infestation and location specific biotic stress. Therefore, gain in productivity can be achieved by solving these three problems. More than 90% of the soybean is produced in the states of Rajasthan, Madhya Pradesh and Maharashtra. Soybean growing areas in Rajasthan, and eastern Madhya Pradesh are low rain fall areas and prone to drought at seed filling stage of the crop. Early maturing varieties can escape the drought occurring at late in the season. However, yields of early maturing varieties (70-80 days) cannot be compared with the full season (115-120 days) varieties. Thus breeding for terminal drought tolerance or providing irrigation at late stage can help in increasing the yield. Breeding for resistance against biotic stress against location specific pest can also result in major yield gain.

An unexplored area is breeding for adaptation to specific latitude. Soybean was domesticated in China in the region between 30° and 45° North Latitude (Hymowitz, 1970). Soybean cultivation was limited to latitude north of 22° till 1970s when the trait "long juvenile period" was introduced in temperate cultivars (Neummaier and James, 1993). Based on the adaptation and maturity duration, genotypes have been classified into thirteen maturity groups (OOO to X). Genotypes belonging to OOO group are early maturing and are adapted to temperate regions whereas the genotypes belonging to maturity group X comprise late maturing types which are adapted to tropical regions (Fehr, 1987). Thus each genotype is adapted to a narrow range of latitude (3° to 5°) and is classified into different maturity group (Zhang *et al.*, 2007). The soybean growing areas in India have been classified into five different zones. Each of these zones spread across more than 7° to 9° latitude and cover locations with different rainfall and biotic stresses. There are few studies on adaptation to specific latitude in Indian context. There is an urgent need to reorganize the zones based on latitude, rainfall, soil type and biotic stress.

REFERENCES

Akkaya, M.S., Shoemake, R.C.R., Specht, J.E., Bhagwat, A.A. and Cregan, P.B. (1995). Integration of simple sequence repeats DNA markers into a soybean linkage map. *Crop Science*, 35: 1439–1445.

Alliprandini, L.F. and Vello, N.A. (2004). Heritability and correlation among traits in four way Soybean crosses. *Euphytica*, 136(1): 81–91.

Anita, R., Kumar, V., Mourya, V., Singh, R.K. and Husain, S.M. (2011). Validation of SSR markers linked to null kunitz tryspin inhibitor allele in indian soybean [*Glycine max* (L.) Merr.] population. *Journal of Plant Biochemistry and Biotechnology*, 20(2): 258–261.

Bachlava, E., Dewey, R.E., Auclair, J., Wang, S., Burton, J.W. and Cardinal, A.J. (2008). Mapping genes encoding microsomal -6 desaturase enzymes and their Co segregation with QTL affecting oleate content in soybean. *Crop Science*, 48: 640–650.

Bachlava, E., Ralph, E.D., Burton, J.W. and Cardinal, A.J. (2009). Mapping and comparison of quantitative trait loci for oleic acid seed content in two segregating soybean populations. *Crop Science*, 49: 433–442.

Bachman, M.S., Tamulonis, J.P., Nickell, C.D. and Ben, A.F. (2001). Molecular markers linked to brown stem rot resistance genes, Rbs_1 and Rbs_2, in Soybean. *Crop Science*, 41: 527–535.

Bailey, M.B., Boerma, H.R. and Parrot, W.A. (1993). Genotypes effects on proliferative embryogenesis and plant regeneration in soybean. *In Vitro Cell Dev. Biol.,* 29P: 102–108.

Beuselinck, P.R., Sleper, D.A. and Bilyeu, K.D. (2006). An assessment of phenotype selection for linolenic acid using genetic markers. *Crop Sci.,* 46: 747–750.

Bilyeu, K.D., Palavalli, K.D., Sleper, D.A. and Beuselinck, P.R. (2003). Three- microsomal omega-3 fatty-acid desaturase genes contribute to soybean linolenic acid levels. *Crop Sci.,* 43: 1833–1838.

Bowers, G.R., Kenty, M.M., Way, M.O., Funderburk, J.E. and Strayer, J.R. (1999). Comparison of three methods for estimating defoliation in soybean breeding programmes. *Agronomy J.,* 91: 242–247.

Bravo, J.J., Fehr, W.R., Welke, G.A., Hammond, E.G. and Cianzio, S.R. (1999). Family and line selection for elevated palmitate of Soybean. *Crop Science*, 39(3): 679–682.

Bubeck, D.M., Fehr, W.R. and Hammond, E.G. (1989). Inheritance of palmitic and stearic acid mutants of soybean. *Crop Sci.,* 29: 652–656.

Burnham, K.D., Dorrance, A.E., VanToai, T.T. and St. Martin, S.K. (2003). Quantitative trait loci for partial resistance to *Phytophthora sojae* in soybean. *Crop Sci.,* 43: 1610–1617.

Byfield, G.E., Xue, H. and Upchurch, R.G. (2006). Two genes from soybean encoding soluble D9 stearoyl-ACP desaturases. *Crop Sci.,* 46: 840–846.

Cardinal, A.J., Burton, J.W., Camacho-Roger, A.M., Yang, J.H., Richard, F.W. and Dewey, R.E. (2007). Molecular analysis of soybean lines with low palmitic acid content in the seed oil. *Crop Sci.,* 47: 304–310.

Cases, S., Smith, S.J., Zheng, Y.W., Myers, H.M., Lear, S.R., Sande, E., Novak, S., Collins, C., Welch, C.B. and Luis, A.J. (1998). Identification of a gene encoding an acyl CoA: Diacylglycerol acyltransferase, a key enzyme in triacylglycerol synthesis. *Proc. Natl. Acad. Sci.,* 95: 13018–13023.

Chapman, A., Pantalone, V.R., Ustun, A., Allen, F.L., Landau-Ellis, D. and Trigiano, R.N. (2003). Quantitative trait loci for agronomic and seed quality traits in an F_2 and $F_{4:6}$ soybean population. *Euphytica*, 129: 387–393.

Charlson, D.V., Bailey, T.B., Silvia, R.C. and Shoemaker, R.C. (2005). Molecular marker Satt481 is associated with iron-deficiency chlorosis resistance in a soybean breeding population. *Crop Science*, 45: 2394–2399.

Chen, H., Cui, S., Fu, S., Gai, J. and Yu, D. (2008). Identification of quantitative trait loci associated with salt tolerance during seedling growth in soybean (*Glycine max* L.). *Crop and Pasture Science*, 59: 1086–1091.

Choi, I.Y, Hyten, D.L., Matukumalli, L.K., Song, Q., Chaky, J.M., Quigley, C.V., Chase, K., Lark, K.G., Reiter, R.S. and Yoon, M.S. (2007). A soybean transcript map: Gene distribution, haplotype and single nucleotide polymorphism analysis. *Genetics*, 176: 685–696.

Christianson, M.L., Warnick, D.A. and Carlson, P.S. (1983). A morpho-genetically competent soybean suspension culture. *Science*, 222: 632–634.

Christou, P., McCabe, D.E. and Swain, W.F. (1988). Stable transformation of soybean callus by DNA-coated gold particles. *Plant Physiology*, 87: 671–674.

Cober, E.R. and Voldeng, D. (2000). Developing high-protein, high-yield soybean populations and lines. *Crop Science*, 40: 39–42.

Costa, M.M., Di, M.A.O., Unêda-Trevisoli, S.H., Arriel, N.H.C., Bárbaro, I.M., Silveira, G.D. and Muniz, F.R.S. (2008). Heritability estimation in early generations of two-way crosses in soybean. *Bragantia*, 67: 101–108.

Destro, D., Bizeti, H.S., Garcia, L.A., Batista, F.I.C.D., Ontalvan, R. and Edison, M. (2003). Comparison between the SPD and the SPDS methods for segregating generation advancement in soybean. *Brazilian Archives of Biology and Technology*, 46: 545–551.

Fehr, W.R., Welke, G.A., Hammond, E.G., Duvick, D.N. and Cianzio, S.R. (1992). Inheritance of reduced linolenic acid content in soybean genotypes A16 and A17. *Crop Sci.,* 32: 903–906.

Gesteria, A.D.S., Schuster, I., Jose, I.C., Piovesan, N.D., Viana, J.M.S., Barros, E.G. and Moreira, M.A. (2003). Biometrical analyses of linolenic acid content of soybean seeds. *Genetics and Molecular Biology*, 26: 65–68.

Gilbert, E.M., Maciei, A.Z. and Holbrook, N.M. (2011). Independent variation in photosynthetic capacity and stomatal conductance leads to differences in intrinsic water use efficiency in 11 soybean genotypes before and during mild drought. *J. of Experimental Botany*, 62: 2875–2887

Gizlice, Z.C., Carter, T.E. and Burton, J.W. (1996). Genetic diversity patterns in North American public soybean cultivars based on coefficient of parentage. *Crop Science*, 36: 753–765.

Gonzalez, J.J.G., Vuong, T.D., Yu, R.Z.O., Lee, J.D., Shannon, G., Ellersieck, M., Nguyen, H.T. and Sleper, D.A. (2011). Major locus and other novel additive and epistatic loci involved in modulation of isoflavone concentration in soybean seeds. *Theor. Appl. Genet.,* 123: 1375–1385.

Gravina, G.A., Sediyama, C.S., Martins, F.S., Moreira, M.A., Barros, E.G. and Cruz, C.D. (2004). Multivariate analysis of combining ability for soybean resistance to *Cercospora sojina* Hara. *Genet. and Mol. Bio.,* 27: 395–399.

Ha, B.K., Hussey, R.S. Hussey and Boerma, H.R. (2007). Development of SNP assays for marker-assisted selection of two southern root-knot nematode resistance QTL in soybean. *Crop Sci.,* 47: S–73–S–82.

Hargreaves, G.H. (1977). World water for agriculture contract AID/ta-C-1103. Utah State University and Agency for International development, Washington, D.C., p. 177.

Hendrika, F., Mienie, C.M.S., Mc Donald, Alex, H. and Waele, D.D. (2008). Identification and validation of genetic markers associated with *Meloidogyne incognita* race 2 resistance in soybean, *Glycine Max* (L.) *Merr. Nematology*, 10: 651–666.

Hinchee, M.A.W., Conner-Ward, D.V., Newell, C.A., McDonnell, R.E., Sato, S.J., Gasser, C.S., Fischhoff, D.A., Re, D.B., Fraley, R.T. and Horsch, R.B. (1988). Production of transgenic soybean plants using Agrobacterium-mediated DNA transfer. *Nat. Biotechnology*, 6: 915–922.

Holt, D.A., Kauffman, H.E., Bellasco, C., Bhatnagar, P.S., Galerani, P., Haun, S. and Roessing, A. (1997). *In*: Proceedings of the conference on meeting the demand for food in the 21st Century; Challenges and Opportunities for Illinois Agriculture.

Hussein, A.H., Lee, G.J., and Boerma, R.H. (2011). Identification of QTL for increased fibrous roots in soybean. *TAG.,* 122: 935–946.

Hymowitz, T. (1970). On the domestication of the soybean. *Econ. Botany*, 24: 408–421.

Hyten, D.L., Song, Q., Zhu, Y., Choi, I.Y., Nelson, R.L., Costa, J.M., Specht, J., Shoemaker, R.C. and Cregan, P.B. (2006). Impact of genetic bottlenecks on soybean genome diversity. *PNAS.,* 103: 16666–16671.

Hyten, D.L., Smith, J.R., Frederick, R.D., Tucker, M.L.,Song, Q. and Cregan, P.B. (2009). Bulked segregant analysis using the golden gate assay to locate the Rpp3 locus that confers resistance to soybean rust in soy. *Crop Sci.,* 49: 265–271.

Ishikawa, G., Yoshitake, T. and Toshiki, N. (2006). A PCR-based method to test for the presence or absence of ß-conglycinin A'- and A-subunits in soybean. *Molecular Breeding: New Strategies in Plant Improvement*, 17: 365–374.

Jackson, E.W., Feng, C., Fenn, P. and Chen, P. (2008). Genetic mapping of resistance to purple seed stain in PI 80837 soybean. *Journal of Heredity*, 99: 319–322.

Jegadeesan, S., Yu, K., Poysa, V., Gawalko, E., Gawalko, Morrison, M.J. Morrison Shi, C. and Cober, E. (2010). Mapping and validation of simple sequence repeat markers linked to a major gene controlling seed cadmium accumulation in soybean [*Glycine max* (L.) Merr]. *Theor. Appl. Genet.,* 121: 283–294.

Kang, S.Y., Lee, K.J., Lee, G.J., Kim, J.B., Chung, S.J., Song, J.Y., Lee, B.M. and D.S. (2010). Development of AFLP and STS markers linked to a water logging tolerance in korean soybean landraces. *Biologia Plantarum*, 54: 61–68.

Kangm S.T. and Mianm M.A.R. (2010). Genetic map of the powdery mildew resistance gene in soybean PI 243540. *Genome*, 53: 400–405.

Karnwal, M.K., Siddhu, P. and Pushpendra. (2009). Early generation selection for yield contributing traits in interspecific crosses of soybean (*Glycine max* (L.) Merrill). *Legume Research,* 32:

Kazi, S., Shultz, J., Afzal, J., Hasmi, R., Jasim, M., Bond, J., Arelli, P.R. and Lightfoot, D.A. (2010). Iso-lines and inbred-lines confirmed loci that underlie resistance from cultivar 'Hartwig' to three soybean cyst nematode populations. *Theoretical and Applied Genetics*, 120(3): 633–644.

Keim, P., Diers, B.W., Olson, T.C. and Shoemaker, R.C. (1990). RFLP mapping in soybean: Association between marker loci and variation in quantitative traits. *Genetics*, 126: 735–742.

Kim, D.H., Kim, K.H., Van, K., Kim, M.Y. and Lee, S.H. (2010). Fine mapping of a resistance gene to bacterial leaf pustule in soybean. *Theor. Appl. Genet.,* 120: 1443–1450.

Kim, K.S., Hill, C.B., Hartman, G.L., Hyten, D.L. and Hudson, M.E. (2010). Fine mapping of the soybean aphid–resistance gene Rag2 in soybean PI 200538. *Theor. Appl. Genet.,* 121: 599–610.

Kim, M.Y., Ha, B.K., Jun, T.H., Hwang, E.Y., Van, K. and Kuk, Y.I. (2004). Single nucleotide polymorphism discovery and linkage mapping of lipoxygenase–2 gene (Lx_2) in soybean. *Euphytica*, 135: 169–177.

Kim, M.S., Park, M.J., Jeong, W.H., Nam, K.C. and Chung, J.Il. (2006). SSR marker tightly linked to the *ti* locus in soybean [*Glycine max* (L.) Merr.]. *Euphytica*, 152: 361–366.

Kim, S., Kim, M.Y., Van, K., Lee, Y.H., Kim, H.S., Cai, C.M., Park, B.S., Seo, H.S., and Lee, S.H. (2011). The Development of a co–dominant marker for the ß–conglycinin a' subunit in soybeans. *Euphytica*, 177: 355–363.

Kita, Y., Nishizawa, K., Takahashi, M., Kitayama, M. and Ishimoto, M. (2007). Genetic improvement of the somatic embryogenesis and regeneration in soybean and transformation of the improved breeding lines. *Plant Cell Reports*, 26: 439–447.

Kumudini, S. (2002). Trials and tribulations: A review of the role of assimilates supply in soybean genetic yield improvement. *Field Crop Research*, 75: 211–222.

Lal, S.K. and Rana, V.K.S. (2000). Genetic enhancement of Soybean. *Ind. J. Genet.,* 60: 483–494.

Lal, S.K., Sapra, R.L. and Talukdar, A. (2008). Soybean ideotype for moisture stress areas of central India. *J. Oilseeds Res.,* 26: 138–140.

Lark, K.G., Weismann, J.M., Matthews, B.F., Palmer, R., Chase, K. and Macalma, T. (1993). A genetic map of soybean (*Glycine max* (L.) using an interspecific cross of two cultivars: 'Minsoy' and 'Noir 1'. *Theor. Appl. Genet.,* 86: 901–906.

Lee, G.J., Boerma, H.R., Villagarcia, M.R., Zhou, X., Carter, T.E., Li Jr, Z. and Gibbs, M.O. (2004). A major QTL conditioning salt tolerance in S–100 soybean and descendent cultivars. *Theor. Appl. Genet.,* 109: 552–561.

Lee, J.D., Shannon, J.G. and Choung, M.G. (2010). Selection for protein content in soybean from single F_2 seed by near infrared reflectance spectroscopy. *Euphytica*, 172: 117–123.

Lee, S.H., Park, K.Y., Lee, H.S, Park, E.H. and Boerma, R.H. (2001). Genetic mapping of QTL's conditioning soybean sprout yield and quality Putative. *Theor. Appl. Genet.,* 103: 702–709.

Li, H., Liu, H., Han, Y., Wu, X., Teng, W., Liu, G. and Li, W. (2010). Identification of QTL underlying Vitamin E contents in soybean seed among multiple environments. *TAG*, 120(7): 1405–1413.

Li, H. and Burton, W.J. (2002). Selecting increased seed density to increase indirectly soybean seed protein concentration. *Crop Sci.,* 42: 393–398.

Li, X., Han, Y., Teng, W., Zhang, S., Yu, K., Poysa, V., Anderson, T., Ding, J. and Li, W. (2010). Pyramided QTL underlying tolerance to *Phytophthora* root rot in mega–environments from Soybean cultivars 'Conrad' and 'Hefeng 25'. *Theor. Appl. Genet.,* 121: 651–658.

Li Y., Hill, C.B., Carlson, S.R., Diers, B.W. and Hartman, G.L. (2007). Soybean aphid resistance genes in the soybean cultivars Dowling and Jackson map to linkage group M. *Mol. Breed. New Strategies in Plant Improvement*, 19: 25–34.

Li, Y., Zhang, C., Gao, Z., Marinus, J.M.S., Zulu, M., Liu, Z., Nan, H.Y., Chang, R.Z. and Qiu, L.J. (2009). Development of SNP markers and haplotype analysis of the candidate gene for *rhg1*, which confers resistance to soybean cyst nematode in soybean. *Molecular Breeding: New Strategies in Plant Improvement*, 24: 63–76.

Manavalan, L.P., Satish, K.G., Tran, L.P. and Nguyen, H.T. (2009). Physiological and molecular approaches to improve drought resistance in soybean. *Plant and Cell Physiol.,* 50: 1260–1276.

Maroof, S.M.A., Jeong, S.C., Gunduz, I., Tucker, D.M., Buss, G.R. and Tolin, S.A. (2008). Pyramiding of soybean mosaic virus resistance genes by marker–assisted selection. *Crop Sci.,* 48: 517–526.

Maupin, L.M., Rosso, M.L. and Rainey, K.M. (2011). Environmental effects on soybean with modified phosphorus and sugar composition. *Crop Sci.,* 51: 642–650.

Mian, M.A.R., Kang, S.T., Scott, E.B. and Ronald, B.H. (2008). Genetic linkage mapping of the soybean aphid resistance gene in Pi 243540. *Theor. Appl. Genet.,* 117: 955–962.

Mian, M.A.R., Wang, T., Phillips, D.V., Jennie, A. and Boe, H.R. (1999). Molecular mapping of the Rcs3 gene for resistance to frogeye leaf spot in soybean. *Crop Sci.,* 39: 1687–1691.

Mian, M.A.R., Ashley, D.A. and Boerma, H.R. (1998). An additional QTL for water use efficiency in soybean. *Crop Sci.,* 38: 390–393.

Mian, M.A.R., Mailey, M.A. and Ashley, P.W.A. (1996). Molecular marker associated with water use efficiency and leaf ash in soybean. *Crop Sci.,* 36: 1252–1257.

Miladinovic, J., Burton, J.W., Tubic, S.B., Miladinovic, D., Dragana Djordjevic, V. and Jukic, V. (2011). Soybean breeding: Comparison of the efficiency of different selection methods. *Turkish J. of Ag. and Forestry Sciences*, 35: 469–480.

Missaoui, A.M, Ha, B.K., Phillips, D.V. and Boerma, H.R. (2007). Single nucleotide polymorphism detection of the rcs3 gene for resistance to frogeye leaf spot in soybean. *Crop Science*, 47: 1681–1690.

Monteros, M.J., Missaoui, A.M., Phillips, D.V., Walker, D.R. and Boerma, H.R. (2007). Mapping and confirmation of the 'Hyuuga' red–brown lesion resistance gene for asian soybean rust. *Crop Science*, 47(2): 829–834.

Monteros, M.J., Burton, J.W. and Boerma, H.R. (2008). Molecular mapping and confirmation of QTLs associated with oleic acid content in N00–3350 soybean. *Crop Science*, 48: 2223–2234.

Morceli, S., Thaiza, G., Trevisoli, S.H., Unêda, M.Jr., Antonio, A.K., Romeu, A.S., Calvo, E.S., Di Mauro, A.O. and Garcia, A. (2008). Identification and validation of microsatellite markers linked to the Rpp5 gene conferring resistance to asian soybean rust. *Pesquisa Agropecuária Brasileira*, 43: 1533–1541.

Morrison, M.J., Harvey, D.V. and Cober, E.R. (1999). Physiological changes from 58 years of genetic improvement of short–season soybean cultivars in Canada. *Agronomy J.,* 91: 685–689.

Narvel, J.M., Jakkula, L.R., Phillips, D.V., Wang, T. and Lee, S.H. (2001). Molecular mapping of *Rxp* conditioning reaction to bacterial pustule in soybean. *Journal of Heredity*, 92: 267–270.

Narvel, J.M., Walker, D.R., Rector, B.G., All, J.N., Parrott, W.A. and Boerma, H.R. (2001a). A retrospective DNA marker assessment of the development of insect resistant soybean. *Crop Sci.,* 41: 1931–1939.

Narvel, J.M., Fehr, W.R. and Welke, G.A. (1998). Agronomic and seed traits of soybean lines lacking seed lipoxygenases. *Crop Sci.,* 38: 926–928.

Neumaier, N. and James, A.T. (1993). Exploiting the long Juvenile trait to improve adaptation of soybeans to the tropics. *Food Legume Newsletter*, 8: 12–14.

Pacheco, R.M., Duarte, J.B., Vencovsky, R., Pinheiro, J.B. and Oliveira, A.B. (2005). Use of supplementary genotypes in AMMI analysis. *Theor. Appl. Genet.,* 110: 812–818.

Panthee, D.R., Kwanyuen, P., Sams, C.E., West, D.R., Saxton, A.M. and Pantalone, V.R. (2004). Quantitative trait loci for β–Conglycinin (7S) and Glycinin (11S) fractions of soybean storage protein. *Journal of the American Oil Chemists' Society*, 81: 1005–1012.

Priolli, R.H.G., Pinheiro, J.B., Zucchi, M.I., Bajay, M.M. and Vello, N.A. (2010). Genetic diversity among Brazilian cultivars based on SSR loci and pedigree data. *Braz. Arch. Biol. Technology*, 53: 519–531.

Qi, Z.M., Sun, Y.N., Wu, Q., Liu, C.Y., Hu, G.H. and Chen, Q.S. (2011). A meta–analysis of seed protein concentration QTL in soybean. *Canadian J. of Plant Sci.*, 91: 221–230.

Rector, B.G., All, J.N., Parrott, W.A. and Boerma, H.R. (1998). Identification of molecular markers linked to quantitative trait loci for soybean resistance to corn earworm. *Theor. Appl. Genet.,* 96: 786–790.

Rector, B.G., All, J.N., Parrott, W.A. and Boerma, H.R. (2000). Quantitative trait loci for antibiosis resistance to corn earworm in soybean. *Crop Sci.,* 40: 233–238.

Reinprecht, Y., Luk–Labey, S.Y., Larsen, J., Poysa, V.W., Yu, K., Rajcan, I., Ablett, G.R. and Pauls, K.P. (2009). Molecular basis of the low linolenic acid trait on soybean EMS mutant line RG10. *Plant Breeding*, 128: 253–258.

Reinprecht, Y., Vaino, W.P., Yu, K., Istvan, R., Albert, G.R. and Pauls, K.P. (2006). Seed and agronomic QTL in low linolenic acid, lipoxygenase–free soybean (*Glycine max* (L.) Merrill) germplasm. *Genome*, 49: 1510–1527.

Richard, A.R. (2004). Physiological traits used in the breeding of new cultivars for water–scarce environments. "New directions for a diverse planet". Proceedings of the 4th International Crop Science Congress, 26 Sep–1 Oct 2004, Brisbane, Australia. Published on CDROM. Web site www.cropscience.org.au

Rotundo, J., Lucas, B., Mark, E.W. and Orf, J.H. (2009). Relationship between assimilate supply per seed during seed filling and soybean seed composition. *Field Crop Research*, 112: 90–96.

Salem, A.M., Kakani, V.G., Koti, S. and Reddy, K.R. (2007). Pollen–based screening of soybean genotypes for high temperatures. *Crop Science*, 47: 219–231.

Sauer, M.L.A., Scott, R.A. and Thomas, M.C. (2008). Marker–assisted selection for low linolenic acid content in soybean. *Journal of Crop Improvement*, 21: 139– 155.

Scaboo, A.M., Pantalone, V.R., Walker, D.R., Boerma, H.R., West, D.R., Walker, F.R. and Sams, C.E. (2009). Confirmation of molecular markers and agronomic traits associated with seed phytate content in two soybean RIL populations. *Crop Sci.,* 49: 426–432.

Schmutz, J., Cannon, S.B., Schlueter, J., Ma, J., Mitros, T., Nelson, W., Hyten, D.L., Song, Q., Thelen, J.J., Cheng, J., Xu, D., Hellsten, U., May, G.D., Yu, Y., Sakurai, T., Umezawa, T., Bhattacharyya, M.K., Sandhu, D., Valliyodan, B., Lindquist, E., Peto, M., Grant, D., Shu, S., Goodstein, D., Barry, K., Futrell–Giggs, M., Abernathy, B., Du, J., Tian, Z., Zhu, L., Gill, N., Joshi, T., Libault, M., Sethuraman, A., Zhang, X.C., Shinozaki, K., Nguyen, H.T., Wing, R.A., Cregan, P., Specht, J., Grimwood, J., Rokhsar, D., Stacey, G., Shoemaker, R.C. and Jackson, S.A. (2010). Genome sequence of the palaeopolyploid soybean. *Nature*, 463:178–183.

Shi, A., Chen, P., Li, D.X., Zheng, C., Hou, A. and Zhang, B. (2008). Genetic confirmation of 2 independent genes for resistance to soybean mosaic virus in J05 soybean using SSR markers. *J. Heredity*, 99: 598–603.

Shi, A., Chen, P., Li, D., Zheng, C., Zhang, B. and Hou, A. (2009). Pyramiding multiple genes for resistance to soybean mosaic virus in soybean using molecular markers. *Molecular Breeding: New Strategies in Plant Improvement*, 23: 113–124.

Shi, A., Chen, P., Vierling, R., Zheng, C., Li, D., Dong, D., Ehsan, S. and Cervantez, I. (2011). Multiplex single nucleotide polymorphism (SNP) assay for detection of soybean mosaic virus resistance genes in soybean. *Theor. Appl. Genet.,* 122: 445–457.

Shoemaker, Specht, Shoemaker, R.C. and Specht, J.E. (1995). Integration of the soybean molecular and classical genetic linkage groups. *Crop Sci.,* 35: 436–446.

Silva, M.F., Schuster, I., Silva, J.F.V., Ferreira, A., Barros, E.G. and Moreira, M.A. (2007). Validation of microsatellite markers for assisted selection of soybean resistance to cyst nematode races 3 and 14. *Pesquisa Agropecuária Brasileira*, 42: 1143–1150.

Sinclair, T.R., Purcell, L.C., King, C.A., Sneller, C.H., Chen, P. and Vadez, V. (2007). Drought tolerance and yield increase of soybean resulting from improved symbiotic N fixation. *Field Crops Research*, 101: 68–71.

Sinclair, T.R., Purcell, L.C., Vadez, V., Serraj, R., King Andy, C. and Nelson, R. (2000). Identification of soybean genotypes with N_2 fixation tolerance to water deficits. *Crop Sci.,* 40: 1803–1809.

Song, Q.J., Marek, L.F., Shoemaker, R.C., Lark, K.G., Concibido, V.C., Delannay, X., Specht, J.E. and Cregan, P.B. (2004). A new integrated genetic linkage map of the soybean. *Theor. Appl. Genet.,* 109: 122–128.

Specht, J.E., Chase, K., Macrander, M., Graef, G.L., Chung, J. and Markwell, J.P. (2001). Soybean response to water: A QTL analysis of drought tolerance. *Crop Sci.,* 41: 493–509.

Spencer, M.M., Pantalone, V.R., Meyer, E.J., Landau-Ellis, D. and Hyten Jr. D.L. (2003). Mapping the *Fas* locus controlling stearic acid content in soybean. *Theor. Appl. Genet.,* 106: 615–619.

Spencer, M.M., Landau-Ellis, D., Meyer, E.J. and Pantalone, V.R. (2004). Molecular markers associated with linolenic acid content in soybean. *Journal of the American Oil Chemists' Society*, 81: 559–562.

St Martin S.K. and Isaias, O.G. (2002). Comparison of three procedures for early generation testing of soybean. *Crop Sci.,* 42: 705–709.

St. Martin, S.K. and Xie, F. (2000). Genetic gain in early stages of a soybean breeding programme. *Crop Sci.,* 40: 1559–1564.

Stefaniak, T.R., Hyten, D.L., Pantalone, V.R., Klarer, A. and Pfeiffer, T.W. (2006). Soybean cultivars resulted from more recombination events than unselected lines in the same population. *Crop Sci.,* 46: 43–51.

Streit, L.G., Walter, R.F., Grace, A.W., Earl, G.H. and Silvia, R.C. (2001a). Family and line selection for reduced palmitate, saturates and linolenate of soybean. *Crop Sci.,* 41: 63–67.

Streit, L.G., Walter, R.F. and Grace, A.W. (2001b). Family and line selection for seed yield of soybean. *Crop Sci.,* 41: 358–362.

Subrahmanyam, D. (2002). Interrelationship between leaf gas-exchange characteristics, area leaf mass and yield in soybean (*Glycine max* L. Merr) *Genotypes. Photosynthetica*, 40: 441–444.

Tuyen, D.D., Lal, S.K. and Xu, D.H. (2010). Identification of a major QTL allele from wild soybean (*Glycine soja* Sieb. and Zucc.) for increasing alkaline salt tolerance in soybean. *Theor. Appl. Genet.,* 121: 229–236.

Ugimoto, T.S., Oshida, S.Y., Watanabe, K., Ino, M.A., Anto, T.K., Aekawa, K.M. and Rie, K.I. (2008). Identification of Ssr markers linked to the *Phytophthora* resistance gene *Rps1*-d in soybean. *Plant Breeding*, 127: 154–159.

Walker, D., Boerma, H.R., All, J. and Wayne, P. (2002). Combining cry1Ac with QTL alleles from PI 229358 to improve soybean resistance to lepidopteron pests. *Molecular Breeding: New Strategies in Plant Improvement*, 9: 43–51.

Wang, K., Li, W., Zhang, L., Liu, C., Zhu, X., Chen, Q. and Hu, G. (2010). Genetic analysis and molecular mapping of two SMV-resistance traits in soybean: Adult-plant resistance and resistance to seed coat mottling. *Agricultural Sciences in China*, 9: 11–18.

Warrag, M.O.A. and Hall, A.E. (1983). Reproductive responses of cowpea to heat stress: Genotypic differences in tolerance to heat at flowering. *Crop Sci.,* 23: 1088–1092.

Warrington, C.V., Zhu, S., Parrott, W.A., All, J.N. and Boerma, H.R. (2008). Seed yield of near–isogenic soybean lines with introgressed quantitative trait loci conditioning resistance to corn earworm (Lepidoptera: Noctuidae) and soybean looper (Lepidoptera: Noctuidae) from PI 229358. *Journal of Economic Entomology*, 101: 1471–1477.

Weikai, Y. and Istvan, R. (2002). Biplot analysis of test sites and trait relations of soybean in Ontario. *Crop Sci.,* 42: 11–20.

Wu, X., Ren, C., Joshi, T., Vuong, T., Xu, D. and Nguyen, H.T. (2010). SNP discovery by high throughput sequencing in soybean. *BMC Genomics*, 11: 469

Zhang, B., Chen, P.C., Shi, A., Hou, A., Tetsuaki, I. and Wang, D. (2009). Putative quantitative trait loci associated with calcium content in soybean seed. *J. Heredity*, 100: 263–269.

Zhang, P., Burton, J.W., Upchurch, R.G., Whittle, E., Shanklin, J. and Dewey, R.E. (2008). Mutations in a D9–stearoyl–ACP–desaturase gene are associated with enhanced stearic acid levels in soybean seeds. *Crop Sci.,* 48: 2305–2313.

Zhang, G., Cuihua G. and Dechun, W. (2009). Molecular mapping of soybean aphid resistance genes in PI 567541B. *Theor. Appl. Genet.,* 118: 473–482.

Zhang, L.X., Kyei–Boahen, S., Zhang, J., Zhang, M.H., Freeland, T.B., Watson, C.E. Jr., and Liu, X.M. (2007). Modifications of optimum adaptation zones for soybean maturity groups in the USA. Online. *Crop Management*, doi:10.1094/CM–2007–0927–01–RS

Zhao, G., Ablett, G.R., Anderson, T.R., Rajcan, I. and Schaafsma, A.W. (2005). Inheritance and genetic mapping of resistance to *Rhizoctonia* Root and Hypocotyl Root in Soybean. *Crop Sci..* 45: 1441–1447.

Zhao, G., Wang, J., Han, Y., Teng, W., Sun, G. and Li, W. (2008). Identification of QTL underlying the resistance of soybean to pod borer, *Leguminivora glycinivorella* (Mats.) Obraztsov and correlations with plant, pod and seed traits. *Euphytica*, 164: 275–282.

12

Oilseed Brassicas

D.K. YADAVA, NAVEEN SINGH, SUJATA VAUSDEV AND K.V. PRABHU

Division of Genetics, Indian Agricultural Research Institute,
New Delhi, India

AREA, PRODUCTION AND PRODUCTIVITY

World Scenario

Rapeseed mustard is the third major oilseed crop at global level, after soybean and palm oil. Worldwide, an incredible increase of more than 75% in production from 33.57 mt (1997-98) to 58.77 mt (2010-11) is attributed to more than 30% expansion in area *i.e.*, from 24.98 mha to 32.63 mha; and more than 34% enhancement in productivity *i.e.*, from 1343 kg/ha to 1800 kg/ha. China, Canada, India, Germany, U.K., France and Australia are the major producers of rapeseed mustard.

Indian Scenario

In India, it is the second most important oilseed crop after soybean. The area, production and productivity of this crop in India has been static with minor seasonal fluctuations over the years: productivity remained at 1145 kg/ha during 2011-12. India's share in the global area planted with rapeseed mustard is 21.45% but it contributes only 11.92% to the global rapeseed mustard production. During 2011-12 there was a production of 6.78 mt from 5.92 mha area under this crop. Rajasthan has emerged as a major mustard growing state with a lion's share of 45–55% over the years. Except Rajasthan and Haryana, there

V.L. Chopra (Ed.): Breeding Field Crops II: Advances, Studium Press (LLC), USA.

is a gradual decline in area under this crop in most of the other states. Uttar Pradesh and Punjab are among the top states with declining trend. *Brassica* has been introduced in three non-traditional states *viz.* Andhra Pradesh, Karnataka and Tamil Nadu during last decade and the short duration varieties *viz.,* Pusa Agrani and Pusa Mahak have given encouraging performance in these states. Among the different oleiferous *Brassica* species grown in India, proportion of area under *Brassica juncea* (Indian mustard) has increased whereas, the area under other species like *B. rapa* cvs. yellow sarson, brown sarson and *B. napus* have declined. With the introduction of short duration (100-110 days) Indian mustard varieties, some area of *B. rapa* cv. *toria* has also been replaced by *B. juncea* in plains as well as in North-Eastern states.

GENETIC RESOURCES

Improvement of Cultivated Species

More than 10000 germplasm lines are being maintained at National Bureau of Plant Genetic Resources, New Delhi, Directorate of Rapeseed Mustard Research, Bharatpur and different AICRP centres in different State Agricultural Universities and ICAR Institutes. Since 1998, a good number of genetic resources including genetic stocks, varieties, germplasm lines for different important traits have been procured, developed, released and registered with NBPGR (Table 1). Good success has been achieved for quality attributes where very stable donors like BJ-1058 and Heera for low glucosinolates; LES-39 (Pusa Karishma), LES-1-27 (Pusa Mustard 21), LET-17 (Pusa Mustard 22), LET-18 (Pusa Mustard 24), LET-36 (Pusa Mustard 29), LES 43 (Pusa Mustard 30), YSRL-9-18-23, TERI (0E) M21 (IC-296684), ELM-079 (RLC-1), ELM-123 (RLC-2), PRQ-2005-1 (IC-546946) for low erucic acid are available. As far as abiotic stresses like salt tolerance [CS-52, CS-54 (CS-614-4-1-4), CS-234-2-2, CS-56, RH-8814 (IC-401570)]; heat tolerance [Pusa Vijay (NPJ-93), Pusa Mustard 25 (NPJ-112), Pusa Mustard 27 (EJ-17), Pusa Mustard 28 (NPJ-124), BPR-541, BPR-543)]; and frost tolerance (RH-819, RGN-48) are concerned a good number of varieties and genetic stocks have been developed and are available in public domain. Well established short duration genotypes *viz.* Pusa Agrani (SEJ-2), Pusa Mahak (JD-6), Pusa Tarak (EJ-9912-13), Pusa Mustard 25 (NPJ-112), Pusa Mustard 27 (EJ-17), Pusa Mustard-28 (NPJ-124), EJ-20, Kanti, NDRE-4, PR-2006-1 and Divya (IC-553910) with 100-120 days maturity have been developed which can be further exploited for breeding early

Table 1: Promising donors identified for different agro-morphological and quality traits

Agronomic attributes	*Species*	*Donor*
Tolerance to *Alternaria* blight	*B. juncea*	EC-129126, EC-399301, PAB-9511, PAB-9534 (IC-556997)#, RC-781
	B. carinata	Kiran (PBC-9921), PC-5, Pusa Swarnim (IGC-01)
	B. napus	GSL-1, HNS-3, PBN-2002, PBN-9501, PBN-9502
Resistance to white rust	*B. juncea*	EC-399300, EC-399301, JMMWR-941-1-2 (JM-1)*, PWR-2001, PWR-9541, Bio-YSR (IC-443623)#, BEC-144, BEC-286, RC-781, NRCDR-515 (IC-555891)#
	B. carinata	JTC-1*, Kiran (PBC-9921)*, PC-5*, PC-5-17, Pusa Gaurav (DLSC-1)*, Pusa Swarnim (IGC-01)*, Pusa Aditya (NPC-9)*
	B. napus	GSL-441, HNS-4, PBN-2001, PBN-2002
Resistance to downy mildew	*B. juncea*	Bio-YSR (IC-443623)#
	B. napus	PBN-2002
Low glucosinolate (<30 micro moles/ g defatted seed meal)	*B. juncea*	NUDHYJ-1, NUDHYJ-2
	B. napus	HNS-99(0E)3, NUDB-9
Low erucic acid (<2%)	*B. juncea*	Pusa Karishma (LES-39)*, LES-1-27 (Pusa Mustard 21)*, LET-17 (Pusa Mustard 22)*, LET-18 (Pusa Mustard 24)*, LET-36, YSRL-9-18-23, TERI (0E) M21 (IC-296684)#, ELM-079*, ELM-123, PRQ-2005-1 ((IC-546946)#
Low erucic acid (<2%) and low glucosinolate (<30 micro moles/ g defatted seed meal)	*B. juncea*	BJ-1058, NUDH-YJ-5 (IC-296507)#, Heera (IC-296501)#
	B. napus	BCN-14, CAN-138, GSC-5*, OCN-3*, NUDB-26-11, TERI (00) R985 (IC-296732)#, TERI (00) R986 (IC-296731)#, TERI (00) R-9903 (IC-404233)*#, PAC-401*, NUDB-38 (IC-296827)#, NUDB-42 (IC-296828)#
High oleic acid (70.1%) with low erucic acid (<2%)	*B. rapa*	PHOP-2-2 (IC-552726)#
	B. napus	TERI (0E) R03 (IC-296685)#, TERI (0E) R09 (IC-296686)#
High oleic and linoleic acid	*B. juncea*	TERI GZ-05 (IC-404233)#
Low linolenic acid (3.03%)	*B. rapa*	PHOT-8-2-11 (IC-552727)#
High oil content (~45%)	*B. juncea*	NDYR-8 (IC-296689)#, NDYR-10 (IC-296690)#
Yellow seeded	*B. juncea*	Pusa Karishma (LES-39)*, Navgold (YRN-6)*, Basanti*, NDYR-8, NUDB-10, TM-1 (IC-296829)#, TM-4, TERI GZ-05, TERI (0E) M21
Early duration (100-120 days)	*B. juncea*	Pusa Agrani (SEJ-2)*, Pusa Mahak (JD-6)*, Pusa Tarak (EJ-9912-13)*, Pusa Mustard 25 (NPJ-112)*, Pusa Mustard 27 (EJ-17)*, Pusa Mustard-28 (NPJ-124)*, EJ-20, Kanti*, NDRE-4*, PR-2006-1*, Divya (IC-553910)#
	B. napus	NUDB-38, NUDB-42
Salt tolerance (Ece upto 10 ds/m)	*B. juncea*	CS-52*, CS-54 (CS-614-4-1-4)*, CS-234-2-2, CS-56*, RH-8814 (IC-401570)#
High temperature tolerance at seedling stage	*B. juncea*	Pusa Vijay (NPJ-93)*, Pusa Mustard 25 (NPJ-112)*, Pusa Mustard 27 (EJ-17)*, Pusa Mustard 28 (NPJ-124)*, BPR-541#, BPR-543#

Table 1: (*Contd....*)

Table 1: *(Contd....)*

Agronomic attributes	*Species*	*Donor*
Tetralocular	*B. juncea*	RB-1 (IC-296741)#, Geeta (RB-9901)*
Bunchy type	*B. juncea*	IC-355399A
Apetalous source	*B. juncea*	RC-199
Cytoplasmic male sterile lines	*B. juncea*	CMS Trachy (IC-296683)#, CMS *B. oxyrrhina* (IC-296701)#, CMS *Moricandia arvensis* (IC-296702)#, CMS *Raphanus* (*ogu*) (IC-296705)#, CMS *M. arvensis* (Corrected) (IC-296706)#, CMS *Diplotaxis erucoides* (IC-296746)#, CMS *D. siifolia* (IC-296749)#
	B. rapa	CMS *B. oxyrrhina* (IC-296707)#
	B. napus	CMS *B. oxyrrhina* (IC-296704)#
Restorer lines	*B. juncea*	CMS Trachy (R) (IC-296682)#, CMS *M. arvensis* (R) (IC-296703)#

* Released varieties, #Registered germplasm. *Source*: Mishra and Kumar (2008), AICRP RM Annual Reports

maturing and high temperature tolerant *B. juncea* varieties. For white rust resistance breeding, Bio-YSR, BEC-144, BEC-286, EC-399299, EC-399301, RC-781, JM-1 etc. have been found to be stable sources and are being used widely in the national programme.

Development of CMS and Restorers

A good number of cytoplasmic male sterile lines in *B. juncea* [CMS *Trachystoma trachy* (IC-296683), CMS *B. oxyrrhina* (IC-296701), CMS *Moricandia arvensis* (IC-296702), CMS *Raphanus* (*ogu*) (IC-296705), CMS *M. arvensis* (Corrected) (IC-296706), CMS *Diplotaxis erucoides* (IC-296746), CMS *D. siifolia* (IC-296749)]; *B. rapa* [CMS *B. oxyrrhina* (IC-296707)] and *B. napus* [CMS *B. oxyrrhina* (IC-296704)] with corresponding restorers like CMS *T. trachy* (R) (IC-296682), CMS *M. arvensis* (R) (IC-296703) have been developed and are being exploited for hybrid development programme in the country (Table 2).

BREEDING OBJECTIVES

Increased Seed and Oil Yield

In oilseed crops, oil content determines the economic yield and market value of the produce. In rapeseed-mustard, the oil content ranges from 35 (*Eruca sativa*/*B. carinata*) to 46 per cent (*B. rapa* cv *yellow sarson*); thus, there is an ample scope for raising its present level in largely cultivated species like *B. juncea* and *B. napus*.

Seed and Oil Quality

Erucic acid and glucosinolates are two components in oil and seed meal cake of *Brassica* varieties being grown in India. Good success has been achieved in the development of low erucic acid varieties in *B. juncea* and double low (low erucic acid and low glucosinolates) types in *B. napus*.

Table 2: Various CMS systems reported in *Brassica* spp.

Cytoplasm donor	***Code***	***Technique used***	***Restoration status***	***Crops***	***Reference(s)***
Diplotaxis siifolia	*sif*	Intergeneric cross	Not available	*B. juncea*	Rao *et al.* (1974)
Raphanus sativus	*ogu*	Interspecific cross, Protoplast fusion	Available	*B. napus* *B. juncea*	Bannerot *et al.* (1974) Kirti *et al.* (1995a)
B. tournefortii	*tour*	Interspecific cross	Unstable, genotype specific partial restoration	*B. napus* *B. juncea*	Banga *et al.* (1995) Rawat and Anand (1979)
B. oxyrrhina	*oxy*	Interspecific cross, protoplast fusion	Not available	*B. juncea*	Prakash and Chopra (1990)
Trychystoma ballii	*trachy*	Protoplast fusion	Incomplete	*B. juncea*	Kirti *et al.* (1995b)
Moricandia arvensis	*mori*	Protoplast fusion	Available	*B. juncea*	Prakash *et al.* (1998)
D. sietiana	*Sie*	Intergeneric cross	Not available	*B. juncea*	Prakash *et al.* (2001)
D. catholica	*cath*	Intergeneric cross	Available	*B. juncea*	Prakash (2001)
Enarthrocarpus lyratus	*lyr*	Intergeneric cross	Available	*B. juncea* *B. napus*	Banga *et al.* (2003) Janeja *et al.* (2003b)
Erucastrum canariense	*can*	Intergeneric cross	Available	*B. juncea* *B. napus*	Prakash *et al.* (2001) Banga *et al.* (2003)
Synthetic *B. napus* ISN-706	126-1	Interspecific cross	Available	*B. juncea*	Sodhi *et al.* (2006)
D. erucoides	*Eru*	Intergeneric cross	Available	*B. juncea*	Bhat *et al.* (2006)
D. berthautii	*bar*	Intergeneric cross	Available	*B. juncea*	Bhat *et al.* (2008)

Development of double low (Canola type) varieties of Indian mustard is a very important objective to make the Indian by-products globally competitive. Low erucic acid with increased oleic acid and balanced content of omega-3 and omega 6 fatty acids along with enhanced oil content (>40%) is desirable in the future varieties.

Early Maturity

Short duration varieties are needed to fit into different cropping systems of the country especially in eastern and north-eastern parts. Toria is low yielder and vulnerable to most biotic and abiotic stresses. Therefore, Indian mustard varieties like Pusa Mustard 25 and Pusa Mustard 28, that mature in about 100 days, are good substitute of toria.

Plant Type

More concerted efforts are needed to identify suitable plant types for different crop geometries and systems. Plant types with determinate growth habit need to be tailored. Efforts are also being made to breed dwarf and basal branching plant types.

Resistance Breeding

Abiotic stresses

High temperature stress at the time of sowing leads to high seedling mortality and thus poor seedling establishment and plant stand. Plants exposed to high temperature stress at reproductive stage show high pollen mortality, increased flower drop, decrease in the number of siliquae and decreased seeds per siliqua. High temperature affects the translocation of photosynthates so that plants exposed to high temperature have shriveled/ small seeds with reduced oil content. The problem is more severe in the late sown crop. Frost is another abiotic stress that occurs every 5 to 7 years and causes huge losses; hence, breeding for frost tolerance needs to be addressed.

Biotic stresses

White rust (*Albugo candida*), *Alternaria* blight and *Sclerotinea* stem rot are major disease of concern and need to be addressed. Likewise, *Orobanche* a parasitic weed, creates havoc to this crop. Breeding for tolerance needs to be undertaken on priority. Mustard aphid (*Lipaphis erysimi*) and painted bug (*Bagrada hilaris*) are insects of economic importance and cause huge losses to the rapeseed mustard crop at various stages of growth. Breeding for resistance to these two insects also needs to be taken on priority.

GENETIC STUDIES OF QUALITY TRAITS

Erucic acid content in *B. juncea* is controlled by two genes, acting in additive manner, each with two alleles (Kirk and Hurlstone, 1983) which were later designated as *FAE1.1* and *FAE1.2*, and mapped to two independent loci that co-segregated with two QTLs (*ea1* and *ea2*) governing the erucic acid trait. Plants having genotypes *e1e1e2e2* possess low erucic acid (<2%) and *E1E1E2E2* possess high erucic acid content (>40%). While the *E1e1E2e2* heterozygote has intermediate phenotype (Gupta *et al.,* 2004).

Genetic analysis from the segregating populations of various crosses involving Bronowski has suggested the involvement of as many as 11 recessive alleles (Kondra and Stefanson, 1970) conditioning the low values of 3-butenyl-4-pentenyl and 2-hydroxy-3-butenyl glucosinolates. The major glucosinolates in *B. juncea* are aliphatics (methionine derivatives) with either 2-propenyl, 3-butenyl or combination of both being dominant types (Love *et al.*, 1990a). Genetic control of total glucosinolates in *B. juncea* has been reported to be under two major genes (Love *et al.,* 1990a), multiple additive alleles at a single locus with maternal effects involved (Love *et al.,* 1990b), six to seven genes (Sodhi *et al.,* 2002) and up to five major quantitative trait loci (QTL) based on molecular mapping information (Mahmood *et al.,* 2003).

BREEDING METHODS EMPLOYED

Being a heterogeneous group of crops in terms of their pollination control, almost all breeding methods are being employed for genetic improvement of different *Brassica* spp. Important varieties developed by using various breeding methods are summarized in Table 3. It is relevant to note here that besides the conventional approaches, tissue culture-based approaches (somaclonal variation) and artificial synthesis of allotetraploids such as *B. juncea* have also yielded successful commercial varieties. For example, the short duration synthetic mustard variety Pusa Agrani has become very popular and spawned several new and improved early varieties. Most of the breeding programmes are directed at achieving improvement using mega varieties. Due to repeated cycles of such breeding, there is great loss of variability and the gene pool becomes narrow. Somaclones are important source of variation when the genetic variability is limited in the germplasm. Somaclonal variation was exploited to develop a strain Bio 902 with improved seed size, and higher yield and was released for commercial cultivation as variety Pusa

Jaikisan. This variety has been extensively cultivated in the major mustard belt of Rajasthan and continues to be in constant demand even after 15 years of its release.

Table 3: Oilseed Brassica cultivars developed using different breeding approaches

Breeding approach	***Varieties developed***
Pureline selection	*B. napus* - GSL-1, Neelam, NUDH-26-11; *B. rapa* yellow sarson -Benoy, Jhumka, Ragini
Mass selection	*Black mustard* - Surya; *Yellow sarson* – YSPb-24; *Taramira* – ITSA, RTM-314, T-27; *Toria* – Agrani, Bhawani, TL-15, M-27, PT-303, RAUTS, ITSA, T-9
Pedigree breeding and its modifications	*B. juncea* - Aravali, RGN-48, RGN-13, NRCDR-2, CS-54, Pusa Mahak, Jagannath, Pusa Agrani, Pusa Bold, RH-781, RH-819, RL-1359, Vasundhra, Vardan, Vaibhav, Ashirwad, Pusa Mustard 24, Pusa Mustard 25, Pusa Mustard 26, Pusa Mustard 27, Pusa Mustard 28, Pusa Mustard 29, Pusa Mustard 30, Pant Rai-19, RH-0406, RGN-229, RGN-236, Raj Vijay Mustard-2; *B. napus* - GSC-5, GSL-2, Sheetal, GSC-5; *B. carinata*-PC-5, Pusa Swarnim, Pusa Aditya; *Yellow sarson* - Subinoy, Gujarat Sarson-1
Recurrent selection	*Toria* – PT-30
Synthetics and Composites	Composites: *Toria* – Jawahar Toria-1, Panchali, PBT-37, PT-507, TH-68; *Synthetics*: *Toria* - Sangam, *Brown sarson* - Pusa Kalyani
Backcross breeding	*B. juncea* - JM-1, JM-2, JM-3, Pusa Karishma, Pusa Mustard 21, Pusa Mustard 22; *B. napus* - OCN-3
Mutation breeding	*B. juncea* - Geeta, RLM-619, RLM-514, TM-2; *Toria* – Anuradha, Parbati; *Yellow sarson* – Narendra Sarson-2
Somaclonal breeding	*B. juncea* - Pusa Jaikisan (Bio 902)
Synthesis of allotetraploids	*B. juncea* - Pusa Agarni, Pusa Mustard 25, Pusa Mustard 26 and Pusa Mustard 28
Heterosis breeding	*B. juncea* - DMH 1, NRCHB 506, PAC 432, PAC 437, DRMRIJ 31

Development of Commercial Hybrids

With the availability of practical and highly effective hybrid seed production mechanism, heterosis is being exploited commercially in oilseed Brassicas. A number of cytosterility sources from *Brassica* coenospecies *viz*., *Diplotaxis siifolia* (Rao *et al.,* 1974), *Raphanus sativus* (Bannerot *et al.,* 1974; Kirti *et al.,* 1995a) *B. tournefortii* (Banga *et al.,* 1994; Rawat and Anand, 1979), *B. oxyrrhina* (Prakash and Chopra, 1990), *Trychystoma ballii* (Kirti *et al.,* 1995b), *Moricandia arvensis* (Prakash *et al.,* 1998) *D. sietiana* (Prakash *et al.,* 2001a), *D. catholica* (Prakash *et al.,* 2001), *Enarthrocarpus lyratus* (Banga *et al.,* 2003a; Janeja *et al.* (2003), *Erucastrum canariense* (Prakash *et al.,* 2001; Banga *et al.,* 2003b), Synthetic *B. napus* ISN-706 (Sodhi *et al.,* 2006), *D. erucoides* (Bhat *et al.,* 2006) and *D. berthautii* (Bhat *et al.,* 2008) were identified and used for development of CMS lines and restorers. Out of these *R. sativus (ogu)*, *M. arvensis (mori)* and Synthetic *B. napus* ISN-706 (126-1) were used for development of commercial hybrids in *Brassica*. Five hybrids of *B. juncea* (NRCHB 506, DMH 1, PAC 432, PAC 437 and DRMRIJ 31) and two of *B. napus* (PGHS 51 and PAC 401) have been released for commercial cultivation in India (Table 4).

Table 4: *Brassica juncea* and *B. napus* hybrids released in India

Name of hybrid	*Developing institution*	*Year of release*	*Area for which released*	*CMS source*
Brassica juncea				
DRMRIJ-31	DRMR, Bharatpur	2013	Rajasthan and Haryana	Moricandia
NRCHB-506	DRMR, Bharatpur	2008	Rajasthan, U.P.	Moricandia
DMH-1	DUSC, Delhi	2008	Punjab, Haryana, Delhi, J&K, Rajasthan, U.P., M.P.	126-1
CORAL-432 (PAC-432)	Advanta India, Bangalore	2010	Punjab, Haryana, Delhi, J&K, Rajasthan	Ogura
CORAL-437 (PAC 437)	Advanta India, Bangalore	2011	Punjab, Haryana, Delhi, J&K, Rajasthan	Ogura
Brassica napus				
PGSH-51	PAU, Ludhiana	1996	Punjab	Tour
Hyola-401 (PAC-401)	Advanta India, Bangalore	1997	Punjab, Haryana, Delhi, J&K, Rajasthan	Ogura

Breeding for Resistance/Tolerance

Biotic stresses

White rust

Jawahar Mustard 1, Jawahar Mustard 2 and Jawahar Mustard 3 have been released as white rust resistant varieties for commercial cultivation.

Abiotic stresses

High temperature tolerance

Short duration *B. juncea* and toria are highly suitable for multiple cropping system and are being grown during early and late seasons. Under early sown conditions, soil temperature crosses 40°C leading to severe seedling mortality and hence poor plant stand. Under late sown conditions there is a severe reduction in the seed size and oil content. Therefore, it becomes imperative to develop temperature resilient genotypes possessing tolerance at seedling and/reproductive stage for wider adaptability and stability of performance. *B. juncea* variety Pusa Mustard 26 developed at IARI, New Delhi possesses tolerance both at seedling and reproductive stages, whereas, Pusa Mustard 25, Pusa Mustard 27 and Pusa Mustard 28 are short duration varieties tolerant to high temperature at seedling stage.

Quality Cultivar Development

With the availability of suitable donors such as ZEM1, ZEM2, Torpe and Bele for low erucic acid and Bronowski for low glucosinolate content coupled with reliable biochemical and/or molecular assays, has led to the development of "0" and "00" varieties of *B. juncea*, *B. napus* and *B. campestris*. Transfer of low erucic acid and/or glucosinolate QTLs/ genes through backcross breeding has been achieved in the past. However, it is a time consuming intensive process and necessitates selfing of every backcross generation and exhaustive phenotyping in the segregating populations through destructive biochemical assays. The use of available DNA markers tightly linked to gene(s) of interest helps in reducing the size of backcross progeny and provides us opportunity for precise transfer of target trait. The gene based markers, identified both for erucic and glucosinolate content, found to be very useful in negative selection for these traits, are being used extensively in quality breeding programmes. Genetic variability for oleic acid is available in the germplasm and is being utilized for developing high oleic acid varieties.

Table 5: Donor sources and breeding methods used in the development of quality varieties

Sl. no.	*Variety*	*Pedigree*	*Breeding method*
B. juncea (Low erucic acid)			
1.	Pusa Karishma	Pusa Barani/ Zem 1	Backcross breeding
2.	Pusa Mustard 21	Pusa Bold / Zem-2	Backcross breeding
3.	Pusa Mustard 22	Pusa Barani/ Zem-2	Backcross method
4.	Pusa Mustard 24	Pusa Bold/ LEB 15//LES 29	Pedigree selection
5.	ELM 079 (RLC-1)	QM-4/ Pusa Bold	Pedigree selection
6.	ELM-123 (RLC-2)	QM-4/ Pusa Bold	Pedigree selection
7.	Pusa Mustard 29	ZEM-2/ Pusa Barani//EC 287711	Pedigree selection
8.	Pusa Mustard 30	Bio-902 × ZEM-1	Pedigree selection
B. napus (Double low)			
9.	GSC 5	Hyola 401 // Agat GSL-8888	Pedigree selection
10.	OCN-3	NECN 13/Tribute//NECN-13	Backcross method
11.	PAC-401	44002A/4154 R	CMS based hybrid
12.	NUDH 26-11	Selection from germplasm	Pure line selection

Directed breeding efforts in India have led to the development and release of eight low erucic varieties of *B. juncea* and four of *B. napus* (Table 5). Efforts are in progress at different National Centres for developing '00' varieties.

Doubled Haploids (DH)

Production of DH lines provides one of the fastest methods of obtaining completely homozygous progenies in a single generation following F_1. This technique facilitates creation of new genetic variability in the form of multitude of homozygous true breeding lines. DH technique is a reliable method for reducing population size in the crosses directed towards assembling a number of alleles/genes in a single genetic background as is required for developing "00" genotypes. It is highly reliable and time saving in the situations where phenotyping is difficult and test crosses are to be made to know the genetic architecture of the segregants. Microspore culture is now routinely used for developing mapping populations, and generating fertility restorers for different cyto-sterility sources.

UTILIZING PLANT BIOTECHNOLOGY TOOLS FOR CULTIVAR DEVELOPMENT

Development of Molecular Markers and Linkage Maps

Discovery of high levels of inter as well as intra-specific DNA polymorphism in RFLP profiles obtained with random genomic DNA clones as probes by Figdore *et al.* (1988) encouraged molecular mapping of *Brassica* genomes and tagging of genes for several useful traits. Different mapping populations at F_2 or later stages (F_3, F_4 or Recombinant Inbred Lines) have been used to construct genetic maps in *B. oleracea* (Slocum *et al.*, 1990; Landry *et al.,* 1992; Kianian and Quiros, 1991), *B. nigra* (Truco and Quiros, 1994), *B. campestris/B. rapa* (Chyi *et al.*, 1992; Kole *et al.*, 1997), *B. napus* (Hoenecke and Chyi, 1991; Landry *et al.*, 1991) and *B. juncea* (Sharma *et al.*, 1994; Mohapatra *et al.*, 2002; Sharma *et al.*, 2002). Backcross mapping population has been used to generate maps for *B. nigra* (Lagercrantz and Lydiate, 1995) and doubled haploid (DH) mapping population has been used to construct maps of *B. oleracea* (Voorrips *et al.*, 1997; Li and Quiros, 2001; Saal *et al.*, 2001), *B. napus* (Ferreira *et al.*, 1994; Uzunova *et al.*, 1995; Foisset *et al.*, 1996) and *B. juncea* (Cheung *et al.*, 1997; Axelson *et al.*, 2000). Apart from these, a combination of mapping populations has also been used. Ramsay *et al.* (1996) and Kearsay *et al.* (1996) have used backcross individuals of DH lines to generate maps of *B. oleracea*. Some maps were also constructed using RAPD and AFLP markers (Sharma *et al.*, 2002; Pradhan *et al.*,

Table 6: Genomic maps developed and genes/QTLs mapped in different *Brassica* spps.

Crop	***Marker/ trait***	***Reference(s)***
i) *Development of molecular markers and genome maps in oilseed crops*		
Brassica spps.	RFLP	Figdore *et al.*, 1988
	Isozyme markers	Arus and Orton, 1983; Chen *et al.,* 1989
	EST Mining	Bhati *et al.*, 2010
B. oleracea	RFLP and RAPD	Slocum *et al.*, 1990; Kianian and Quiros, 1991; Landry *et al.*, 1992; Kearsay *et al.*, 1996; Ramsay *et al.*, 1996; Voorrips *et al.,* 1997; Li and Quiros, 2001; Saal *et al.*, 2001
B. rapa	RFLP and RAPD	Song *et al.*, 1991 Chyi *et al.,* 1992; Kole *et al.*, 1997
	SSR, IP	Li *et al.*, 2010
B. nigra	RFLP and RAPD	Truco and Quiros, 1994; Lagercrantz and Lydiate, 1995
B. napus	RFLP and RAPD	Hoenecke and Chyi, 1991; Landry *et al.*, 1991; Ferreira *et al.,* 1994; Uzunova *et al.*, 1995; Foisset *et al.*, 1996
	SSR	Lydiate and Sharpe, 2003
	SRAP	Sun *et al.*, 2007
B. juncea	RFLP, AFLP, RAPD	Sharma *et al.,* 1994; Cheung *et al.*, 1997; Axelson *et al.*, 2000; Mohapatra *et al.*, 2002; Sharma *et al.*, 2002; Pradhan *et al.*, 2003; Mahmood *et al.*, 2005, Kalita *et al.*, 2007
	SSR	Koundal *et al.*, 2008; Parida *et al.,* 2010; Yadava *et al.*, 2009; Pradhan *et al.*, 2011
	IP	Panjabi *et al.*, 2008
ii) *Mapping and tagging QTLs*		
B. oleracea	*Plasmodiophora brassicae*	Figdore *et al.*, 1993; Voorrips *et al.,* 1997
	Xanthomonas campestris	Camarago *et al.,* 1995
B. rapa	Club root	Saito *et al.*, 2006; Werner *et al.*, 2008;
	Xanthomonas campestris	Soengas *et al.*, 2007
	Albugo candida	Kole *et al.*, 1996
	Fatty acids	Teutonico and Osborn, 1994; Tanhuanpaa *et al.,* 1996; Tanhuanpaa *et al.,* 1998; Jagannath *et al.* (2011)
	Seed coat colour	Teutonico and Osborn, 1994; Chen *et al.*, 1997; Rahman *et al.,* 2007
B. nigra	Flowering time	Lagercrantz *et al.*, 1996
B. napus	*Leptosphaeria maculans*	Dion *et al.*, 1995; Ferreira *et al.*, 1995a; Leflon *et al.*, 2007
	Turnip mosaic virus	Walsh *et al.*, 1999
	Sclerotinia sclerotiorum	Zhao and Meng, 2003

Table 6: (*Contd....*)

Table 6: (*Contd....*)

Crop	*Marker/ trait*	*Reference(s)*
	Verticillium wilt	Happstadius *et al.*, 2003
	Albugo candida	Ferreira *et al.*, 1995c
	Vernalization requirement	Ferreira *et al.*, 1995b; Teutonico and Osborn, 1995; Camarago and Osborn, 1996
	Oil content, protein, fatty acid	Arondel *et al.*, 1992; Ecke *et al.*, 1995; Hu *et al.*, 1995; Tanhuanpaa *et al.*, 1995; Jourden *et al.*, 1996a, 1996b; Jourden *et al.*, 1996c; Thormann *et al.*, 1996; Barret *et al.*, 1998b; Fourmann *et al.*, 1998; Hu *et al.*, 1999; Schierholt *et al.*, 2000; Zhao *et al.*, 2006; Delourme *et al.*, 2006; Qiu *et al.*, 2006; Rahman *et al.*, 2008; Nath and Goswami, 2009
	Glucosinolates	Uzunova *et al.*, 1995; Toroser *et al.*, 1995; De Quiroz and Mithen, 1996; Hasan *et al.*, 2008
	Seed coat colour	Van Deynze *et al.*, 1995
	Male sterility/ fertility restorer genes	Delourme *et al.*, 1994, Jean *et al.*, 1997; Delourme *et al.*, 1998; Yi *et al.*, 2006; Huang *et al.*, 2007; He *et al.*, 2008
	Yield	Shi *et al.*, 2009
B. juncea	*Albugo candida*	Cheung *et al.*, 1998; Prabhu *et al.*, 1998; Mukherjee *et al.*, 2001; Varshney *et al.*, 2004; Panjabi-Masand *et al.*, 2010
	Seed coat colour	Upadhyay *et al.*, 1996; Negi *et al.*, 2000; Li *et al.*, 2010
	Oil content	Sharma *et al.*, 1999; Sharma *et al.*, 2002
	Erucic acid	Gupta *et al.*, 2004
	Glucosinolates	Stringam and Thiagarajah, 1995; Good *et al.*, 2003; Mahmood *et al.*, 2003; Ripley and Roslinsky, 2005; Ramchiary *et al.*, 2007; Bisht *et al.*, 2009
	Moricandia arvensis	Ashutosh *et al.*, 2007
iii) *Marker assisted selection (MAS)*		
B. juncea	Erucic acid	Singh, 2010; Singh *et al.*, 2011; Naresh *et al.*, 2012
B. juncea	Glucosinolate	Yadava *et al.*, 2012

2003). Pradhan *et al.* (2003) constructed a high-density genetic linkage map of *B. juncea* (2n = 36) using AFLP and RFLP markers in a F_1-derived doubled-haploid population. Now, the number of publicly available *Brassica* microsatellite primers is increasing as a result of publicly funded international initiatives (http://www.brassica.info/ssr/SSRinfo.htm). The release of a set of robust, highly polymorphic, mapped SSR markers spanning the entire *B. napus* genome (Lydiate and Sharpe, 2003) into the public domain greatly assisted in genome mapping and gene tagging. To enrich the SSR resource further, Federico *et al.* (2008) developed 587 new primer pairs flanking SSRs using sequence information from 3,500 genomic clones mainly from *B. oleracea* to identify di, tri, tetra and penta-nucleotide repeats. Yadava *et al.* (2009) evaluated the cross-transferability and polymorphic potential of the genomic SSRs to assess their utility across *Brassica* species and related genera. The study clearly revealed that the available SSR markers can be used effectively for monitoring gene introgression. However, due to low level of intra-specific polymorphism, a large set of markers would need to be used (Koundal *et al.*, 2008).

Sun *et al.* (2007) used sequence related amplified polymorphism (SRAP) to construct an ultra dense genetic recombination map for a doubled haploid population in *B. napus*. Gao *et al.* (2007) constructed a high-density genetic map of *B. oleracea* adding over 1,000 new markers to *Brassica* molecular tools. Panjabi *et al.* (2008) designed and tested the efficacy of PCR-based Intron Polymorphism (IP) markers to analyze genome-wide synteny between the oilseed *B. juncea* and *A. thaliana* and analyzed the arrangement of 24 genomic block segments in the A, B and C *Brassica* genomes to study the evolutionary events that contributed to karyotype variations. These recently reported SRAP and IP markers should find application in tagging of useful genes in oilseed *Brassica* crops. Particulars of molecular maps developed and QTLs identified are summarized in Table 6.

Mapping and Tagging of Genes/ QTL

Disease resistance

Molecular markers have been generated for the genes conferring resistance to *Leptosphaeria maculans, Plasmodiophora brassicae, Xanthomonas campestris, Sclerotinia sclerotiorum, Verticillium* wilt, clubroot, turnip mosaic virus and white rust resistance in different *Brassica* spps. Cheung *et al.* (1998) identified one co-segregating dominant RFLP marker x140a, mapped to LG7 in *B. juncea* and designated as *Acr*. In the same year, white rust resistance locus $Ac2_1$ present in an Eastern European source that was effective against a Canadian isolate of the pathogen was mapped using RAPD markers (Prabhu *et al.*, 1998). Mukherjee *et al.* (2001) mapped a locus designated as *Ac*2(t) that is effective against Indian isolate of the white rust pathogen. Varshney *et al.* (2004) identified a more tightly linked AFLP marker and a PCR based cleaved amplification polymorphic sequence (CAPS) marker for closely linked RAPD marker has been developed. Panjabi-Masand *et al.* (2010) have reported two independent loci for conferring resistance to *Albugo candida* in the east European germplasm of Indian mustard. A *B. rapa* linkage map comprising ten linkage groups (A01–A10) with a total map distance of 664 cM was produced, based on 223 AFLP and 23 microsatellite markers (Soengas *et al.*, 2007).

Oil content and fatty acid composition

Both oil quantity and quality traits have been studied using molecular markers. Ecke *et al.* (1995) mapped three loci on different linkage groups in *B. napus* using RFLP markers. In *B. juncea*, Sharma *et al.* (1999) identified three loci based on segregation of RAPD markers in a recombinant inbred population. Zhao *et al.* (2006) reported mapping of QTLs for oil and protein content in rapeseed. Delourme *et al.* (2006) reported that additive effects are the main factor contributing to variation in oil content. Two RAPD markers, K-01_{1100} and 25a, were generated and linked to the linolenic acid concentration (Hu *et al.*, 1995; Tanhuanpää *et al.*, 1995). RAPD markers linked to oleic, linolenic and linoleic acid were identified in *B. napus* (Hu *et al.*, 1999). Markers linked to genomic regions controlling linolenic acid concentration in *B. napus* corresponding to *fad3* (omega-3-desaturase) gene in *A. thaliana* (Arondel *et al.*, 1992) were also identified (Jourden *et al.*, 1996 a,b; Thormann *et al.*, 1996). In another study, a single QTL containing 6 markers associated with oleic, palmitic and linoleic acid content was detected in *B. rapa* (Tanhuanpää *et al.*, 1996). In *B. juncea*, Sharma *et al.* (2002) mapped two major QTLs influencing oleic acid level using both single factor analysis of variance and interval mapping. Erucic acid loci have been linked to molecular markers by Ecke *et al.* (1995), Thormann *et al.* (1996), Jourden *et al.* (1996c) and Barret *et al.* (1998b) using RFLP analysis in *B. napus*. Two QTLs underlying the variation of seed erucic acid content were assigned to two linkage groups of *B. juncea* map using double haploid mapping populations derived from high × high erucic acid hybrid (Gupta *et al.*, 2004). Qiu *et al.* (2006) constructed a genetic linkage map consisting of 277 loci and identified reproducible QTL for seed oil and erucic acid content. The SRAP markers based on *B. napus FAE1* gene recently developed by Rahman *et al.* (2008) are expected to be highly useful in MAS for erucic acid content.

Seed glucosinolate content

Ripley and Roslinsky (2005) reported an ISSR marker tightly linked to high 2-propenyl glucosinolate. It has a good potential to be used for MAS for low glucosinolate in the Canola Breeding programme.

Ramchiary *et al.* (2007) used a recurrent backcross selection (RSB) method with a doubled haploid (DH) generation interspersing backcross generations for the introgression of low glucosinolate alleles from an east European *B. juncea* line, Heera into an Indian variety, Varuna. A validation study on a population of low glucosinolate DH lines derived from all the backcross generations of the RSB breeding programme revealed that the QTL detected in BC(4)DH were the 'true' QTL. This study was extended by Bisht *et al.* (2008) who mapped a total of six QTLs for this trait. Hasan *et al.* (2008) reported four genes involved in the biosynthesis of indole, aliphatic and aromatic glucosinolates that might be associated with known quantitative trait loci for total seed glucosinolate content in *B. napus*.

Seed coat colour

RFLP markers linked to seed coat colour in *B. napus* were identified using the BSA approach (Van Deynze *et al.*, 1995). Seed coat colour trait in *B. campestris* was tagged with RAPD markers using *B. campestris-oleracea* addition lines (Chen *et al.*, 1997). Two RFLP markers flanking one of the interacting loci were identified. Negi *et al.* (2000) tagged the seed coat colour trait using a combined approach of BSA and AFLP in *B. juncea*.

Genetic male sterility and fertility restorer genes

Genetic male sterility is useful in breeding two-line hybrids. Yi *et al.* (2006) mapped *Bnms1* gene using AFLP markers converted to SCAR markers in *B. napus*, for use in MAS. Huang *et al.* (2007) identified 17 AFLP markers linked to the *Bnms3* gene for recessive genic male sterility. He *et al.* (2008) mapped the *Bnms3*, a recessive gene controlling genetic male sterility in *B. napus,* to a region of 0.56 cM, with 0.15 cM from marker SEP8 and 0.41 from marker SEP4, respectively. Four RAPD markers $OPC\text{-}02_{1150}$, $OPD\text{-}02_{1000}$, $OPF\text{-}06_{1200}$ and $OPG\text{-}02_{700}$ were linked to the fertility restorer gene for "Ogura" radish cms system (Delourme *et al.*, 1994, 1998). The *Rfp1* locus responsible for fertility restoration in "Polima" CMS was linked to 10 RFLP and one RAPD marker. Ashutosh *et al.* (2007) mapped the fertility restorer gene for a cytoplasmic male sterile (CMS) line of *B. juncea* developed earlier through somatic hybridization with *Moricandia arvensis* using AFLP markers.

Marker assisted selection (MAS)

In India, these markers are being used in MAS/backcrossing for improvement of quality traits such as erucic acid (Singh 2010; Singh *et al.,* 2011; Naresh *et al.*, 2012) and glucosinolate (Yadava *et al.*, 2012). Presently Delhi University South Campus, New Delhi; IARI, New Delhi; Punjab Agricultural University, Ludhiana and TERI, New Delhi are using the markers reported by Bisht *et al.* (2008) for introgression of low glucosinlates and erucic acid (Gupta *et al.,* 2004).

Mapping populations

A large number of mapping populations including recombinant inbred lines (RILs) and doubled haploids are being used to tag and map the genes of inetrest in *B. juncea*. Twelve mapping populations (RILs) involving indigenous and exotic germplasm lines are being developed at IARI, New Delhi. These include the populations for white rust, glucosinolate, erucic acid, other important traits like seed colour, plant type, seed size, oil content etc. Delhi University South Campus has developed a doubled haploid population involving Varuna and Heera and it has been used widely for molecular map of *B. juncea*, tagging the genes for white rust, erucic acid and glucosinolates. Punjab Agricultural University, Ludhiana and TERI, Delhi have also developed the mapping population for quality and other traits of economic importance.

Indian Efforts Towards Development of Transgenic Brassicas

Recombinant DNA technology is a potential modern tool to transfer desirable traits across species and genomes. When no source of resistance is available in the crossable germplasm, the transgenic approach may be explored for introgression of the resistance from unrelated species and genera. Efforts have been made for standardizing protocols for incorporating transgenes governing resistance/ tolerance against different biotic and abiotic stresses and for development of male sterility in *Brassica* species in India.

Herbicide resistance

The first effort to create transgenics in *B. juncea* was for developing phosphinothricin (PPT) herbicide resistant lines using *Agrobacterium tumefaciens*. The developed constructs contained the bar gene encoding the enzyme phosphinothricin-acetyl-transferase which inactivates PPT by acetylating it (Mehra *et al.*, 2000). Bisth *et al.* (2004) developed Indian mustard transgenic lines resistant to the herbicide 2, 4- dichlorophenox (2, 4-D). The constructs contained the tfdA gene encoding the enzyme 2,4-D monooxygenase cloned downstream to 35S promoter along with a leader sequence from RNA4 of alfalfa mosaic virus. Wild type plants were affected by as low as 10 mg per litre and were completely killed at 50 mg per litre concentration, whereas the four transgenic lines were tolerant upto 500 mg per litre of 2,4-D.

Aphid resistance

Kanrar *et al.* (2002) transferred a cDNA encoding wheat germ agglutinin (WGA) to Indian mustard cv RLM-198 through *Agrobacterium*-mediated transformation. The bioassays using leaf discs showed that feeding on transgenics induced high mortality. Fecundity of aphids was reduced significantly.

Alternaria blight resistance

Kanrar *et al.* (2001) demonstrated that the expression of hevin, a rubber tree lectin, in Indian mustard confers resistance against *Alternaria brassicae*. Mondal *et al.* (2007) reported expression of a class I basic glucanase gene, under control of CaMV 35S promoter, in Indian mustard and its resistance against *Alternaria* leaf spot. In an *in vitro* antifungal assay, transgenics arrested hyphal growth of *Alternaria brassicae* by 15-54%. Under pathogen challenged condition in poly-house, trangenics showed restricted number, size and spread of lesions caused by *Alternaria brassicae*.

Abiotic stress tolerance

Rajwanshi *et al.* (2007) have reported a marker free approach for developing abiotic stress tolerant transgenic Indian mustard to avoid the extra burden of marker on plant that also makes transgenic plants unacceptable to the consumers. Saxena *et al.* (2011) have reported over-expression of Glyoxalase II gene for enhanced salinity tolerance in Indian mustard.

Other important traits

Studies on other traits have revealed that increased oil content in transgenics are associated with differential expression of novel *Brassica*-specific transcripts (Sharma *et al.,* 2012).

VARIETIES

Varieties of Different Cultivated Oil seed *Brassicas* Released After 1998

More than 75 varieties/hybrids of different *Brassica* spps. and *Eruca sativa* have been developed after 1998 which include 48 varieties/ hybrids of *Brassica juncea*, eight of *Brassica napus*, seven of *B. rapa*, four of *B. carinata*, two of *Eruca sativa* and one of *B. nigra*. Among these varieties/ hybrids, the historical achievement is the development of low erucic acid (0) varieties [LES-39 (Pusa Karishma), LES-1-27 (Pusa Mustard 21), LET 18 (Pusa Mustard 24), LET-17 (Pusa Mustard 22), ELM 079 (RLC-1), ELM-123 (RLC-2), LET-36 (Pusa Mustard 29) and LES-43 (Pusa Mustard 30)] of *B. juncea*, low erucic acid and low glucosinolate (00) varieties of *B. napus* and hybrids in case of *B. juncea* (Table 7).

Table 7: Different varieties of Rapeseed Mustard released from 1998-2013

Variety	*Release year*	*Developing institute*	*Area of adaptation*
Brassica campestris* cv. *Brown sarson			
Shalimar arson 1	2012	SKAUST, Srinagar	Recommended for J&K.
KBS-1	2001	ORS, HPKV, Kangra	Resistant to white rust & tolerant to cold. Recommended for HP
Brassica campestris* cv. *Yellow sarson			
Ragini	2001	Mahyco, Jalna	Irrigated conditions of Maharashtra
YSH 0401	2009	HAU, Hisar	For whole mustard growing areas of the country under irrigated conditions
NRCYS-05-02	2009	DRMR, Bharatpur	For whole mustard growing areas of the country under irrigated conditions
Brassica campestris* cv. *Toria			
TS-46	2000	AAU, Shilongani	Suitable for late sown irrigated conditions of Assam
Parbati	2001	OUAT, Bhubaneswar	Orissa
Anuradha	2002	OUAT, Bhubaneswar	Rainfed and irrigated in Orissa
Brassica carinata			
PC-5-17	2001	PAU, Ludhiana	Rainfed and irrigated conditions of Punjab, Haryana, Delhi and Himachal Pradesh

Table 7: (*Contd....*)

Table 7: (*Contd....*)

Variety	*Release year*	*Developing institute*	*Area of adaptation*
JTC-1	2001	ZARS, Tikamgarh	Rainfed conditions of Punjab, Haryana, Delhi and Himachal Pradesh
Pusa Swarnim (IGC-01)	2003	IARI, New Delhi	Rainfed & irrigated areas of Haryana, Punjab, UP, HP, J&K and Rajasthan
Pusa Aditya (NPC-9)	2006	,,	Rainfed and marginal lands of NCR Delhi
Brassica juncea			
Pusa Karishma (LES-39)	2005	IARI, New Delhi	First quality (Zero erucic acid) variety from Public sector. Irrigated conditions of NCR, Delhi
Pusa Mahak (JD-6)	2005	,,	Early maturing variety recommended for irrigated conditions of Zone V. Substitute for toria
Pusa Mustard 21 (LES-1-27)	2007	,,	Second quality (Zero erucic acid) variety from Public sector for Zone II
Pusa Mustard -22 (LET-17)	2008	,,	Quality (Zero erucic acid) variety for NCR, Delhi
Pusa Mustard -24 (LET-18)	2009	,,	Quality (Zero erucic acid) variety for Zone II
Pusa Vijay (NPJ-93)	2008	,,	NCR, Delhi, High temperature tolerance
Pusa Tarak (EJ-13)	2009	,,	Early maturing variety for NCR, Delhi. Substitute for toria
Pusa Mustard 25 (NPJ-112)	2010	,,	Early maturing variety. Suitable for zone II. Potential substitute for toria
Pusa Mustard 26 (NPJ-113)	2011	,,	Late sown variety. Suitable for Zone II.
Pusa Mustard 27 (EJ-17)	2012	,,	Early maturing variety. Suitable for Zone III. Potential substitute for toria
Pusa Mustard 28 (NPJ-124)	2011	,,	Early maturing variety. Suitable for Zone II. Potential substitute for toria
Pusa Mustard -29 (LET-36)	2013	,,	Quality (Zero erucic acid) variety for Zone II
Pusa Mustard -30 (LES-43)	2013	,,	Quality (Zero erucic acid) variety Zone III
Basanti	2001	CSAU&T, Kanpur	Yellow seeded, for UP
Urvashi	2001	,,	Irrigated areas of UP
Kanti	2003	,,	High temperature tolerant at seedling stage, early maturing, UP
Maya	2003	,,	Irrigated conditions of Zone-III
Ashirwad	2005	,,	For late sowing in Zone-III
PR-2006-1	2012	GBPUAT, Pantnagar	Early maturing variety. Suitable for zone II.
Narendra Ageti Rai-4	2001	NDUAT, Faizabad	Early maturing variety for UP
Narendra Swarna Rai-8	2005	,,	Suitable for irrigated conditions of UP, Yellow seeded variety
CS-54	2005	CSSRI, Karnal	Sodic and saline soils of Zone II
CS-56	2008	,,	,,

Table 7: (*Contd....*)

Table 7: (*Contd....*)

Variety	*Release year*	*Developing institute*	*Area of adaptation*
JM-2	2005	ZARS, JLNKVV, Morena	Resistant to white rust, MP & Chhattisgarh
JM-3	2005	,,	MP
Raj Vijay Mustard 2	2013	Rajmata Vijaya Raje Sciendia Krishi Vishwa Vidyalaya, Gwalior	Delhi, Haryana, Jammu, Punjab and Northern Rajasthan under late sown conditions
PBR-210	2007	PAU, RRS, Bathinda	Irrigated conditions of Punjab
PBR-357	2012	,,	Irrigated conditions of Zone II
RL-99-27	2004	PAU, Ludhiana	Rainfed situations of Zone II
ELM-079	2007	,,	Low erucic acid variety for Punjab
ELM-123	2012	,,	Low erucic acid variety for Zone II
NRCDR-02	2006	DRMR, Bharatpur	Irrigated conditions of Zone II
NRCHB-506	2009	,,	Irrigated conditions of UP, Rajasthan, MP (Hybrid)
NRCHB-101	2009	,,	Late sown conditions of Zone III
NRCDR-601	2010	,,	Irrigated conditions of Zone II
DRMRIJ-31	2013	,,	Irrigated conditions of Zone II (Hybrid)
DMH-1	2009	Delhi Uni. South Campus (DHARA)	Irrigated conditions of Zone II & III (Hybrid)
PAC-432	2010	Advanta India	Irrigated conditions of Zone II (Hybrid)
PAC-437	2011	,,	Irrigated conditions of Zone II (Hybrid)
RCC-4	2001	ORS, HPKVV, Kangra	HP
RGN-13	2003	RAU, ARS, Sriganganagar	Highly resistant to frost & tolerant to high temperture. Irrigated in Rajasthan
RGN-48	2005	,,	Rainfed, Zone II. Tolerant to frost
RGN-73	2006	,,	Irrigated areas of Zone III and Rajasthan
RGN-145	2008	,,	Late sown irrigated for Zone - II
RGN-229	2013	,,	Rainfed conditions of Zone - II
RGN-236	2013	,,	Late sown conditions of Zone - II
Arawali	2001	RAU, RRS, Navgaon	Rainfed areas of Zone II
RN-505	2005	,,	Late sown areas in Rajasthan
YRN-6 (Nav Gold)	2006	,,	Late sown, Zone II, Yellow seeded
Geeta	2003	HAU, RRS, Bawal	Rainfed in Zone II, Tetralocualar variety
RB-50	2008	,,	Rainfed in Zone II,
Swarnjyoti	2003	HAU, Hisar	Late sown conditions of Zone III and Haryana
Vasundhara	2003	,,	Irrigated conditions of Zone III
RH-0406	2013	,,	Rainfed conditions of Zone II
Shivani	2005	BAU, Kanke, Ranchi	Rainfed & low fertility, Jharkhand
Brassica napus			
GSC-5	2005	PAU, Ludhiana	Double low variety for Punjab
OCN-3	2008	,,	Double low variety for Punjab
PAC-401	2001	Advanta India Ltd., Bangalore	Double low hybrid suitable for irrigated conditions of Zone I and II

Table 7: (*Contd....*)

Table 7: (*Contd....*)

Variety	*Release year*	*Developing institute*	*Area of adaptation*
Neelam Kangra	2001	ORS, HPKVV,	Rainfed and irrigated areas of HP
TERI Uttam Jawahar	2007	TERI, New Delhi	Irrigated conditions of MP
NUDB-26-11	2008	Dhara	Irrigated conditions of J&K, HP and Uttarakhand
Brassica nigra			
Surya	2003	RAURS, Lam (Guntur) AP	Upland rice fellow and intercrop with black gram in AP
Eruca sativa			
RTM-314	2001	SKN College of Agriculture, Jobner	Rainfed situations of Haryana, Punjab, Rajasthan and Uttar Pradesh
RTM-2002	2007	SKN College of Agriculture, Jobner	Rainfed conditions of Rajasthan

ALL INDIA COORDINATED RESEARCH PROJECT ON RAPESEED MUSTARD

Systematic research work on rapeseed mustard was initiated after the constitution of Indian Central Oilseed Committee. This committee was designated as Oilseed Development Council in 1966. In 1967 an All India Coordinated Research Project on Oilseeds was initiated under the leadership of Project Coordinator which included five crops *viz.* groundnut, rapeseed mustard, sesame, linseed and castor. In 1981 a separate All India Coordinated Research Project on Rapeseed and Mustard was established at HAU, Hisar and 14 research centres were established for research work on rapeseed and mustard. On October 20, 1993, National Research Centre on Rapeseed Mustard was established at Sewar, Bharatpur in Rajasthan. The Indian Council of Agricultural Research raised the NRC on Rapeseed Mustard to Directorate of Rapeseed Mustard Research in February 2009. Presently there are eleven main and twelve sub-centres under the AICRP Rapeseed Mustard. In addition there are twenty two verification centres also for evaluation of the advance material in different zones under AICRP trials. The country has been divided into six agro-climatic zones for evaluation and release of the material (Table 8).

Future thrust areas

- Enhancing level of heterosis of hybrids through proper utilization of genetic resources and focused breeding options for their better commercial adoption

Table 8: National level zonalization for rapeseed mustard evaluation

Zone	***States***	***Centres***
Zone-I	Jammu & Kashmir, Himachal Pradesh	Khudwani, Kangra, Dhaula Kuan
Zone –II	Jammu & Kashmir, Punjab, Haryana, Rajasthan, Delhi	RS Pura (Chatha), Ludhiana, Bathinda, Hisar, Bawal, Karnal, Alwar (Navgaon), Bikaner, Sriganganagar, Delhi (IARI)
Zone-III	Uttar Pradesh, Uttarakhand, Madhya Pradesh, Rajasthan	Kanpur, Faizabad, Varanasi, Lucknow, Agra, Pantnagar, Morena, Kota, Bharatpur
Zone-IV	Gujarat, Maharastra, Rajasthan	Jaipur (Jobner), Jodhpur, Junagarh, S.K. Nagar, Jalgaon, Nagpur
Zone-V	Chhattisgarh, Bihar, Jharkhand, WB, Orissa, Assam, NE hill States	Jagdalpur, Dholi, Kanke (Ranchi), Berhampore, Bhubaneswar, Shillongani, Imphal
Zone-VI	AP, TN, Karnataka	Hyderabad, Gunter, Bangalore, Dharwad, Bijapur, Coimbatore, Wellington

- Developing early maturing, dwarf and determinate type varieties for enhancing cropping intensity and suitability for different cropping systems.
- Development of varieties resistant to biotic stresses especially white rust, *Alternaria* blight, *Sclerotonia* stem rot, aphids and *Orobanche* for stabilizing the productivity.
- Development of varieties with enhanced water and nutrient use efficiency for optimum utilization of the available soil moisture and nutrients.
- Tailoring varieties tolerant to abiotic stresses such as drought, high temperatures at sowing and/or maturity, salinity and frost in rapeseed mustard.
- Developing double zero varieties (erucic acid <2 % and glucosinolate < 30 micromoles/ g of defatted seed meal cake) with high oleic acid.
- Improving the oil content of present day varieties from about 40 percent to 45 percent

REFERENCES

AICRP-RM. Annual Progress Reports of All India Coordinated Research Project on Rapeseed and Mustard, Directorate of Rapeseed-Mustard Research, Sewar, Bharatpur.

Arondel, V., Lemieux, B., Hwang, T., Gibson, S., Goodman, H.M. and Somerville, C.R. (1992). Map-based cloning of a gene controlling omega-3 fatty acid desaturation in *Arabidopsis*. *Science*, 258: 1352–1355.

Axelson, T., Bowman, C.M., Sharpe, A.G., Lydiate, D.J. and Lagercrantz, U. (2000). Amphidiploid *Brassica juncea* contains conserved progenitor genomes. *Genome,* 43: 679–688.

Arus, P. and Orton, T.J. (1983). Inheritance and linkage relationships of isozyme loci in *Brassica oleracea*. *J. of Hered.*, 74: 405–412.

Ashutosh, S., P.C., Prakash, S. and Bhat, S.R. (2007). Identification of AFLP markers linked to the male fertility restorer gene of CMS (*Moricandia arvensis*) *Brassica juncea* and conversion to SCAR marker. *Theor App Genet.,* 114 (2): 385–392.

Banga, S.S., Banga, S.K. and Sandha, G.S. (1994). Hybrids in oilseed rape based on *tour* cytoplasm. *Cruciferae Newlett.,* 16: 73–74.

Banga, S.S., Deol, J.S. and Banga, S.K. (2003a). Alloplasmic male sterile *Brassica juncea* with *Enarthrocarpus lyratus* cytoplasm and introgression of gene(s) for fertility restoration from cytoplasmic donor species. *Theor. App. Genet.,* 106: 1390–1395.

Banga, S.S., Banga, S.K. and Bhaskar, P.B. (2003b). Alloplasmic line of *Brassica napus* L. with *Erucstrum canariense* cytoplasm is male sterile. Proc GCIRC 11th International Rapeseed Congr, pp. 324–325.

Bannerrot, H., Boulidard, L., Couderon, Y. and Temple, J. (1974). Transfer of cytoplasmic male sterility from *Raphanus sativus* to *Brassica oleracea*. *In*: Wills, A.B. and North, C. (*Eds.*) Proc. Eucarpia Meet Cruciferae. Scotish Hortic Res Inst, Invergavrie, UK, pp. 52–54.

Barret, P., Delourme, R., Renard, M., Domergue, F., Lessire, L., Delseny, M. and Roscoe, T.J. (1998). A rapeseed *FAE1* gene is linked to the E1 locus associated with variation in the content of erucic acid. *Theor Appl Genet.*, 96: 177–186.

Bhat, S.R., Vijayan, P., Ashutosh Dwivedi, K.K. and Prakash, S. (2006). *Diplotaxis erucoides*-induced cytoplasmic male sterility in *Brassica juncea* is rescued by the *Moricandia arvensis* restorer: Genetic and molecular analyses. *Pl. Breed*, 125: 150–155.

Bhat, S.R., Kumar, P. and Prakash, S. (2008). An improved cytoplasmic male sterile (*Diplotaxis berthautii*) *Brassica juncea*: Identification of restorer and molecular characterization. *Euphytica*, 159: 145–152.

Bhati, J., Sonah, H., Jhang, T., Singh, N.K. and Sharma, T.R. (2010). Comparative analysis and EST mining reveals high degree of conservation among five Brassicaceae species. Comparative and Functional Genomics, Article ID 520238, p. 13. doi:10.1155/2010/520238

Bisht, N.C., Gupta, V., Ramachiary, N., Sodhi, Y.S., Mukhopadhyay, A., Arumungam, N., Pental, D. and Pradhan, A.K. (2009). Fine mapping of loci involved with glucosinolate biosynthesis in oilseed mustard (*Brassica juncea*) using genomic information from allied species. *Theor Appl Genet.,* 11: 413–421.

Bisht, N.C., Burma, P.K. and Pental, D. (2004). Development of 2,4-D resistant transgenics in Indian oilseed mustard (*Brassica juncea*). *Curr. Sci.*, 87: 367–370.

Camarago, L.E.A. and Osborn, T.C. (1996). Mapping loci controlling flowering time in *Brassica oleracea*. *Theor Appl Genet.*, 92: 610–616.

Camarago, L.E.A., Williams, P.H. and Osborn, T.C. (1995). Mapping of quantitative trait loci controlling resistance of *Brassica oleracea* to *Xanthomonas campestris* pv. *campestris* in the field and green house. *Phytopathology*, 85: 1296–1300.

Chen, B.Y., Heneen, W.K. and Simonsen, V. (1989). Comparative and genetic studies of isozymes in re-synthesized and cultivated *Brassica napus* L., *B. campestris* and *B. alboglabra* Bailey. *Theor Appl Genet.*, 77: 673–679.

Chen, B.Y., Jorgensen, R.B., Cheng, B.F. and Heneen, W.K. (1997). Identification and chromosomal assignment of RAPD marker linked with a gene for seed coat colour in a *Brassica campestris-alboglabra* addition line. *Hereditas*, 126: 133–138.

Cheung, W.Y., Friesen, L., Rakow, G.F.W., Seguin Swartz, G. and Landry, B.S. (1997). An RFLP based linkage map of mustard [*Brassica juncea* (L) Czern and Coss]. *Theor. Appl. Genet*, 94: 841–851.

Cheung, W.Y., Gugel, R.K. and Landry, B.S. (1998). Identification of RFLP markers linked to the white rust resistance gene (*Acr*) in mustard [*Brassica juncea* (L.) Czern. and Coss.]. *Genome*, 41: 626–628.

Chyi, Y.S., Hoeneke, M.E. and Sernyk, J.L. (1992). A genetic linkage map of restriction fragment length polymorphism loci for *Brassica rapa* (syn *campestris*). *Genome*, 35: 746–757.

De Quiroz, H.C. and Mithen, R. (1996). Molecular markers for low glucosinolate alleles in oilseed rape (*Brassica napus* L.). *Mol. Breed.*, 2: 277–281.

Delourme, R., Bouchereau, A., Hubert, N., Renard, M. and Landry, B.S. (1994). Identification of RAPD markers linked to a fertility restorer gene for the Ogura radish cytoplasmic male sterility of rapeseed (*Brassica napus* L). *Theor Appl Genet.*, 88: 741–748.

Delourme, R., Falentin, C., Huteau, V., Clouet, V., Horvais, R., Gandon, B., Specel, S., Hanneton, L., Dheu, J.E., Deschamps, M., Margale, E., Vincourt, P. and Renard. M. (2006). Genetic control of oil content in oilseed rape (*Brassica napus* L.). *Theor Appl Genet.*, 113(7): 1331–1345.

Delourme, R., Foisset, N., Horvais, R., Barret, P., Champagne, G., Cheung, W.Y., Landry, B.S. and Renard, M. (1998). Characterisation of the radish introgression carrying the *Rfo* restorer gene for the *Ogu*-INRA cytoplasmic male sterility in rapeseed (*Brassica napus* L). *Theor Appl Genet.*, 97: 129–134.

Dion, Y., Gugel, R.K., Rakow, G.F.W., Seguin Swartz, G. and Landry, B.S. (1995). RFLP mapping of resistance to the blackleg disease [casual agent, *Leptosphaeria maculans* (Desm) Ces et de Not.] in canola (*Brassica napus* L). *Theor Appl Genet.,* 91: 1190–1194.

Downey, R.K. (1964). A selection of *Brassica campestris* containing no erucic acid in its seed oil. *Can. J. Plant Sci.,* 44: 295.

Downey, R.K., Craig, B.M. and Young, C.G. (1969). Breeding rapeseed for oil and meal quality. *J Am oil Chem Soc.,* 46: 121–123

Ecke, W., Uzunova, M. and Weissleder, K. (1995). Mapping the genome of rapeseed (*Brassica napus* L) II. Localization of genes controlling erucic acid synthesis and oil content. *Theor Appl Genet.*, 91: 972–977.

Federico, L. Iniguez Luy, Voort Amy, V. and Osborn Thomas, C. (2008). Development of a set of public SSR markers derived from genomic sequence of a rapid cycling *Brassica oleracea* L. genotype. *Theor Appl Genet.,* 117(6): 977–985.

Ferreira, M.E., Williams, P.H. and Osborn, T.C. (1994). RFLP mapping of *Brassica napus* using doubled haploid lines. *Theor Appl Genet.*, 89: 615–621.

Ferreira, M.E., Rimmer, S.R., Williams, P.H. and Osborn, T.C. (1995a). Mapping loci controlling *Brassica napus* resistance to *Leptosphaeria maculans* under different screening conditions. *Genetics*, 85: 213–217.

Ferreira, M.E., Satagopan, J., Yandell, B.S., Williams, P.H. and Osborn, T.C. (1995b). Mapping loci controlling vernalization requirement and flowering time in *Brassica napus*. *Theor Appl Genet.*, 90: 727–732.

Ferreira, M.E., Williams, P.H. and Osborn, T.C. (1995c). Mapping loci controlling resistance to *Albugo candida* in *Brassica napus* using molecular markers. *Phytopathology*, 85: 218–220.

Figdore, S.S., Kennard, W.C., Song, K.M., Slocum, M.K. and Osborn, T.C. (1988). Assessment of the degree of restriction fragment length polymorphism in *Brassica*. *Theor Appl Genet.*, 75: 833–840.

Figdore, S.S., Ferreira, M.E., Slocum, M.K. and Williams, P.H. (1993). Association of RFLP markers with trait loci affecting clubroot resistance and morphological characters in *Brassica oleracea* L. *Euphytica*, 69: 33–44.

Foisset, N., Delourme, R., Barret, P., Hubert, N., Landry, B.S. and Renard, M. (1996). Molecular mapping analysis of *Brassica napus* using isozyme, RAPD and RFLP markers on doubled haploid progeny. *Theor. Appl. Genet.*, 93: 1017–1025.

Fourmann, M., Barret, P., Renard, M., Pelletier, G., Delourme, R. and Brunel, D. (1998). The two genes homologous to *Arabidopsis FAE1* co–segregate with the two loci governing erucic acid content in *Brassica napus*. *Theor. Appl. Genet.*, 96: 852–858.

Good, A.G., Stringam, G.R., Mahmood, T., Ekuere, U. and Yeh, F. (2003). Molecular mapping of seed aliphatic glucosinolates in *Brassica juncea*. *Genome*, 46(5): 753–760.

Gupta, V., Mukhopadhaya A., Arumugam N., Sodhi Y.S., Pental D. and Pradhan A.K. (2004). Molecular tagging of erucic acid trait in oilseed mustard (*Brassica juncea*) by QTL mapping and single nucleotide polymorphism in *FAE1* gene. *Theor. Appl. Genet.,* 108: 743–749.

Happstadius, I., Ljungberg, A., Kristiansson, B. and Dixelius, C. (2003). Identification of *Brassica oleracea* germplasm with improved resistance to *Verticillium* wilt. *Plant Breed*, 122: 30–34.

Hasan, M., Friedt, W., Kühnemann Pons, J., Freitag, N.M., Link, K. and Snowdon, R.J. (2008). Association of gene–linked SSR markers to seed glucosinolate content in oilseed rape (*Brassica napus* ssp. *napus*). *Theor. Appl. Genet.*, 116(8): 1035–1049.

He, J., Ke, L., Hong, D., Xie, Y., Wang, G., Liu, P. and Yang, G. (2008). Fine mapping of a recessive genic male sterility gene (*Bnms3*) in rapeseed (*Brassica napus*) with AFLP and *Arabidopsis* derived PCR markers. *Theor. Appl. Genet.*, 117(1): 11–18.

Hoenecke M. and Chyi, Y.S. (1991). Comparison of *Brassica napus* and *B. rapa* genomes based on restriction fragment length polymorphism mapping. Proc. 8th Intl. Rapeseed Cong., *Saskatchewan,* 7: 1102–1107.

Hu, J., Li, G., Struss, D. and Quiros, C.F. (1999). SCAR and RAPD markers associated with 18–carbon fatty acids in rapeseed, *Brassica napus*. *Plant Breed.*, 118: 145–150.

Hu, J., Quiros, C., Arus, P., Struss, D. and Robbelen, G. (1995). Mapping of a gene determining linolenic acid concentration in rapeseed with DNA–based markers. *Theor. Appl. Genet.*, 90: 258–262.

Huang, Z., Chen, Y., Yi, B., Xiao, L., Ma, C., Tu, J. and Fu, T. (2007). Fine mapping of the recessive genic male sterility gene (*Bnms3*) in *Brassica napus* L. *Theor. Appl. Genet.*, 115(1): 113–118.

Jagannath, A., Sodhi, Y.S., Gupta, V., Mukhopadhyay, A., Arumungam, N., Singh, I., Rogtagi, S., Burma, P.K., Pradhan, A.K. and Pental, D. (2011). Eliminating expression of erucic acid–encoding loci allows the identification of "hidden" QTL contributing to oil quality fractions and oil content in *Brassica juncea* (Indian mustard). *Theor. Appl. Genet.,* 122: 1091–1103.

Janeja, H.S., Banga, S.K., Bhaskar, P.B. and Banga, S.S. (2003). Alloplasmic male sterile *Brassica napus* with *Enarthrocarpus lyratus* cytoplasm: Introgression and molecular mapping of an *E. lyratus* chromosome segment carrying a fertility restoring gene. *Genome,* 46: 792–797.

Jean, M., Brown, G.G. and Landry, B.S. (1997). Genetic mapping of nuclear fertility restorer genes for the Polima cytoplasmic male sterility in canola (*Brassica napus* L) using DNA markers. *Theor. Appl. Genet.*, 95: 321–328.

Jonsson, R. (1977). Erucic acid heredity in rapeseed (*Brassica napus* L.) and *Brassica campestris*. *Hereditas,* 86: 159–170.

Jonsson, R. and Uppstrom, B. (1986). Quality breeding in rapeseed. *In*: Sval of 1886–1986. Research and results in *Plant Breeding*. pp 173–184.

Josefsson, E. and Appleqvist, L.A. (1968). Glucosinolates in seed of rape and turnip rape as affected by variety and environment. *J. Sci. Fd. Agric.*, 19: 564–570.

Jourden, C., Barret, P., Brunel, D., Delourme, R. and Renard, M. (1996a). Specific molecular marker of genes controlling linolenic acid content in rapeseed. *Theor. Appl. Genet.*, 93: 512–518.

Jourden, C., Barret, P., Horvais, R., Delourme, R. and Renard, M. (1996b). Identification of RAPD marker linked to linolenic acid genes in rapeseed. *Euphytica*, 90: 351–357.

Jourden, C., Barret, P., Horvais, R., Foisset, N., Delourme, R. and Renard, M. (1996c). Identification of RAPD markers linked to loci controlling erucic acid level in rapeseed. *Mol Breed.*, 2: 61–71.

Kalita, M.C., Mohapatra, T., Dhandapani, A., Yadava, D.K., Srinivasan, K, Mukherjee, A.K. and Sharma, R.P. (2007). Comparative evaluation of RAPD, ISSR and Anchored–SSR markers in assessment of genetic diversity and fingerprinting of oilseed *Brassica* genotypes. *J. Plant Biochem. Biotech.,* 15(1): 41–48.

Kanrar, S., Venkateshwari, J.C., Kirti, P.B. and Chopra, V.L. (2001). Transgenic expression of hevein, the rubber tree lectin, in Indian mustard confers protection against *Alternaria brassicae*. *Plant Sci.,* 162: 441–448.

Kanrar, S., Venkateswari, J., Kirti, P.B. and Chopra, V.L. (2002). Transgenic Indian mustard (*Brassica juncea*) with resistance to mustard aphid (*Lipaphis erysimi*). *Pl. Cell Reporter,* 20: 976–981.

Kearsay, M.J., Ramsay, L.D., Jennings, D.E., Lydiate, D.J., Bohoun, E.J.R. and Marshall, D.F. (1996). Higher recombination frequency in female compared to male meiosis in *Brassica oleracea*. *Theor. Appl. Genet.*, 92: 363–367.

Kianian, S.F. and Quiros, C.F. (1991). Genetic analysis of major multigene families in *Brassica oleracea* and related species. *Genome*, 35: 516–527.

Kirk, J.T.O. and Oram, R.N. (1981). Isolation of erucic acid free lines of *Brassica juncea*. *J. Aust. Inst. Agric. Sci.,* 47: 51–52

Kirti, P.B., Mohapatra, T., Baldev, A., Prakash, S. and Chopra, V.L. (1995a). A stable cytoplasmic male sterile line of *Brassica juncea* carrying restructured organelle genomes from somatic hybrid *Trachystoma ballii* + *Brassica juncea*. *Plant Breed.,* 114: 434–438.

Kirti P.B., Banga, S.S., Prakash, S. and Chopra, V.L. (1995b). Transfer of *ogu* cytoplasmic male sterility to *Brassica juncea* and improvement of the male sterile line through somatic cell fusion. *Theor. Appl. Genet.,* 91: 517–521.

Kirti, P.B., Baldev, A., Gaikwad, K., Bhat, S.R., Dinesh, K. V., Prakash, S. and Chopra, V.L. (1997). Introgression of a gene restoring fertility to CMS (*Trachystoma) Brassica juncea* and genetics of restoration. *Plant Breed.,* 116: 259–262.

Kole, C., Teutonico, R., Williams, P.H. and Osborn, T.C. (1996). Molecular mapping of a locus controlling resistance to *Albugo candida* in *Brassica rapa*. *Phytopathology*, 86: 367–369.

Kole, C., Kole, P., Vogelzang, R. and Osborn, T.C. (1997). Genetic linkage map of a *Brassica rapa* recombinant inbred population. *J. Heredity*, 88: 553–557.

Koundal, V., Parida, S.K., Yadava, D.K., Ali, A., Koundal, K.R. and Mohapatra, T. (2008). Evaluation of microsattelite markers for genome mapping in Indian mustard (*Brassica juncea* L.). *J. Plant Biochem.Biotech.,* 17(1): 69–72.

Landry, B.S., Hubert, N., Crete, R., Chiang, M.S., Lincoln, S.E. and Etoh, T. (1992). A genetic map for *Brassica oleracea* based on RFLP markers detected with expressed DNA sequences and mapping of resistance genes to race 2 of *Plasmodiophora brassicae* (Woronin). *Genome*, 35: 409–420.

Landry, B.S., Hubert, N., Etoh, T., Harada, J.J. and Lincoln, S.E. (1991). A genetic map for *Brassica napus* based on restriction fragment length polymorphisms detected with expressed DNA sequences. *Genome*, 34: 543–552,

Leflon, M., Brun, H., Eber, F., Delourme, R., Lucas, M.O., Vallee, P., Ermel, M., Balesdent, M.H., and Chèvre, A.M. (2007). Detection, introgression and localization of genes conferring specific resistance to *Leptosphaeria maculans* from *Brassica rapa* into *B. napus*. *Theor Appl Genet.,* 115(7): 897–906.

Lagercrantz, U. and Lydiate, D.J. (1995). RFLP mapping in *Brassica nigra* indicates differing recombination rates in male and female meioses. *Genome,* 38: 255–264.

Lagercrantz, U., Putterill, J., Coupland, G. and Lydiate, D. (1996). Comparative mapping of *Arabidpsis* and *Brassica*, fine scale genome colinearity and congruence of genes controlling flowering time. *The Plant J.*, 9: 13–20.

Li, G. and Quiros, C.F. (2001). Sequence–related amplified polymorphism (SRAP), a new marker system based on a simple PCR reaction: Its application to mapping and gene tagging in *Brassica*. *Theor. Appl. Genet.*, 103(2–3): 455–461.

Li, X., Ramchiary, N., Choi, S.R., Nguyen, D.V., Hossain, M.J., Yang, H.K., and Lim, Y.P. (2010). Development of a high density integrated reference genetic linkage map for the multinational *Brassica rapa* Genome Sequencing Project. *Genome,* 53(11): 939–947.

Lydiate, D. and Sharpe, A. (2003). Aligning genetic maps of *Brassica napus* using microsatellite markers. Plant and Animal Genome XI, San Diego, USA.

Mehra, S., Pareek, A., Bandopadhyay, P., Sharma, P., Burma, P.K. and Pental, D. (2000). Development of transgenics in Indian oilseed mustard (*Brassica juncea*) resistant to herbicide phosphinothricin. *Curr. Sci,.* 78: 1358–1364.

Mishra, S. and Manchanda, S.C. (2012). Cooking oils for heart health. *J. Preventive Cardiology,* 1: 123–131.

Misra, A.K. and Kumar, A. (2008). Rapeseed mustard genetic resources and its utilization. National Research Centre on Rapeseed and Mustard (ICAR), Sewar, Bharatpur – 321 303 (Rajasthan). p (viii+48).

Mohapatra, T., Upadhyay, A., Sharma, A. and Sharma, R.P. (2002). Detection and mapping of duplicate loci in *Brassica juncea*. *J. Plant Biochem. Biotech.*, 11: 37–42.

Mohapatra, T. and Yadava, D.K. (2008). Marker assisted selection for oilseed improvement. *In:* Vegetable oils scenario: Approaches to meet the growing demands (*Ed.* Hegde. D.M.). ISOR, DOR, Hyderabad. pp. 115–150.

Mondal, K.K., Bhattacharya, R.C., Koundal, K.R. and Chatterjee, S.C. (2007). Transgenic Indian mustard (*Brassica juncea*) expressing tomato glucanase leads to arrested growth of *Alternaria Brassicae*. *Plant Cell Reporter,* 26: 247–253.

Mukherjee, A.K., Mohapatra, T., Varshney, A., Sharma, R. and Sharma, R.P. (2001). Molecular mapping of a locus controlling resistance to *Albugo candida* in *Brassica juncea*. *Plant Breed.*, 120(6): 483–487.

Naresh, S., Vasudev, S., Yadava, D.K., Singh, N., Dharitree, C., Jain, N., Singh, S., Singh J., Giri, S.C., Dass, B., Singh, R., Shekhawat, G.S., Mohapatra, T. and Prabhu, K.V. (2012). Identification of new functional molecular markers for erucic acid genes in ZEM–2 derived *Brassica juncea* variety LES–1–27 (Ab). International Conference on Plant Biotechnology for Food Security: New Frontiers organized by Society for Plant Biochemistry and Biotechnology, NRC on Plant Biotechnology and IARI, New Delhi from 21.24 Feb. 2012 at New Delhi. MB 79: 88–89.

Nath, U.K. and Goswami, G. (2009). Inheritance and variation of erucic acid content in a transgenic rapeseed (*Brassica napus* L.) doubled haploid population. *Mol Breed.*, 23: 125–138.

Negi, M.S., Devic, M., Delseny, M. and Lakshmikumaran, M. (2000). Identification of AFLP fragments linked to seed coat colour in *Brassica juncea* and conversion to SCAR marker for rapid selection. *Theor Appl Genet.*, 101: 146–152.

Panjabi, M. P., Yadava, S.K., Sharma, P., Kaur, A., Kumar, A., Arumugam, N., Sodhi, Y.S., Mukhopadhyay, A., Gupta, V., Pradhan, A.K. and Pental, D. (2010). Molecular mapping reveals two independent loci conferring resistance to *Albugo candida* in the east European germplasm of oilseed mustard *Brassica juncea. Theor Appl Genet.,* 121: 137–145.

Panjabi, P., Jagannath, A., Bisht, N.C., Padmaja, L.K., Sharma, S., Gupta, V., Pradhan, A.K. and Pental, D. (2008). Comparative mapping of *Brassica juncea* and *Arabidopsis thaliana* using Intron Polymorphism (IP) markers: Homoeologous relationships, diversification and evolution of the A, B and C. *Brassica* genomes. *BMC Genomics,* 9: 113.

Parida, S., Yadava, D.K. and Mohapatra, T. (2010). Microsatellites in *Brassica* unigenes: Relative abundance, marker design and use in comparative physical mapping and genome analysis. *Genome*, 53: 55–67.

Prabhu, K.V., Somers, D.J., Rakow, G. and Gugel, R.K. (1998). Molecular markers linked to white rust resistance in mustard *Brassica juncea*. *Theor. Appl. Genet.*, 97: 865–870.

Pradhan, A., Gupta, V., Mukhopadhyay, A., Arumugam, N., Sodhi, Y. and Pental, D. (2003). A high–density linkage map in *Brassica juncea* (Indian mustard) using AFLP and RFLP markers. *Theor. Appl. Genet.*, 106(4): 607–614.

Pradhan, A., Nelson, M.N., Plummer, J.A., Cowling, W.A. and Yan, G. (2011). Characterization of *Brassica nigra* collections using simple sequence repeat markers reveals distinct groups associated with geographical location, and frequent mislabelling of species identity. *Genome*, 54(1): 50–63.

Prakash, S. and Chopra, V.L. (1990). Male sterility caused by cytoplasm of *Brassica oxyrrhina* in *B. campestris* and *B. juncea*. *Theor. Appl. Genet.*, 79: 285–287.

Prakash, S., Kirti, P.B., Bhat, S.R., Gaikwad, K., Kumar, D. and Chopra, V.L. (1998). A *Moricandia arvensis* based cytoplasmic male sterility and fertility restoration system in *Brassica juncea*. *Theor. Appl. Genet.*, 97: 488–492.

Prakash, S. (2001). Utilization of wild germplasm of *Brassica* allies in developing male sterility–fertility restoration systems in Indian mustard–*Brassica juncea*, *In*: Houli, L. and Fu, T.D. (*Eds*.), Proc. Int. Symp. Rapeseed Science, Huazhong Agricultural University, China, Science Press, New York. pp. 73–78.

Prakash, S., Ahuja, I., Upreti, H.C., Kumar, D., Bhat, S.R., Kirti, P.B. and Chopra, V.L. (2001). Expression of male sterility in alloplasmic *Brassica juncea* with *Erucastrum canariense* cytoplasm and development of a fertility restoration system. *Plant Breed.*, 120: 178–182.

Qiu, D., Morgan, C., Shi, J., Long, Y., Liu, J., Li, R., Zhuang, X., Wang, Y., Tan, X., Dietrich, E., Weihmann, T., Everett, C., Vanstraelen, S., Beckett, P., Fraser, F., Trick, M., Barnes, S., Wilmer, J., Schmidt, R., Li, J., Li, D., Meng, J. and Bancroft, I. (2006). A comparative linkage map of oilseed rape and its use for QTL analysis of seed oil and erucic acid content. *Theor. Appl. Genet.*, 114(1): 67–80.

Rahman, M., McVetty, P.B.E. and Li, G. (2007). Development of SRAP, SNP and Multiplexed SCAR molecular markers for the major seed coat color gene in *Brassica rapa L. Theor. Appl. Genet.*, 115(8): 1101–1107.

Rahman, M., Sun, Z., McVetty, P.B.E. and Li, G. (2008). High throughput genome–specific and gene–specific molecular markers for erucic acid genes in *Brassica napus* (L.) for marker–assisted selection in plant breeding. *Theor. Appl. Genet.,* 117(6): 895–904.

Rajwanshi, R., Roy, S. D., Pooggin, M., Hohn, T. and Bhall–Sarin, N. (2007). Marker free approach for developing abiotic stress tolerant transgenic *Brassica juncea* (Indian mustard). *In*: Proceedings of 2007 WSEAS Conference on Celluar and Molecualr Biology –Biophysics and Bioengineering, Athens, Greece , August 26–28, 2007.

Rakow, G. (1973). Selection for content of linoleic and linolenic acids in rapeseed after mutagenic treatment *Z. Pflanzenzuecht,* 69: 62–68.

Ramchiary, N., Bisth, N.C., Gupta, V., Mukhopadhaya, A., Arunmugam, N. Sodhi, Y.S., Pental, D. and Pradhan, A.K. (2007). QTL analysis reveals context–dependent loci for seed glucosinolate trait in the oilseed *Brassica juncea*: Importance of recurrent selection backcross scheme for identification of QTL. *Theor Appl. Genet.,* 116: 77–85.

Ramsay, L.D., Jennings, D.E., Bohuon, E.J.R., Arthur, A.E., Lydiate, D.J., Kearsey, M.J. and Marshall, D.F. (1996). The construction of a substitution library of recombinant backcross lines in *Brassica oleracea* for the precision mapping of quantitative loci. *Genome*, 39: 558–567.

Rao, G.V., Batra–Swarup, V., Prakash, S. and Shivanna, K.R. (1974). Development of a new cytoplasmic male sterile system in *Brassica juncea* through wide hybridization. *Plant Breed.*, 112: 171–174.

Rawat, D.S. and Anand, I.J. (1979). Male sterility in Indian mustard. *Ind. J. Genet.*, 39: 412–415.

Ripley, V.L. and Roslinsky, V. (2005). Identification of an ISSR marker for 2–propenyl glucosinolate content in *Brassica juncea* and conversion to SCAR marker. *Mol. Breed.*, 16: 57–66.

Roy, N.N. and Tarr, A.W. (1986). Development of Near–zero linolenic acid (18:3) lines of Rapeseed (*Brassica napus* L.) *Z Pflanzenzuecht*, 96: 218–223.

Saal, B., Plieske, J., Hu, J., Quiros, C.F. and Struss, D. (2001). Microsatellite markers for genome analysis in *Brassica*. II. Assignment of rapeseed microsatellites to the A and C genomes and genetic mapping in *Brassica oleracea* L. *Theor. Appl. Genet.*, 102: 695–699.

Saito, M., Kubo, N., Matsumoto, S., Suwabe, K., Tsukada, M. and Hirai, M. (2006). Fine mapping of the clubroot resistance gene, *Crr3*, in *Brassica rapa*. *Theor Appl Genet.*, 114(1): 81–91.

Saxena M. Roy, S. Deb, Singla–Pareek, S.L., Sopory, S.K. and Bhalla–Sarin, N. (2011). Overexpression of glyoxalase II gene leads to enhanced salinity tolerance in *Brassica juncea*. *The Open Plant Sci. J.*, 5: 23–28.

Schierholt, A., Becker, H.C. and Ecke, W. (2000). Mapping a high oleic acid mutation in winter oilseed rape (*Brassica napus* L.). *Theor Appl Genet.*, 101: 897–901.

Sharma, A., Mohopatra, T. and Sharma, R.P. (1994). Molecular mapping and character tagging in *Brassica juncea*. I Degree, nature and linkage relationship of RFLPs and their association with quantitative traits. *J. Pl. Biochem and Biotech.*, 3: 85–89.

Sharma, N., Anderson, M., Kumar, A., Zhang, Y., Giblin, M.E., Abrams, S.R., Zaharia, L.I., Taylor, D.C. and Fobert, P.R. (2008). Transgenic increases seed oil content are associated with differential expression of novel *Brassica*–specific transcripts. *BMC Genomics,* 9: doi 10–1186/1471–2164–9–619.

Sharma, R., Aggarwal, R.A.K., Kumar, R., Mohapatra, T. and Sharma, R.P. (2002). Construction of RAPD linkage map and localization of QTLs for oleic acid level using recombinant inbreds in mustard. *Genome*, 45 (3): 467–472.

Sharma, R., Mohapatra, T., Mukherjee, A.K., Krishanpal and Sharma, R.P. (1999). Molecular markers for seed oil content in Indian mustard. *J. Pl. Biochem. Biotech.*, 8: 61–64.

Shi, J., Li, R., Qui, D., Jiang, C., Long, Y., Morgan, C., Bancroft, I., Zhao, J., and Meng, J. (2009). Unraveling the complex trait of crop yield with quantitative trait loci mapping in *B. napus*. *Genetics,* 182: 851–861.

Singh. J. (2010). Validation and use of molecular markers tagged to low erucic acid trait for marker assisted selection in Indian mustard (*Brassica juncea* L.). M.Sc. thesis submitted to Post Graduate School, Indian Agricultural Research Institute, New Delhi.

Singh, J., Yadava, D.K., Sugandh, N., Sujata, V., Jain, N., Singh, N., Singh, S, Mohapatra, T. and Prabhu, K.V. (2011). Validation of SNPs for low erucic acid in *Brassica juncea* (Ab). National Seminar on Contemporary approaches to crop improvement organized by Indian Society of Genetics and Plant Breeding at UAS Bangalore from 22–23 April, 2011. PS–I: 71.

Slocum, M.K., Figdore, S.S., Kenard, W.C., Suzuki, J.Y. and Osborn, T.C. (1990). Linkage arrangement of restriction fragment length polymorphic loci in *Brassica oleracea*. *Theor Appl Genet.*, 80: 75–64.

Snowdon R.J. and Iniguez Luy, F.L. (2012). Potential to improve oilseed rape and canola breeding in the genomics era. *Pant. Breed.,* 131: 351–360.

Sodhi, Y. S., Chandra, A., Verma, J.K., Arumugam, N., Mukhopadhyay, A., Gupta, V. Pental, D. and Pradhan, A.K. (2006). A new cytoplasmic male sterility system for hybrid seed propuduction in Indian oilseed mustard *Brassica juncea*. *Theor. Appl. Genet.*, 114: 93–99.

Soengas, P., Hand, P., Vicente, J.G., Pole, J.M. and Pink, D.A.C. (2007). Identification of quantitative trait loci for resistance to *Xanthomonas campestris* pv. *campestris* in *Brassica rapa*. *Theor. Appl. Genet.*, 114(4): 637–645.

Song, K.M., Suzuki, J.Y., Slocum, M.K., Williams, P.H. and Osborn, T.C. (1991). A linkage map of *Brassica rapa* (syn *campestris*) based on restriction fragment length polymorphism loci. *Theor. Appl. Genet.*, 82: 296–304.

Stefansson, B.R., Hougen, F.W. and Downey, R.K. (1961). Note on the isolation of rape plants with seed oil free from erucic acid. *Can. J. Plant Sci.*, 41: 218–219.

Stringam, G.R. and Thiagarajah, M.R. (1995). Inheritance of alkenyl glucosinolate in traditional and microspore–derived doubled haploid populations of *Brassica juncea* L. Czern and Coss. *In*: Proceedings of 9th International Rapeseed Congress – Rapeseed today and tomorrow, Cambridge, U.K. pp. 804–806.

Sun, Z., Wang, Z., Tu, J., Zhang, J., Yu, F., McVetty, P.B.E. and Li, G. (2007). An ultradense genetic recombination map for *Brassica napus*, consisting of 13551 SRAP markers. *Theor. Appl. Genet.*, 114(8): 1305–1317.

Tanhuanpaa, P., Vilkki, J. and Vihinen, M. (1998). Mapping and cloning of *FAD2* gene to develop allele–specific PCR for oleic acid in spring turnip rape (*Brassica rapa* ssp. oleifera). *Mol. Breed.*, 4: 543–550.

Tanhuanpaa, P.K., Vilkki, J.P. and Vilkki, H.J. (1995). Association of a RAPD marker with linolenic acid concentration in the seed oil of rapeseed (*Brassica napus* L). *Genome*, 38: 414–416.

Tanhuanpaa, P.K., Vikki, J.P. and Vikki, H.J. (1996). Mapping of a QTL for oleic acid concentration in spring turnip rape (*Brassica rapa* ssp *oleifera*). *Theor. Appl. Genet.*, 92: 952–956.

Teutonico, R.A. and Osborn, T.C. (1994). Mapping of RFLP and quantitative trait loci in *Brassica rapa* and comparison to the linkage maps of *B. napus, B. oleracea* and *Arabidopsis thaliana*. *Theor. Appl. Genet.*, 89: 885–894.

Teutonico, R.A. and Osborn, T.C. (1995). Mapping loci controlling vernalization requirement in *Brassica rapa*. *Theor. Appl. Genet.*, 91: 1279–1283.

Thormann, C.E., Romero, J., Mantet, J. and Osborn, T.C. (1996). Mapping loci controlling the concentration of erucic and linolenic acids in seed oil of *Brassica napus* L. *Theor Appl Genet.*, 93: 282–286.

Toroser, D., Thormann, C.E., Osborn, T.C. and Mithen, R. (1995). RFLP mapping of quantitative trait loci controlling seed aliphatic glucosinolate content in oilseed rape (*Brassica napus* L). *Theor. Appl. Genet.*, 91: 802–808.

Truco, M.J. and Quiros, C.F. (1994). Structure and organization of the B genome based on a linkage map in *Brassica nigra*. *Theor. Appl. Genet.*, 89: 590–598.

Upadhyay, A., Mohapatra, T., Pai, R.A. and Sharma, R.P. (1996). Molecular mapping and character tagging in Indian mustard (*Brassica juncea*) II. RFLP marker association with seed coat colour and quantitative traits. *J. Plant Biochem. Biotech.*, 5: 17–22.

Uppstrom, B. and Johansson, M. (1985). Determination of sinapine in rapeseed. *Sver Utsaedestoeren Tidskr*, 95: 123–128.

Uzunova, M., Ecke, W., Weissleder, K. and Robbelen, G. (1995). Mapping the genome of rapeseed (*Brassica napus* L) I Construction of an RFLP linkage map and localization of QTLs for seed glucosinolate content. *Theor. Appl. Genet.*, 90: 194–204.

Van Deynze, A.E., Landry, B.S. and Pauls, K.P. (1995). The identification of restriction fragment length polymorphisms linked to seed colour genes in *Brassica napus*. *Genome*, 38: 534–542.

Van, R.L. and Roslinsky, V. (2005). Identification of an ISSR marker for 2–propenyl glucosinolate content in *Brassica juncea* and conversion to SCAR marker. *Mol. Breed.,* 16: 57–66.

Varshney, A., Mohapatra, T. and Sharma, R.P. (2004). Development and validation of CAPS and AFLP markers for white rust resistance gene in *Brassica juncea*. *Theor. Appl. Genet.,* 109: 153–159.

Voorrips, R.E., Jongerius, M.C. and Kanne, H.J. (1997). Mapping of two genes for resistance to clubroot (*Plasmodiophora brassicae*) in a population of doubled haploid lines of *Brassica oleracea* by means of RFLP and AFLP markers. *Theor. Appl. Genet.*, 94: 75–82.

Walsh, J.A., Sharpe, A.G., Jenner, C.E. and Lydiate, D.J. (1999). Characterisation of resistance to turnip mosaic virus in oilseed rape (*Brassica napus*) and genetic mapping of TuRB01. *Theor. Appl. Genet.,* 99: 1149–1154.

Werner, S., Diederichsen, E., Frauen, M., Schondelmaier, J. and Jung, C. (2008). Genetic mapping of clubroot resistance genes in oilseed rape. *Theor. Appl. Genet.,* 116(3): 363–372.

Yadava, D.K., Parida, S.K., Dwivedi, V.K., Varshney, A., Ghazi Irfan, A., Sujata V. and Mohapatra, T. (2009). Cross–transferability and polymorphic potential of genomic STMS markers of *Brassica species. J. Plant Biochem. Biotech.*, 18(1): 29–36.

Yadava, D.K., Naresh, S., Jain, N., Singh, N., Vasudev, S., Singh, S., Giri, S.C., Dass, B., Singh, R., Mohapatra, T. and Prabhu, K.V. (2012). Transferring low glucosinolate trait in low erucic acid *Brassica juncea* genotypes through MABB (Ab). International Conference on Plant Biotechnology for Food Security: New Frontiers organized by Society for Plant Biochemistry and Biotechnology, NRC on Plant Biotechnology and IARI, New Delhi from 21.24 Feb. 2012 at New Delhi. *MB,* 80: 89.

Yadava, D.K., Vasudev, S., Singh, N., Mohapatra, T. and Prabhu, K.V. (2011). Breeding major oilcrops: Present status and future research needs. Technical innovations inmajor Oilseed crops, Vol. 1 Breeding (Ed. S.K. Gupta). Springer. pp. 17–52.

Yadava, D.K., Vasudev, S., Singh, N., Naresh, S., Singh, S., Yadav, S., Jain, N., Mohapatra, T. and Prabhu, K.V. (2012). Broadening genetic diversity for breeding high value oilseed *Brassica* cultivars using conventional and molecular approaches. Souviner 1st National Brassica Conference on Production barriers and technological options in oilseed *Brassica* organized by CCS HAU, Hisar, Society for Rapeseed Mustard Research, Bharatpur and ICAR from 2–3 March, 2012 at CCS HAU, Hisar (Eds. Kumar A., Singh D., Meena P.D. and Kumar V. 2012). p. 10.

Yi, B., Chen, Y., Lei, S., Tu, J. and Fu, T. (2006). Fine mapping of the recessive genic male–sterile gene (*Bnms1*) in *Brassica napus* L. *Theor. Appl. Genet.,* 113(4): 643–650.

Zhao, J., and Meng, J. (2003). Genetic analysis of loci associated with partial resistance to *Sclerotinia sclerotiorum* in rapeseed (*Brassica napus* L.). *Theor. Appl. Genet.,* 106: 759–764.

Zhao, J., Becker, H.C., Zhang, D., Zhang, Y. and Ecke, W. (2006). Conditional QTL mapping of oil content in rapeseed with respect to protein content and traits related to plant development and grain yield. *Theor. Appl. Genet.*, 113(1): 33–38.

13

Sunflower

P. Kadirvel and M. Sujatha

Directorate of Oilseeds Research, Hyderabad, India

Sunflower (*Helianthus annuus* L.) is an important oilseed crop which competes in the "world oilseed complex" with three other major oilseeds *viz.*, soybean, groundnut and rapeseed. In 2010, global area under sunflower cultivation was about 23 million hectares and the production was about 30 million tonnes with an average productivity of 1322.6 kg/ ha (Table 1; FAOSTAT). An introduced crop in India in early 1970s, sunflower has become an important oilseed crop with 1 million hectares under its cultivation (Table 2). In 2000-01, Karnataka (45%), Maharashtra (31%) and Andhra Pradesh (18%) were the leading states, which covered almost 94 percent of total acreage of sunflower in India. In contrast, during 2009-10, Karnataka accounted for 54 percent followed by Andhra Pradesh with 24 percent (Table 3). A substantial reduction in sunflower area has occurred in Maharashtra between 2001 (31%) and 2009 (15%). The analysis of growth rates of sunflower during the past decade in the country indicates clear signals of deceleration in the area and production (Vision 2030, DOR; Nayak *et al.*, 2010). The crop, hitherto confined to southern parts of India, has made inroads into non-traditional areas such as Punjab, Haryana, Uttar Pradesh as spring/summer crop during the last decade.

Sunflower is primarily used for extracting cooking oil but also finds use in other sectors such as an annual ornamental, confectionary, animal feed, cosmetic and pharmaceutical industry. The seeds contain 40–46

V.L. Chopra (Ed.): Breeding Field Crops II: Advances, Studium Press (LLC), USA.

percent edible oil. Traditional sunflower oil is composed of saturated - C16:0 palmitic (7%) and C18:0 stearic (4%), monounsaturated - C18:1 oleic (20%) and polyunsaturated - C18:2 linoleic (69%) fatty acids. Sunflower oil is valued as a vegetable oil because of high level of polyunsaturated linoleic acid, which can reduce the risk of

Table 1: Area, production and productivity of sunflower in major growing countries

Country	*Area (m ha)*	*Production (m t)*	*Productivity (kg/ha)*
Ukraine	4.526	6.771	1496
Russian Federation	5.575	5.345	958
Argentina	1.489	2.221	1491
France	0.695	1.633	2350
Bulgaria	0.700	1.596	2280
Turkey	0.641	1.320	2058
Romania	0.786	1.263	1606
China	0.970	1.710	1763
Hungary	0.501	0.970	1934
United States of America	0.758	1.241	1636
Spain	0.698	0.887	1271
India	1.000	0.650	650
World-Total	23.104	30.559	1323

FAOSTAT | © FAO Statistics Division 2012 | Accessed on 10 June 2012

Table 2: Area, production and productivity of sunflower in India over the past five decades

Year	*Area (m ha)*	*Production (m t)*	*Productivity (kg/ha)*
2010	1.000	0.650	650.0
2000	1.074	0.646	602.0
1990	1.633	0.873	534.6
1980	0.119	0.066	555.3
1970	0.117	0.076	649.6

FAOSTAT | © FAO Statistics Division 2012 | Accessed on 10 June 2012

Table 3: Area, production and productivity of sunflower in major growing states of India during 2009-10

State	*Area (ha)*	*Production (t)*	*Productivity (kg/ha)*
Karnataka	794000	304000	383
Andhra Pradesh	350000	270000	771
Maharashtra	210000	114000	520
Bihar	22600	31700	1403
Odisha	17700	16000	904
Tamil Nadu	14500	19000	1329

http://www.krishisewa.com/articles/2012/sunflower3.html

cardiovascular disease. In addition, the oil contains tocopherols, which are antioxidants that provide oxidative stability to the oil. More details can be found in Seiler and Jan (2010).

Genetic Resources

Sunflower is domesticated from wild *H. annuus* in North America. It was introduced into Europe during early 16th century. Russia was the first country to recognize it as a source of oil. In 1969, sunflower was introduced as an oil seed crop in India. Taxonomically, sunflower belongs to the family Compositae (Asteraceae), subfamily Asteroideae, tribe Heliantheae, subtribe Helianthineae. As of now, 51 *Helianthus* species have been described which include 14 annuals and 37 perennials. Chromosome number (n) of annual species is 17 and perennials have n = 17, 34 or 51 (Seiler and Jan, 2010). The North Central Regional Plant Introduction Station, Ames, Iowa of the USDA-ARS maintains the largest sunflower germplasm collection in the world. USDA-GRIN database reports maintenance of 4155 accessions of which 2853 accessions belong to *H. annuus* collected from 59 countries (USDA-GRIN database as on 10 June 2012). Distribution of USDA germplasm has largely contributed to research programmes in Argentina, France, Italy, Spain, Germany, Bulgaria, Romania, Czech Republic, Hungary, Russia, Serbia, India, China and Mexico. In India, more than 2000 accessions are maintained at the Directorate of Oilseeds Research, Hyderabad and distributed through the All India Coordinated Research Programmes (AICRP) for sunflower improvement in the country. The Directorate of Oilseeds Research also maintains 43 wild *Helianthus* species for genetic enhancement of sunflower through interspecific gene transfer.

Hybrids

Earlier, sunflower varieties were open pollinated populations. VNIMK 6540, VNIMK 8931 and Peredovik of Russia are the best known examples. However, the open pollinated varieties had problems of uniformity and therefore efforts were made to develop hybrids. Genic male sterility system (GMS) was first explored to develop hybrids. GMS-based hybrids include INRA6501 and Airelle, which were popular until 1980. Later on GMS based hybrids were abandoned due to the problems associated with maintenance and hybrid seed production. Leclercq (1969) discovered cytoplasmic male sterility (CMS) source, known as PET1, from a cross between *H. petiolaris* and *H. annuus*, which spearheaded the development of commercial hybrids throughout the world. More than 70 CMS sources have been reported in sunflower (Horn *et al.*, 2002). Since introduction of the first sunflower hybrid BSH-1 (CMS234A × RHA274) in 1980, several varieties and hybrids from both public and private sectors have been released for cultivation in India (Tables 4, 5). More than 95% of the area under sunflower cultivation is

Table 4: Public sector varieties in India

Variety	*Year of release*	*Releasing centre*	*Yield potential (kg/ha)*	*Oil content (%)*	*Recommended states/regions situations*	*Salient features/ traits*
DRSF-108	2004	DOR, Hyderabad	900-1800	36-39	All India	High oil
CO-5	2005	TNAU Coimbatore	1000-1700	39-42	Tamil Nadu	-
TAS-82	2005	PDKV Akola	800-1200	40-42	Maharashtra	Black seeded variety
LSF-8	2006	Latur, Maharashtra	100-1400	36-39	Maharashtra	Tolerant to downy mildew, rust and *Alternaria*
DRSF-113	2007	DOR, Hyderabad	1000-1500	36-39	All India	High yield
Phule Raviraj	2009	MPKV, Rahuri	1795	34	Western Maharashtra	Tolerant to necrosis, *Alternaria* and Capitulum borer

Source: Recent varieties and hybirds of annual oilseeds recommended for different states (2001-2010), Directorate of Oilseeds Research

occupied by hybrids. Of these, the hybrids from private sector dominate (@95%) of the market. Breeding methods such as backcross or recurrent selection or pedigree selection are commonly used to develop parental lines for the hybrid programmes (Vear, 2010).

MAJOR BREEDING OBJECTIVES AND ACCOMPLISHMENTS

Primary focus of both public and private sunflower breeding programs in India is oriented towards enhancing seed yield, improving oil content and introducing disease resistance (*Alternaria helianthi,* downy mildew, sunflower necrosis disease, powdery mildew). In other countries, breeding focus is on development of high oleic lines, resistance to major diseases (*Sclerotinia sclerotiorum*, downy mildew, orobanche) and dual purpose types – as confectionery and oil-type hybrids. For spring cultivation in India, development of cultivars that combine short vegetative period with high productivity remains the main challenge.

Table 5: Public sector hybrids in India

Hybrid	*Year of release*	*Releasing centre*	*Yield potential (kg/ha)*	*Oil content (%)*	*Recommended states/regions situations*	*Salient features/ traits*
KBSH-41	2002	UAS, Bangalore	1300-1500	39-41	Karnataka	Tolerant to moisture stress
KBSH-42	2002	UAS, Bangalore	1300-1500	38-41	Karnataka	Tolerant to moisture stress
PSFH-118	2002	PAU, Ludhiana	1400	40	Punjab	Resistant to stem and head rot
KBSH-44	2003	UAS, Dharwad	1400-1600	36-38	All India	Resistant to downy mildew
LSFH-35 (Maruti)	2003	Latur, Maha-rashtra	1400-1500	39-41	Maharashtra	Resistant to downy mildew
NDSH-1 (NDSH-15)	2003	ANGRAU, Nandyal	1400	40	Andhra Pradesh	Early maturing hybrid
RSFH-1	2004	UAS, Raichur	1300-1500	39-41	Karnataka (Zones 2 and 3)	High oleic hybrid
HSFH-848	2005	HAU, Hisar	1800-2400	41-42	Haryana	Dwarf hybrid
DRSH-1	2006	DOR, Hyderabad	1300-1600	42-44	All India	High oil
KBSH-53	2009	UAS, Bangalore	1700-2700	42-44	Karnataka	Resistant to powdery mildew
PSFH-569	2009	PAU Ludhiana	2232	40	Punjab	High oil, early hybrid

Source: Recent varieties and hybirds of annual oilseeds recommended for different states (2001-2010), Directorate of Oilseeds Research

Seed Yield

Seed yield *per se* is a low heritablity trait that is highly influenced by environment and is controlled by multiple genes. Number of seeds per head and weight of 1000 seeds are the primary traits that determine yield in sunflower. These traits may have higher heritability compared to yield *per se* but may be negatively correlated. Despite complications, seed yield of sunflower hybrids has substantially improved over the past 40 years due to breeding efforts. For instance, in the year 2000, some of the sunflower varieties in France showed 140% gain as compared to during 1965-1975 (Vear, 2010).

Oil Content and Quality

Oil content is considered to be a highly heritable trait. Therefore, selection can be made on individual plants from early generations of a breeding programme. But breeding for high oil content *per se* is complicated because seed yield and oil content are negatively correlated. There was no significant improvement in oil content *per se* but the gain for oil yield per hectare was 148% from 1970 to 2000, which may be attributed to the balanced improvements in seed yield and oil content (Vear, 2010). During the last 30 years, fatty acid components have been extensively modified in sunflower. A variety of specialty oil types, which have different levels and/or combinations of fatty acid components as well as tocopherols were bred. Genetics of fatty acid and tocopherol traits have been well worked out. Genes that control oil quality traits include *Ol1* and several modifier genes for high levels of oleic acid content; *P1*, *P2* and *P3* for high palmitic acid; *Es1*, *Es2*, *Es3* for high stearic acid content; *Tph1* and *Tph2* for altered tocopherol composition (Fernández-Martínez *et al.*, 2007). Breeding for enhanced oil quality has been successful because of the smaller number of genes controlling the traits.

Earliness

Cultivars with different maturity periods are needed to suit the specific requirements of different regions. In general, hybrids that mature in less than 100 days from planting to maturity are classified as early, 100-120 days as medium and 120-140 days as late. Kaya *et al.* (2004) recommended a standard set of hybrids for different maturity groups to classify maturity of new lines and hybrids. Conventional breeding for earliness involves development of inbred lines with acceptable flowering onset for the region followed by selection among lines based on seed moisture content at harvest of the hybrids compared with their seed yield (Vear, 2010).

Lodging Resistance and Dwarfness

Dwarf or semidwarf cultivars are needed to avoid yield loss due to lodging and stalk breakage caused by excessive growth. In India, Morden (Cerenianka-66) is the only dwarf (66–110 cm) and early maturing variety available for cultivation. Other dwarf sources include DDR (90.3 cm), Donsky (65.5 cm), and Donskoi 47 (79.8 cm) and a mutant derived from Morden (35.7 cm) (Jagadeesan *et al.*, 2007).

Resistance to Diseases

Breeding for resistance (or at least tolerance) to pathogens is always an aim in sunflower breeding. Brief account of some of the important diseases is given below.

Alternaria blight *(Alternaria helianthi)*

Lack of sources of resistance to *A. helianthi* has been a major constraint in sunflower breeding. Resistance appears to be quantitative; hence the usefulness of such levels of resistance may not be effective under epidemic conditions. However, resistance to *A. helianthi* has been reported in the perennial species like *H. maximiliani, H. mollis, H. divaricatus, H. simulans* and *H. occidentalis* (diploids), *H. pauciflorus* and *H. decapetalus* (tetraploids) and *H. resinosus* and *H. tuberosus* (hexaploids); subsequently, interspecific cross derivatives were developed (Sujatha *et al.*, 1997; Sujatha and Prabakaran, 2006). Reddy *et al.* (2006) reported that RHA 587 had high level of resistance (1-5% infection) against *Alternaria* in both field and laboratory conditions, which may be a good source for breeding programmes.

Rust (Puccinia helianthi)

P. helianthi is a dynamic pathogen and has several races worldwide. Excellent sources of resistance to rust have been identified in cultivated sunflower and some of this resistance has been transferred into released germplasm (Miller and Gulya, 2001). Several genes conferring resistance to rust have also been identified in sunflower that include *R1*, *R2*, *R3*, *R4*, *R5*, *R11*, *Pu6*, and R_{adv} (Qi *et al.*, 2011). In India, Sujatha *et al.* (2003) reported that interspecific cross derivatives PS 1089 (derived from *H. argophyllus* x cultivated sunflower) and PS 2011 and PS 2032 (derived from *H. petiolaris* x cultivated sunflower) were immune to rust isolate at DOR, Hyderabad.

Powdery mildew (Golovinomyces cichoracearum)

Jan and Chandler (1985) incorporated powdery mildew resistance, which was incompletely dominant, into a cultivated background. This was released as germplasm PM1 (Jan and Chandler, 1988). The species *H. argophyllus* and *H. debilis* were identified to possess at least two genes. The action of major genes in powdery mildew resistance from *H. debilis* ssp. *debilis* has been investigated (Jan and Chandler, 1985). In India, powdery mildew has become an important disease of sunflower only recently. The following sources of resistance to powdery mildew have been identified at DOR, Hyderabad - three annual wild species (*H. argophyllus, H. agrestis, H. debilis*), six perennials (*H. angustifolius, H. atrorubens, H. rigidus, H. salicifolius* Dietr., *H. pauciflorus* Nutt and *H. resinosus* Small), two interspecific derivatives (ID-16, ID-25), and two exotic lines with multiple resistance (TX16R, EC-537925) (Prathap Reddy *et al.*, 2013).

Downy mildew (Plasmopara halstedii)

Downy mildew is also a very dynamic pathogen, which has several races around the world. At least 24 downy mildew resistance genes (*Pl1–16, Plv, Plw, Plx-z, Mw, Mx* and *PlArg*) have been identified so far in cultivated and wild sunflowers. Of these, only the *Pl8*, *PlArg* and *Pl15* genes are reported to be resistant to all *P. halstedii* races (Liu *et al.*, 2012). Genetic stocks/varieties with multiple resistance genes have been developed by backcross breeding programmes (Hulke *et al.*, 2010; Jocić *et al.*, 2010). In India, downy mildew was reported around 1986 in Marathwada region of Maharashtra State where sunflower is extensively grown. Hybrids resistant to downy mildew were released for commercial cultivation, which substantially reduced the downy mildew incidence in India (Shirshikar, 2005).

Sunflower Necrosis Disease (SND)

SND was first observed in India during 1997 and caused significant yield losses. No reliable sources of resistance to SND in cultivated sunflower have been identified yet. Artificial screening experiments indicated that wild species, *H. occidentalis* and *H. maximiliani* show immune type of reaction against SND (M. Sujatha, unpublished). Transgenics conferring resistance to necrosis disease through deployment of the TSV coat protein gene have been developed at the Directorate of Oilseeds Research.

More information on the global scenario of the progress made in breeding for resistance to diseases in sunflower is available in the Proceedings of the International Symposium on “Sunflower Breeding on Resistance to Diseases" organized by All-Russia Research Institute of Oil Crops by V.S. Pustovoit (VNIIMK) and The International Sunflower Association (ISA) at Krasnodar, Russia June 23-24, 2010.

Resistance to Insect Pests

For sunflower in India, only a few insect species have become economically important for causing yield losses. These include *Helicoverpa armigera* and the tobacco caterpillar (*Spodoptera litura* (Fabr.). Genetic variability for resistance to *S. litura* is limited in the cultivar germplasm of sunflower. Sujatha and Lakshminarayana (2007) reported that *H. occidentalis* and *H. argophyllus* were immune with no leaf damage under field conditions. Laboratory bioassays confirmed resistance in eight species *viz., H. occidentalis, H. argophyllus, H. tuberosus, H. maximiliani, H. mollis, H. simulans, H. divaricatus* and *H. hirsutus*.

Tolerance to Abiotic Stresses

Škorić (2009) reviewed the progress made in sunflower breeding for tolerance to abiotic stresses.

Sunflower is considered to be a moderately sensitive crop to drought stress. Developing high-yielding cultivars that flower and mature before soil water conditions become limiting may be a useful approach. The genetics of drought tolerance has not been studied thoroughly in sunflower. Sources of drought tolerance were found in *H. argophyllus* and *H. deserticola*. Breeding for stay green has been useful for drought tolerance. Some of the stay green inbred lines include HA-48, HA-22, CMS-1-50, PH-BC-2-91, PR-ST-3, RHA-SES and RHA-583. Use of *H. argophyllus* in breeding for drought tolerance has been challenging because it was difficult to preserve the original structure (pubescence) of *H. argophyllus* leaves in the backcrosses progenies. Other species such as *H. deserticola, H. hirsutus, H. maximiliani,* and *H. tuberosus* may also be explored (Škorić, 2009).

Private Sector Hybrids in India

In India, private seed sector played a leading role in hybrid development and seed production in sunflower. In the year 1990-91, 22,000 tonnes of private hybrid seeds and about 4900 tonnes of varieties were marketed. No public bred hybrid seed was available (Gadwal, 2003). Now, more than 95% of hybrid seeds in the country are produced by private sector as against only about 5% from the public sector (Seed Section, DOR).

Molecular Breeding

Genomic resources

Genetic linkage maps of varying densities have been constructed in wild and cultivated sunflowers using RFLP, RAPD, AFLP and SSR markers. As of now, about 2,040 SSR markers have been developed (as reviewed by Sujatha and Sujatha, 2012). Lai *et al.* (2005) mined more than 67,000 ESTs in sunflower for SNP discovery and mapped 243 SNP markers. Bachlava *et al.* (2012) developed an array of 10,640 SNPs through transcriptome re-sequencing. Burke *et al.* (2012) developed a consensus genetic linkage map containing *ca.* 8500 SNPs along with *ca.* 1500 previously mapped SSRs. This SNP array was used to construct genetic linkage maps of the *H. argophyllus* and *H. niveus* spp. *tephroides*, and to association mapping within the sunflower gene pool. Through National Sunflower Association (NSA) funded initiative, a SNP chip consisting of 8,700 SNP markers has been developed and validated in a diverse panel of 1,200 sunflower accessions (Venkatramana *et al.*, 2012). The NSA, in conjunction with the USDA and sunflower seed companies, has established the NSA Sunflower SNP Consortium to help sunflower breeders. Availability of fully sequenced sunflower genome would expedite large scale discovery of robust markers such as SSR and SNP for breeding applications. Significant progress has also been made towards developing a reference genome sequence in sunflower. A physical map that covers more than 85% of the genome has been developed (Kane *et al.*, 2011). A TILLING population for high throughput identification of EMS-induced point mutations in sunflower genome has been developed, which may facilitate a better understanding of gene function in sunflower (Sabetta *et al.*, 2011).

Molecular genetic diversity analysis

Marker-based genetic diversity analysis in germplasm collections can be helpful for reliable classification of accessions and identification of core accessions for specific breeding purposes. In sunflower, molecular markers are widely used for identification of inbred lines, cultivars and wild species (Liu and Burke, 2006), for analysis of genetic diversity and relationships (Yue *et al.*, 2009), population structure (Mandel *et al.*, 2011) and linkage disequilibrium (Fusari *et al.*, 2008).

Genetic mapping of agronomic traits

A wide range of molecular markers such as RFLP, AFLP, SSR and TRAP developed in sunflower have led to mapping of several major genes and QTLs for agronomically important traits. So far, progress has been made in mapping of major genes for the traits - resistance to downy mildew disease – *Pl1, Pl2, Pl5, Pl6, Pl7, Pl8, Pl13, Pl16,* Pl_{ARG}; resistance to Orobanche – *Or5*, resistance to chlorotic mottle virus- *Rcmo1*, resistance to rust disease - *R1, R2, R4,* R_5, *R11* and *Radv*; tocopherol content – *Tph1* (beta), *Tph2* (gamma); stearic acid content - *Es1, Es2 and Es3*; fertility restoration – *Rf1, Rf3, Rf4, Rf5*; oleic acid – *Ol*; nuclear male sterility - *Ms10, ms11*; branching – *b1*; branching and pericarp pigment – *Hyp*; Lemon ray flower color *Yf, Yf1* and chlorophyll deficiency – *Yl*. QTLs have been reported for + traits such as photoperiod response and flowering, oil content, resistance to downy mildew, black stem, mid stalk rot and basal stem rot, drought, salinity or chilling stresses, early domestication, germination and seedling development, plant height, self-incompatibility, self-pollination and seed dormancy and seed quality (Schuppert *et al.*, 2006; Davar *et al.*, 2010; Wieckhorst *et al.*, 2010; Yue *et al.*, 2010; Lawson *et al.*, 2011; Qi *et al.*, 2011; Liu *et al.*, 2012; Qi *et al.*, 2012a; 2012b and as reviewed by Sujatha and Sujatha, 2012).

Marker assisted selection (MAS)

A few examples highlight MAS having been successfully practiced in sunflower. Lawson *et al.* (1998) applied SCAR markers to select for rust resistance genes *R1* and R_{adv}. Jociæ *et al.* (2010) applied marker-assisted backcrossing to incorporate *Pl6* and *Pl7* genes into commercial sunflower lines using co-dominant CAPS markers developed by Pankovic *et al.* (2007). Marker-assisted gene introgression or pyramiding of desirable genes appears to be very promising in sunflower given the diversity of genes that have been genetically mapped for various traits.

Genetic seed purity testing

Seed quality is of great importance for enhancing productivity, especially in cross-pollinated crops like sunflower where most of its cultivated area is with hybrids. In sunflower, seed quality problems are often encountered due to spurious parental lines multiplication and production of certified hybrid seed. Research at various Centres under AICRP (Sunflower) resulted in development of appropriate seed production technologies for the maintenance of parental lines as well as for certified seed production of hybrids/varieties. Systematic procedures with respect to season, row ratio of female and male plants, isolation distance for breeder, foundation and certified seed production were developed.

Molecular markers proved to be useful in circumventing the difficulties associated with traditional grow-out-test (GOT) for genetic purity testing of hybrid seed lots. In sunflower, SSR markers have been used to distinguish inbred lines and hybrids (Antonova *et al.*, 2006; Pallavi *et al.*, 2011), which in turn could be used for assessing the genetic purity in seed lots.

In Vitro Techniques

Homozygous lines assume great importance in commercial hybrid production for fixing heterozygosity and also in development of dihaploid lines for mapping desirable traits. Anther and microspore cultures allow acceleration of breeding programmes by providing homozygous doubled haploids within a comparatively short time. In sunflower, anther culture still needs considerable improvement as the regeneration rates reported till date are very low (@ 10%) and the anther culture response is strongly influenced by physical, nutritional, physiological and genetic factors.

A prerequisite for genetic transformation is the availability of a reproducible and highly efficient protocol of shoot regeneration. During the past three decades, there have been several reports of plant regeneration from sunflower through direct and callus-mediated organogenesis and somatic embryogenesis from zygotic embryos and seedling tissues. The regeneration protocols described thus far are plagued with problems of genotype dependence, low rate of plant regeneration and reproducibility, precocious flowering, vitrification and poor rooting, lengthy culture time, abnormal morphogenesis, etc. A genotype-independent regeneration system through which transgenic plants could be produced would overcome a major bottleneck in sunflower. Owing to difficulties in adventitious shoot regeneration, most of the transformation experiments have relied on the use of shoot apices and cotyledonary nodes. Problems that need to be overcome include lack of an efficient adventitious shoot regeneration system, low *Agrobacterium* virulence, low transformation rates, lack of a stringent selection system for transformants, unusual sensitivity to antibiotics, genotype dependence of regeneration efficiency, gene instability and low expression levels of transgenes that lead to development of chimeras. Till date there are no GM varieties authorized for commercial use in sunflower.

Current Challenges in Sunflower Breeding

A major challenge would be to develop superior hybrids and populations which can surpass the presently grown hybrids in both seed yield and oil content. These hybrids should combine high seed yield, high oil content, bold seeds with high 100 seed weight, resistance to *Alternaria* leaf spot, powdery mildew, necrosis and downy mildew. There is also an immediate need to develop early maturing hybrids of 75-80 days duration superior to Morden. Most of the sunflower breeding research in USA and European countries is supported by private sector and complemented with molecular breeding tools. There is a need to have such initiatives in Indian sunflower programmes to exploit the biotechnological tools for accelerating the breeding processes using marker assisted breeding and in transferring genes from alien sources through wide hybridization and transgenic approach.

REFERENCES

Antonova, T.S., Guchetl, S.Z., Tchelustnikova, T.A. and Ramasanova, S.A. (2006). Development of marker system for identification and certification of sunflower lines and hybrids on the basis of SSR-analysis. *Helia*, 29: 63–72.

Bachlava, E., Taylor, C.A., Tang, S., Bowers, J.E., Mandel, J.R., Burke, J.M. and Knapp, S.J. (2012). SNP discovery and development of a high-density genotyping array for sunflower. *PLoS ONE* 7(1): e29814. doi:10.1371/journal.pone.0029814.

Burke J.M., Mandel, J.R., Bowers, J.E., Bachlava, E., Marek, L., Nambeesan, S., Taylor, C.A., Tang, S., Rieseberg, L.H. and Knapp, S.J. (2012). SNP diversity, genome mapping, and association genetics in sunflower. *In*: Proceedings of the 18th International Sunflower Conference, February 27 - March 1 / 2012, Argentina.

Davar, R., Darvishzadeh, R., Majd, A., Ghosta, Y. and Sarrafi, A. (2010). QTL mapping of partial resistance to basal stem rot in sunflower using recombinant inbred lines. *Phytopathol. Mediterr.*, 49: 330–341.

Fernández-Martínez, J.M., Pérez-Vich, B., Velasco, L. and Domínguez, J. (2007). Breeding for specialty oil types in sunflower. *Helia*, 30(46): 75–84.

Fernández-Martínez, J.M., Domínguez, J., Pérez-Vich, B. and Velasco, L. (2010). Update on breeding for resistance to sunflower broomrape. *Helia*, 33: 1–12.

Fusari, C.M., Lia, V.V., Hopp, H.E., Heinz, R.A. and Paniego, N.B. (2008). Identification of single nucleotide polymorphisms and analysis of linkage disequilibrium in sunflower elite inbred lines using the candidate gene approach. *BMC Plant Biol.*, 8:7 doi:10.1186/1471-2229-8-7.

Gadwal, V.R. (2003). The Indian seed industry: Its history, current status and future. *Current Sci.*, 84: 399–406.

Horn, R., Kusterer, B., Lazarescu, E., Prüfe, M., Özdemir, N. and Friedt, W. (2002). Molecular diversity of CMS sources and fertility restoration in the genus. *Helianthus. Helia*, 25: 29–40.

Hulke B.S., Miller, J.F., Gulya, T.J. and Vick, B.A. (2010). Registration of the oilseed sunflower genetic stocks HA 458, HA 459, and HA 460 possessing genes for resistance to downy mildew. *J. Plant Reg.*, 4: 1.

Jagadeesan, S., Kandasamy, G., Manivannan, N. and Muralidharan, V. (2008). A valuable sunflower dwarf mutant. *Helia*, 31: 79–82.

Jan, C.C. and Chandler, J.M. (1985). Transfer of powdery mildew resistance from *Helianthus debilis* Nutt. to cultivated sunflower. *Crop Sci.,* 25: 664–666.

Jan, C.C. and Chandler, J.M. (1988). Registration of powdery mildew resistant sunflower germplasm pool, PM 1. *Crop Sci.,* 28: 1040.

Jociæ S., Cvejiæ, S., Hladni, N., Miladinoviæ, D. and Miklie̊, V. (2010). Development of sunflower genotypes resistant to downy mildew. *Helia*, 33: 173–180.

Kane, N.C, Gill, N., King, M.G., Bowers, J.E., Berges, H., Gouzy, J., Bachlava, E., Langlade, N.B., Lai, Z., Stewart, M., Burke, J.M., Vincourt, P., Knapp, S.J. and Rieseberg, L.H. (2011). Progress towards a reference genome for sunflower. *Botany*, 89: 429–437.

Kaya Y., Baltensperger, D., Nelson, L. and Miller, J. (2004). Maturity grouping in sunflower (*Helianthus annuus* L.). *Helia*, 27: 257–270.

Kolkman, J.M., Berry, S.T., Leon, A.J., Slabaugh, M.B., Tang, S., Gao, W., Shintani, D.K., Burke, J.M. and Knapp, S.J. Single nucleotide polymorphisms and linkage disequilibrium in sunflower. *Genetics*, 177: 457–468.

Lai, Z., Livingstone, K., Zou, Y., Church, S.A., Knapp, S.J., Andrews, J. and Rieseberg, L.H. (2005). Identification and mapping of SNPs from ESTs in sunflower. *Theor. Appl. Genet.,* 111: 1532–1544.

Lawson, W.R., Jan, C.C., Shatte, T., Smith, L., Kong, G.A. and Kochman, J.K. (2011). DNA markers linked to the R_2 rust resistance gene in sunflower (*Helianthus annuus* L.) facilitate anticipatory breeding for this disease variant. *Mol. Breed.,* 28: 569–576.

Lawson, W.R., Goulter, K.C., Henry, R.J., Kong, G.A. and Kochman, J.K. (1998). Marker assisted selection for two rust resistance genes in sunflower. *Mol. Breed.,* 4: 227–234.

Leclercq, P. (1969). Une stérilité mâle chez le tournesol. *Ann. Amélior Plantes*, 19: 99–106.

Liu, A. and Burke, J.M. (2006). Patterns of nucleotide diversity in wild and cultivated sunflower. *Genetics*, 173: 321–330.

Liu, Z., Mulpuri, S., Feng, J., Vick, B.A. and Jan, C.C. (2012). Molecular mapping of the Rf_3 fertility restoration gene to facilitate its utilization in breeding confection sunflower. *Mol. Breed.,* 29: 275–284.

Mandel, J.R., Dechaine, J.M., Marek, L.F. and Burke, J.M. (2011). Genetic diversity and population structure in cultivated sunflower and a comparison to its wild progenitor, *Helianthus annuus* L. *Theor. Appl. Genet.,* 123: 693–704.

Miller, J.F. and Gulya, T.J. (2001). Registration of three rust resistant sunflower germplasm populations. *Crop Sci.,* 41: 601.

Nayak A., Gracy, C.P., Nagashree, N. and Girisha, K. (2010). An overview of sunflower in India. http://www.krishisewa.com/articles/2012/sunflower3.html

Pallavi, H.M., Gowda, R., Shadakshari, Y.G., Bhanuprakash, K. and Vishwanath, K. (2011). Identification of SSR markers for hybridity and seed genetic purity testing in sunflower (*Helianthus annuus* L.). Helia, 34(54): 59–66.

Pankovic, D., Radovanovic, N., Jocic, S., Satovic, Z. and Škoric, D. (2007). Development of codominant amplified polymorphic sequence markers for resistance of sunflower to downy mildew race 730. *Plant Breed.,* 126: 440–444.

Qi, L.L., Hulke, B.S., Vick, B.A. and Gulya, T.J. (2011). Molecular mapping of the rust resistance gene R_4 to a large NBS–LRR cluster on linkage group 13 of sunflower. *Theor. Appl. Genet.,* 3: 351–358.

Qi, L.L., Gulya, T.J., Hulke, B.S. and Vick, B.A. (2012a). Chromosome location, DNA markers and rust resistance of the sunflower gene R_5. *Mol. Breed.,* 30: 745–756.

Qi, L.L., Seiler, G.J., Vick, B.A. and Gulya, T.J. (2012b). Molecular mapping of $R_{(11)}$ gene conferring resistance to recently emerged rust races, tightly linked to male fertility restoration, in sunflower (*Helianthus annuus* L.). *Theor. Appl. Genet.,* 125: 921–932.

Prathap Reddy, K., Chander Rao, S., Kirti, P.B. and Sujatha, M. (2013). Development of a scoring scale for powdery mildew (*Golovinomyces cichoracearum* (DC.) V.P. Heluta) disease and identification of resistance sources in cultivated and wild sunflowers. *Euphytica* 190: 385–399.

Reddy, C.V.C.M., Reddy, A.V.V., Sinha, B. and Shanta Lakshmi, M. (2006). Screening of sunflower genotypes for resistance against *Alternaria* Blight. *Asian J. Plant Sci.,* 5: 511–515.

Sabetta W., Alba, V., Blanco, A. and Montemurro, C. (2011). sunTILL: A TILLING resource for gene function analysis in sunflower. *Plant Methods,* 7: 20. http://www.plantmethods.com/content/7/1/20

Schuppert, G., Tang, S., Slabaugh, M. and Knapp, S.J. (2006). The sunflower high–oleic mutant *Ol* carries variable tandem repeats of *FAD2–1*, a seed–specific oleoyl–phosphatidyl choline desaturase. *Mol. Breed.,* 17: 241–256.

Seiler, G. and Jan, C.C. (2010). Basic information. *In*: Hu, J., Seiler, G. and Kole, C. (*Eds.*) Genetics, Genomics and Breeding of Sunflower. Science Publishers, USA, pp. 1–50.

Shirshikar, S.P. (2005). Present status of sunflower downy mildew disease in India. *Helia*, 28: 153–158.

Škorić, D. (2009). Sunflower breeding for resistance to abiotic stresses. *Helia*, 32: 1–16.

Sujatha, M., Prabakaran, A.J. and Chattopadhyay, C. (1997). Reaction of wild sunflowers and certain interspecific hybrids to *Alternaria helianthi*. *Helia*, 20: 15–24.

Sujatha, M., Prabakaran, A.J., Sudhakara Babu, S.N. and Rama Devi, R. (2003). Differential reaction of recombinant interspecific inbred lines of sunflower to red rust incited by *Puccinia helianthi*. *Helia*, 26: 25–36.

Sujatha, M. and Prabakaran, A.J. (2006). Ploidy manipulation and introgression of resistance to *Alternaria helianthi* from wild hexaploid *Helianthus* species to cultivated sunflower (*H. annuus* L.) aided by anther culture. *Euphytica*, 152: 201–215.

Sujatha, M. and Lakshminarayana, M. (2007). Resistance to *Spodoptera litura* (Fabr.) in *Helianthus* species and backcross derived inbred lines from crosses involving diploid species. *Euphytica*, 155: 205–213.

Sujatha, K. and Sujatha, M. (2012). Present status and future prospects of molecular markers and marker assisted selection in sunflower (*Helianthus annuus* L.) *The J. Plant Sci. Res.,* 28: 7–26.

Vear, F. (2010). Classic genetics and breeding. *In*: Genetics, Genomics and Breeding of Sunflower. *In:* Hu, J., Seiler, G. and Kole, C. (*Eds.*) Genetics, Science Publishers, USA, pp. 51–77.

Venkatramana, P., Hulke, B., Qi, L. and Schultz, Q. (2012). SNP assisted breeding for sunflower improvement. *In*: Proceedings of the 18th International Sunflower Conference, February 27 – March 1/2012, Argentina.

Wieckhorst, S., Bachlava, E., Dußle, C.M., Tang, S., Gao, W., Saski, C., Knapp, S.J., Schön, C.C., Hahn, V. and Bauer, E. (2010). Fine mapping of the sunflower resistance locus Pl_{ARG} introduced from the wild species *Helianthus argophyllus*. *Theor. Appl. Genet.,* 121: 1633–1644.

Yue, B., Vick, B.A., Cai, X. and Hu, J. (2010). Genetic mapping for the Rf_1 (fertility restoration) gene in sunflower (*Helianthus annuus* L.) by SSR and TRAP markers. *Plant Breed.,* 129: 24–28.

Yue, B., Cai, X., Vick, B.A. and Hu, J. (2009). Genetic diversity and relationships among 177 public sunflower inbred lines assessed by TRAP markers. *Crop Sci.,* 49: 1242–1249.

14

Groundnut

A. Bandyopadhyay[1] and P. Manivel[2]

[1] National Fund for Basic Strategic and Frontier Research in Agriculture, Krishi Anusandhan Bhawan, New Delhi, India
[2] Directorate of Medicinal and Aromatic Plants Research, Anand, India

There has been a declining trend in area under groundnut. In 2001 the area was about 8 million ha but in 2008 the area has been estimated to be about 6.41 m ha with a production of about 9.36 m tonnes (DAC, 2008). Groundnut accounts for about 24.2% of area and 38% of production under oil seeds. Various socioeconomic factors have contributed to this declined area and production but lack of concerted efforts to apply scientific advances for groundnut improvement has definitely been a major factor.

The decennial annual compound growth rate of productivity in India during 1991-2000 had been 6.64% (Singh and Basu, 2004). A study of changes that occurred in the yield components of cultivars of groundnut released from 1905 to 1995, shows a trend of pod yield improvement of cultivars released after 1985 than those released earlier (Manivel *et al.*, 2000). Though as many as 69 cultivars have been notified from 1995 to 2010 (possibly the highest rate per year), their productivity does not establish a definite upward trend (Table 1). There has been little shift in breeding strategy for groundnut improvement. Even now, in most of the high yielding groundnut varieties released globally , productivity increase has resulted from higher harvest index obtained through reduction of total biomass (Zhu *et al.*, 2010). Breeding for high biomass combined with higher harvest index should have been the strategic shift of breeding methodology for achieving higher groundnut yields. This strategy necessitates the development of genotypes with resistance to stresses.

V.L. Chopra (Ed.): Breeding Field Crops II: Advances, Studium Press (LLC), USA.

Table 1: List of recently released groundnut varieties

Name of cultivar	*Habit**	*Pedigree/ Parentage*	*Developing centre*	*Year of release*	*Area of adoption*	*Pod yield (tons/ha)*
Smruti (OG 52-1)	SB	Pollachi × RS 60	OUA&T, Chiplima	1995	Orissa	1.98 (kharif); 3.62 (rabi-summer)
Tirupati 4 (TCGS 30)	SB	JL 24 × Ah.316/S	ANGRAU, Tirupati	1995	Andhra Pradesh	1.6 (kharif); 2.49 (rabi-summer)
R 9251 (KRG-3)	SB	JLM 1 × TG 23	UAS, Raichur	1997	Karnataka	2.35
Kadiri 4 (K 150)	SB	Dh 3-30 × NCAC -2230	ANGRAU, Kadiri	1997	Andhra Pradesh	2.30
JGN 3 (Jawahar groundnut 3)	SB	Jyoti x JL 24	JNKVV, Khargone	1997	Madhya Pradesh	1.53
ALR 3 (ALG 63)	SB	(R-33-1 × ICG (FDRS) 68 × (NcAc 17090 × ALR 1)	TANU, Aliyar Nagar	1999	Tamil Nadu	1.93
Co 3 (TNAU 256)	SB	VRI 3 × JL 24	TNAU, Coimbatore	1999	Tamil Nadu	1.75 (kharif); 2.10 (rabi-summer)
GG 5	SB	27-5-1 × JL24	JAU, Junagadh	1999	Gujarat	1.27
Manjra (LGN 2)	VB	R 33-1 × TMV 10	MAU, Latur	2000	Gujarat & Southern Rajasthan	1.50-2.00
VRI (Gn) 5	SB	CG 26 × ICGS 44	TNAU, Vriddhachalam	2001	Tamil Nadu	2.13 (kharif); 2.38 (rabi-summer)
Co (Gn) 4 (TNAU 269)	SB	TMV 10 × ICGS 82	TNAU, Coimbatore	2001	Tamil Nadu	1.50 (kharif); 1.95 (rabi-summer)
GG 7 (J-38)	SB	S 206 × FESR 8 1-1-9-B-B	JAU, Junagadh	2001	Gujarat & Southern Rajasthan	2.15
AK 159	SB	JL 24 × CGC-4018	PKV, Akola	2002	Maharashtra and M.P.	1.60
Sneha	VB	IES 883 × JL 24	KAU, Vellayani	2002	Kerala	3.82
Snigdha	VB	Dh (E) 32 × JL 24	KAU, Vellayani	2002	Kerala	3.55
GG 6	SB	CGC 3 × FESR-5-P6-b1-b1-1-1-1-B-B	JAU, Junagadh	2003	Gujarat	2.78
GG 14 (JSP 28)	VR	GG11 × R 33-1-90-215-1-B-2-B-B	JAU, Junagadh	2003	Northern Rajasthan, Punjab, Haryana & U.P.	2.16

Table 1: (*Contd....*)

Table 1: *(Contd....)*

Name of cultivar	***Recommended seasons***	***No. of days to maturity***	***Shelling (%)***	***Oil (%)***	***HKW (g)***	***Reaction to major pests and diseases***	***Special characters***
Smruti (OG 52-1)	Kharif and rabi-summer	110	70.0	48.0	–	Resistant to collar rot, stem rot, rust and leaf spots	Suitable for residual moisture situation
Tirupati 4 (TCGS 30)	Kharif and rabi-summer	100-105	72-76	48.8	42-47	–	Early maturity
R 9251 (KRG-3)	Kharif	90-105	71.0	49.0	35.0	Tolerant of PBND	Early maturing
Kadiri 4 (K 150)	Rabi-summer	110-115	74.0	48.0	32.0	Tolerant of LLS and PBND	–
JGN 3 (Jawahar groundnut 3)	Kharif	104	72.0	49.6	32.4	Tolerant of sucking pests and foliar diseases.	Drought tolerant
ALR 3 (ALG 63)	Kharif	110-115	69.0	50.0	36.0	Tolerant of LLS and resistant to rust	Suitable for early sowing of south-west monsoon
Co 3 (TNAU 256)	Kharif and rabi-summer	105-110	71.4	49.2	61.0	Tolerant of PBND	Bold kernel
GG 5	Kharif	110	73.7	49.0	44.0	–	Drought tolerant; leaflets stay green at maturity
Manjra (LGN 2)	Kharif	115-120	65-70	49.0	43-48	–	–
VRI (Gn) 5	Kharif and rabi-summer	105-110	74.0	50.0	40.0	Resistant to rust & LLS	–
Co (Gn) 4 (TNAU 269)	Kharif and rabi-summer	105-110	73.0	53.0	52.0	Resistant to rust & LLS	High oil content
GG 7 (J-38)	Kharif	100	76.0	49.0	48.0	Tolerant of LLS	Early maturity
AK 159	Kharif	106	66-68	51.0	34.0	–	High oil content
Sneha	Kharif	115				–	Early maturity
Snigdha	Kharif	115				Resistant to pest and diseases	–
GG 6	Rabi-summer	100	73.0	50.0	45.0	–	Early maturity
GG 14 (JSP 28)	Kharif	123	65.0	52.0	49.0	Tolerant of thrips, *Spodoptera*, leaf miner	–

Table 1: *(Contd....)*

Table 1: *(Contd....)*

Name of cultivar	*Habit**	*Pedigree/ Parentage*	*Developing centre*	*Year of release*	*Area of adoption*	*Pod yield (tons/ha)*
TPG 41	SB	TG 28 A × TG 22	BARC, Trombay	2004	All India	2.08
TG 37A	SB	TG 25 × TG 26	BARC, Trombay	2004	Gujarat, Rajasthan, U.P., Punjab, Orissa, W.B., Bihar and Assam	2.08 (kharif); 2.84 (rabi-summer)
Vikas (GPBD 4)	SB	KRG 1 × ICGV 86855	UAS, Dharwad	2004	Maharashtra, Karnataka, A.P. & T.N.	2.20
TLG 45	SB	TG 19 × TAG 24	MAU, Latur	2004	Maharashtra	1.50
SG 99	SB	ICGV 86829 × ICGV 87160 (ICG (FDRS) 10	PAU, Ludhiana	2004	Punjab	2.2 (kharif); 2.50 (rabi-summer)
Phule Unap (JL 286)	SB	JL 86 × NCAC 343	MPKV, Jalgaon	2004	Maharashtra	2.23
Prutha (Dh 86)	SB	Dh 40 × Dh 8	UAS, Dharwad	2005	Gujarat, Southern Rajasthan, Maharashtra, Orissa and W.B.	4.02
Kadiri 5	SB	JL 24 × VG 55-7	ANGRAU, Kadiri	2005	Andhra Pradesh	2.20
Kadiri 6 (K 1240)	SB	JL 24 × Ah 316/S	ANGRAU, Kadiri	2005	Andhra Pradesh	3.100-3.800
Pratap Mugphali 2 (ICUG 92195)	SB	ICGV 86055 × ICG (FDRS) 10	MPUA&T, Udaipur	2005	Rajasthan	1.80-2.80
Pratap Mugphali 1 (ICUG 92035)	SB	ICGV 86033 × ICG 2214	MPUA&T, Udaipur	2005	Rajasthan	2.50-3.00
Ratneshwar (LGN 1)	SB	JL 24 × NCAC 17090	MAU, Latur	2005	Maharashtra	1.49
Co(GN) 5	VB	(Co 2 × ICGS 86010) × (Co 2 × VG 119 × ICGS 50)	TNAU, Vriddhachalam	2005	Tamil Nadu	1.58

Table 1: *(Contd....)*

Table 1: (*Contd....*)

Name of cultivar	***Recommended seasons***	***No. of days to maturity***	***Shelling (%)***	***Oil (%)***	***HKW (g)***	***Reaction to major pests and diseases***	***Special characters***
TPG 41	Rabi-summer	122	69.0	49.0	65.0	Tolerant of rust	Large seeds (>60 g/100 kernel); high O/L ratio (3.27); possesses fresh seed dormancy (up to 25 days)
TG 37A	Kharif and rabi-summer	105-110	64.0	48.0	39.0	Tolerant to collar rot, rust and LLS	Possesses fresh seed dormancy (up to 15 days)
Vikas (GPBD 4)	Kharif	107	68.0	49.0	36.0	Resistant to LLS and rust	-
TLG 45	Kharif	114	66.0	51.0	59.0	–	Large seeded
SG 99	Kharif and rabi-summer	102	65.0	52.3	42.0	Tolerant of PBND	-
Phule Unap (JL 286)	Rabi-summer	93-95	72-75	49-50	37-38	Tolerant of LLS, rust and Stem rot, tolerant of thrips, leaf miner & *Spodoptera litura*	Early maturity
Prutha (Dh 86)	Rabi-summer	125-127	68.0	48.0	43.0	Tolerant of tikka and sucking pests	Semi dwarf with high harvest index (HI)
Kadiri 5	Kharif	100-110	70.0	48.0	44.0	Tolerant of LLS	Drought tolerant
Kadiri 6 (K 1240)	Rabi-summer	100-105	70.0	49.0	39.0	Tolerant of leaf spots	
Pratap Mugphali 2 (ICUG 92195)	Kharif	105	69	48-50	38	Tolerant of ELS, LLS, PBND, *S. litura*, leaf miner and thrips	–
Pratap Mugphali 1 (ICUG 92035)	Kharif	108	62.0	48-50	40.0	Tolerant of ELS, LLS, PBND, *S. litura*, leaf miner and thrips	–
Ratneshwar (LGN 1)	Kharif	105	69.0	51.2	32.0	Tolerant of LLS, stem rot, rust and PBND	–
Co(GN) 5	Kharif	125	73.8	53.8	46.0	Tolerant of rust and PBND, leaf miner and *Spodoptera*	–

Table 1: (*Contd....*)

Table 1: (*Contd....*)

Name of cultivar	*Habit**	*Pedigree/ Parentage*	*Developing centre*	*Year of release*	*Area of adoption*	*Pod yield (tons/ha)*
Utkarsh (CSMG 9510)	VR	MA 10 × ICG 7894	CSAUA&T, Mainpuri	2005	U.P., Punjab, & Northern Rajasthan	2.19
GG 21 (JSSP 15)	VB	Somnath × NCAC-2232 90-167-1-B-1-B-B-B	JAU, Junagadh	2005	U.P., Punjab, Northern Rajasthan	1.84
RG 382 (Durga)	VR	BG 2 × Tiftran	RAU Durgapura	2005	Rajasthan	2.20
GG 8 (J 53)	SB	27-5-1 × JL 24 30-3-2-B-B	JAU, Junagadh	2006	Northern Maharashtra and Madhya Pradesh	1.71
TG 38B (TG 38)	SB	Gamma ray induced mutation from F1 seeds of Girnar 1 × TG 26	BARC, Trombay	2006	Orissa, West Bengal, Bihar & North Eastern states	2.77
Prasuna (TCGS 341)	SB	TCG- 1717 × TCG-1518	ANGRAU, Tirupati	2006	Andhra Pradesh	2.0 (kharif); 4.00 (rabi-summer)
Abhaya (TPT 25)	SB	K 134 × TAG 24	ANGRAU, Tirupati	2006	Andhra Pradesh	2.30 (kharif); 3.76 (rabi-summer)
TMV (Gn) 13	SB	Selection from Pollachi red local	TNAU, Coimbatore	2006	Tamil Nadu	2.58
GG 16 (JSP 39)	VR	JSP 14 × JSSP 4 S-94-15-B-10-1-B-B	JAU, Junagadh	2006	Tamil Nadu, A.P, Kerala & Southern Maharashtra	2.06
Vasundhara (Dh 101)	SB	R 9241 × Dh 51-2	UAS, Dharwad	2007	W.B, Orissa, Jharkhand & Assam	2.87
ICGV 91114	SB	ICGV 86055 × ICGV 86533) F2-P8-B1-B1-B1-B1	ICRISAT, Hyderabad	2007	Andhra Pradesh	2.00
Kalahasti (TCGS 320)	SB	TCG- 1709 × TCG-1518	ANGRAU, Tirupati	2007	Andhra Pradesh	1.40 (kharif); 3.76 (rabi-summer)

Table 1: (*Contd....*)

Table 1: *(Contd....)*

Name of cultivar	***Recommended seasons***	***No. of days to maturity***	***Shelling (%)***	***Oil (%)***	***HKW (g)***	***Reaction to major pests and diseases***	***Special characters***
Utkarsh (CSMG 9510)	Kharif	125	66.0	49.0	54.0	Resistant to rust	Possesses fresh seed dormancy (up to 40-45 days)
GG 21 (JSSP 15)	Kharif	123	64.0	53.0	53.0	–	–
RG 382 (Durga)	Kharif	129	64.0	55.0	59.0	Tolerant of ELS and rust, jassids, leaf miner and thrips.	Drought tolerant
GG 8 (J 53)	Kharif	104-107	69.0	46.0	35.0	Tolerant of PBND, collar rot and rot diseases	–
TG 38B (TG 38)	Rabi-summer	115-125	71.0	48.0	44.0	Tolerant to stem rot	–
Prasuna (TCGS 341)	Kharif and rabi-summer	110	74-77	50.0	45-50	Tolerant of Kalahasti malady	–
Abhaya (TPT 25)	Kharif and rabi-summer	106	76.0	52.3	38-40	Tolerant of LLS, sucking insects (jassids and thrips) and *Spodoptera litura*	Tolerant of early and mid season moisture stress conditions
TMV (Gn) 13	Kharif	105	75.0	50.0	40-42	Tolerant of LLS, rust and bud necrosis	Tolerant of early and mid season moisture stress conditions
GG 16 (JSP 39)	Kharif	119	63.0	46.0	43.0	Tolerant of PBND, root rot, thrips, *Spodoptera*, Leaf miner	–
Vasundhara (Dh 101)		120-130	67.0	50.0	35.0	Tolerant of stem rot, PBND, thrips and *Spodoptera litura*	–
ICGV 91114	Kharif	90-95	75.0	48.0	41.0	Tolerant of rust and LLS	Early maturity
Kalahasti (TCGS 320)	Kharif and rabi-summer	110	74-76	52.3	50-60	Tolerant of bud necrosis, resistant to jassids	Suitable for rabi situation for kalahasti malady endemic areas and also for kharif in North coastal and north Telangana of Andhra Pradesh

Table 1: *(Contd....)*

Table 1: *(Contd....)*

Name of cultivar	*Habit**	*Pedigree/ Parentage*	*Developing centre*	*Year of release*	*Area of adoption*	*Pod yield (tons/ha)*
Narayani (TCGS 29)	SB	JL 24 × Ah 316/S	ANGRAU, Tirupati	2007	Andhra Pradesh	2.25 (kharif); 3.76 (rabi-summer)
AK 265	VB	ICGS 11 × US 63	PDKV, Akola	2007	Southern Maharashtra, Karnataka, A.P., & T.N.	1.90
M 548	VR	M 37 × Blanco Puro White	PAU, Ludhiana	2007	Punjab	2.18
TBG 39 (TG 39)	VB	TAG 24 × TG 19	RAU, Bikaner	2007	Rajasthan	3.15
AK 303	VB	(ICGV 88384 × JL 24) × (ICGV 88438 × ICG 5240)	PDKV, Akola	2007	Maharashtra	2.1
TG 51	SB	TG 26 × Chico	BARC, Trombay	2008	W.B., Orissa, Jharkhand and Assam	2.67
Ajeya (R 2001-3)	SB	ICGS 11 × ICG 4728	UAS, Raichur	2008	Southern Maharashtra, Karnataka, A.P., & T.N.	2.44
VL- Moongphali-1	SB	Selection from ICGV 86590 (X 14-4-B-19-8 × PI 239747	VPKAS, Almora	2008	Uttarakhand	1.94
Girnar 2 (PBS-24030)	VB	Selection from the cross M 13 × R 33-1)	NRCG, Junagadh	2008	U.P., Punjab, Northern Rajasthan	2.91
ICGV 00348	VB	ICGV 87290 × ICGV 87846	TNAU, Vriddhachalam	2008	Southern Maharashtra, Karnataka, A.P., and T.N.	2.01
VRI (Gn) 7	VB	TMV 1 × JL 24	Vriddhachalam	2008	Tamil Nadu	1.86
VRI (Gn) 6 (VG 9816)	SB	ALR 2 × VG 9513	TNAU, Vriddhachalam	2009	Southern Maharashtra, Karnataka, A.P., & T.N.	2.26
Jawahar Groundnut 23 (JGN 23)	SB	JL 24 × ICG 7886	JNKVV, Khargone	2009	Madhya Pradesh	1.63
Kadiri 9	SB	Kadiri 4 × Vemana	ANGRAU, Kadiri	2009	Andhra Pradesh	2.50-3.00

Table 1: *(Contd....)*

Table 1: (*Contd....*)

Name of cultivar	*Recommended seasons*	*No. of days to maturity*	*Shelling (%)*	*Oil (%)*	*HKW (g)*	*Reaction to major pests and diseases*	*Special characters*
Narayani (TCGS 29)	Kharif and rabi-summer	100	72-76	47.5	40-45	–	Early maturity, tolerant of mid season moisture stress conditions
AK 265	Kharif	120	68.0	47.0	42.0	Resistant to foliar diseases	Drought tolerant
M 548	Kharif	123	69.0	51.4	61.0	Tolerant of leaf spots and collar rot	Large seeded, high protein content
TBG 39 (TG 39)	Kharif	116	64.5	50.0	66.2	–	Large seeded, high yielding
AK 303	Kharif	125	72.0	49.0	80.0	–	Large seeded
TG 51	Rabi-summer	124	68.0	49.0	39.0	Tolerant of stem rot and root rot	–
Ajeya (R 2001-3)	Kharif	105-120	68.0	46-48	30-36	Resistant to PBND	Drought tolerant, wider adaptability
VL- Moongphali-1	Kharif	125-130			42.0	Resistant to LLS and root rot diseases	–
Girnar 2 (PBS-24030)	Kharif	130	69.0	51.0	61.0	Torelant of rust, LLS, PSND & sucking pests	Large seeded, stay green leaves at harvest
ICGV 00348	Kharif	124	66.0	47.0	41.0	Tolerant of leaf spots and rust disease	–
VRI (Gn) 7	Kharif	120-125	72.0	48.0	46.0	Tolerant of leaf miner, LLS and rust	–
VRI (Gn) 6 (VG 9816)	Kharif	100-106	66.0	47.0	29.0	Tolerant of LLS, rust, PBND	–
Jawahar Groundnut 23 (JGN 23)	Kharif	104	70.0	49.0	36.0	Tolerant of ELS & LLS	Drought tolerant
Kadiri 9	Kharif	105-110	65-70		31-35	–	Tolerant of early and late season drought

Table 1: (*Contd....*)

Table 1: *(Contd....)*

Name of cultivar	***Habit****	***Pedigree/ Parentage***	***Developing centre***	***Year of release***	***Area of adoption***	***Pod yield (tons/ha)***
Greeshma	SB	TIR 46 × JUG 37	ANGRAU, Tirupati	2009	Andhra Pradesh	2.10 (kharif); 4.00 (rabi-summer)
Kadiri 7	VB	{ICGV 86522 × ICG (FDRS) 10} × ICGV 91172	ANGRAU, Kadiri	2009	Andhra Pradesh	1.64
Kadiri 8	VB	{ICGV 86522 × ICG (FDRS) 10} × ICGV 91172	ANGRAU, Kadiri	2009	Andhra Pradesh	1.52
Mallika (ICHG 00440)	VB	(ICGV 88386 × ASHFORD) × ICGV 95172 F2-SSD-B1-B1-B3-B1	RAU, Hanumangarh	2009	All India	2.58
TGLPS 3 (TDG-39)	VB		UAS, Dharwad	2009	Karnataka	2.50-3.00
JL 501	SB	Selection from TAG 24	MPKV, Jalgaon	2010	Gujarat and Southern Rajasthan	1.66
Vijetha (R 2001-2)	SB	ICGS 11 × ICG 4728	UAS, Raichur	2010	W.B., Orissa, Jharkhand, Assam, Maharashtra, Karnataka, A.P. &, T.N.	1.630 (Zone IV); 2.522 (Zone V)
Girnar 3 (PBS 12160)	SB	Girnar 1× ICGS 11	DGR, Junagadh	2010	W.B, Orissa, Manipur	1.52
Kadiri Haritandhra (K 1319)	SB	91/57-2 × PI 476177	ANGRAU, Kadiri	2010	Karnataka and Maharashtra	3.72
GPBD-5	SB	TG 49 × GPBD 4	UAS, Dharwad	2010	Jharkhand and Manipur	1.5
HNG 69	VB	CSMG 84-1 × PG 1	RAU, Hanumangarh	2010	U.P., Punjab and Northern Rajasthan	2.80
GJG-HPS-1 (JSP-HPS-44)	VR	JSP-21 × VG -5	JAU, Junagadh	2010	Gujarat	2.12
Phule Vyas (JL 220)	SB	JL 80 × VG 77	MPKV, Jalgaon	1998-99	Maharashtra	2.00

Table 1: *(Contd....)*

Table 1: (*Contd....*)

Name of cultivar	***Recommended seasons***	***No. of days to maturity***	***Shelling (%)***	***Oil (%)***	***HKW (g)***	***Reaction to major pests and diseases***	***Special characters***
Greeshma	Kharif and rabi-summer	95-100	75-76	48-50	42-45	Tolerant of LLS	Tolerant of drought, high temperature and aftatoxin; early maturing
Kadiri 7	Kharif	120-125	68.0	47.0	65-75	Tolerant of sucking pest and leaf spots	Large seeded
Kadiri 8	Kharif	120-125	69.0	47.0	65-75	Tolerant of sucking pest and leaf spots	Large seeded
Mallika (ICHG 00440)	Kharif	125-130	66-70	48.0	73.0	Resistant to collar rot and PBND	Large seeded
TGLPS 3 (TDG-39)	Kharif	115-120					High harvest index (HI), bold kernel with less aflotoxin and confectionary type
JL 501	Kharif	102	67.0	48.0	36.0	–	–
Vijetha (R 2001-2)	Kharif	105-120	68.0	46-48	30-36	Resistant to PBND	–
Girnar 3 (PBS 12160)	Kharif	104-111	70.0	45.0	34.0	Tolerent of leaf miner and thrips	–
Kadiri Haritandhra (K 1319)	Rabi-summer	122	68.0	48.0	41.0	Multiple diseases (rust, ELS, LLs, stem rot, PBND) and insect pests (thrips, *Spodoptera litura*, jassid, Helicoverpa) resistant	Possesses fresh seed dormancy (upto 20days)
GPBD-5	Kharif	105-110	68.0	46.0	37.0	Resistant to LLS and rust	–
HNG 69	Kharif	131	66.0	50.0	51.0	Tolerant to collar rot, stem rot and ELS	–
GJG-HPS-1 (JSP-HPS-44)	Kharif	116	68.0	49.0	56.0	–	–
Phule Vyas (JL 220)	Kharif	90-95	68.0	52.0	46.0	–	Early maturity; high oil content

*SB = Spanish bunch, VB = Virginia bunch, VR = Virginia runner

Breeding for Tolerance to Abiotic Stresses

Efforts for breeding cultivars tolerant to abiotic stresses, the major risk element in groundnut cultivation, have yet to deliver results. Efforts have nevertheless been made to understand the nature of tolerance to moisture stress and developing criteria of selection for tolerance. Three physiological traits: (1) specific leaf area (SLA, $g.cm^{-2}$) which measures leaf thickness, (2) SPAD Chlorophyll Meter Reading (SCMR), which measures the chlorophyll concentration in the leaf and (3) Carbon Isotope Discrimination ($\Delta^{13}C$, expressed as % of the carbon isotope, ^{13}C of total C in plant dry matter which primarily is isotope ^{12}C) which indirectly measures transpiration efficiency, have been found associated with water use efficiency (WUE), and advocated as surrogate traits for WUE (Farquhar *et al.*, 1989; Wright *et al.*, 1988, 1994; Nageswara Rao and Wright, 1994). $\Delta^{13}C$ is negatively correlated with WUE but the strength of correlation like any other surrogate trait is influenced by the species and other genetic and environmental conditions. Measurement of this trait based on mass spectrometry is expensive too. Drought tolerance is not a single trait because it varies from season to season and is influenced by the stage of the plant life. Also at any stage the tolerance *per se* is a manifestation of interplay of several anatomical, morphological, physiological and biochemical factors with environment. A major reason for less than expected advance in understanding drought tolerance is that the root system of the plant has received very little focus while trying to decipher the mechanism of drought tolerance. The quantification and selection for this trait in the segregating generations remain to be a challenge to the breeder. It should also be kept in mind that the nature of drought tolerance for the survival of a plant as an organism for reproductive success and that required by the humans for getting the desired plant product may not be the same. Hence many of the physiological measures of drought tolerance may not be meaningful from the product point of view. Some of the varieties with some degrees of drought tolerance, released in India are: K 134, ICGS 1 (ICGV 87119), ICGS 5 (ICGV 87121), ICGS 37 (ICGV 87187) ICGS 76 (ICGV 87141), ICGV 91114, and ICGV 00350, though these have not been specifically bred for drought tolerance.

Although drought and high temperature go hand in hand, attempts to understand the effects of high temperature stress have been few and far between. Post-rainy season groundnut in India which is known to give much higher yield than the rainy season crop do face the problem of high temperature at the time of peak pod formation even when the crop is mostly irrigated. This high temperature coupled with high humidity at the time of harvest drastically decreases the viability of the seed harvested from the crop. It is thus not possible to use the seed from one post rainy-season crop to sow the next season. Breeding for tolerance to high temperature in post-rainy season cultivars is therefore an urgent requirement.

According to studies on quantitative inheritance of SCMR and SLA (Upadhyaya *et al.*, 2011), for SCMR at 80 days after sowing (DAS), additive effects were important, whereas dominance effects with duplicate epistasis were important for SCMR at 60 DAS. SLA is important at both stages. The use of such information for breeding and selection is yet to be put in practice and translating all this knowledge into practised technology still remains a challenge.

Efforts are on to identify molecular markers for different components of drought tolerance and to use genes like DREB1 associated with drought tolerance for developing transgenics.

Many countries in Asia (including India, especially the north-eastern hills and some red lateritic areas) and Africa have acid soils which are deficient in calcium. Groundnut yields in these areas are low and pod filling is poor (*i.e.,* occurrence of pops or empty pods). Aluminium toxicity, fixation of phosphates etc. are associated problems and make screening for resistance very complex. Soil acidity is one stress that limits large scale adoption of groundnut cultivation in the north-eastern hills of India though it is known that the yield and quality of groundnut there are much better than in the plains. Progress in this direction has been tardy.

Photoperiod is not important in groundnut in the classical sense but studies indicate a quantitative influence on reproductive efficiency which implies that cultivars with less sensitivity would be required for wider adaptation and for fitting into cropping sequences. The large differences in floral efficiency (fruit: flower ratio) and yield depend on short or long day lengths. In India, two crops (*kharif* and *rabi*-summer) are grown in a year. As selection in breeding material is frequently practiced in both seasons, most resulting genotypes are relatively insensitive to photoperiodic effects. A few cultivars released in India like, ICGS 11, ICGS 44 and ICGS 37 have been found to be relatively photoperiod insensitive (Nigam *et al.*, 1991). Dwivedi *et al.* (2000) reported both seasonal and genotypic differences for most of the traits studied. Photoperiodic differences (short, long and day neutral conditions) influenced only the shelling outturn among yield components and eicosenoic and palmitic acids among fatty acids but not oil content, oleic (O) and linoleic acid (L) and O/L ratio. Studies to understand the mechanism and inheritance of photoperiod sensitivity and effective laboratory screening procedures need to be developed and standardized for obtaining desired results through breeding.

Breeding for Resistance to Biotic Stresses

The expectation of faster progress in breeding for disease resistant groundnut has not been fully met though varieties with some degree of resistance to different diseases, especially the foliar fungal disease, do trickle out occasionally. These have been in most cases classified as tolerant (Table 1) meaning uncertain resistance. This is the situation because efforts to understand the pathogens and the mechanism through which it operates have not received the required attention. Very exciting fields of study like genetics of susceptibility and quantitative resistance have so far been ignored in groundnut.

Whereas the importance of the major foliar diseases like Late Leaf Spot (LLS) (*Phaeoisariopsis personata* and *Mycosphaerella berkeleyi*), Early Leaf Spot (ELS) (*Cercospora arachidicola*, *Mycospharella arachis*) and rust (*Puccinia arachidis*) have remained undiminished the soil-borne disease, stem rot caused by the fungus *Sclerotium rolfsii* has become very important almost in all groundnut growing areas in India. The most likely reason has been the fast depleting organic matter content in soil. Practically no sources of resistance in the cultivated species are known. Breeding for stem rot resistance is technically difficult by classical means mainly because of the facultative nature of the fungus and lack of availability of any selection criterion applicable in the segregating generations. In fact the very definition of resistance to the diseases is in terms of plants affected in a population and so that it can not be applied in a segregating generation. Criteria for measuring resistance based on individual plants should be developed. Once that is done the quantitative variation in resistance which is visible in the field can probably be accumulated in a few genotypes by cyclical crossing and selection. If successful, such resistance is expected to be more durable. Molecular methods may be able to aid this process in addition to the conventional efforts in breeding based on major genes.

An epidemic of peanut stem necrosis disease (PSND) caused by Tobacco streak virus occurred during monsoon season of 2000 for the first time in Anantapur district of Andhra Pradesh, India. The virus is known to cause up to 50% yield losses in groundnut. The virus is known to attack many other crops and weed species and is transmitted through pollen assisted by thrips (*Thysanoptera: Thripidae*). In groundnut it is not seed-transmitted. None of the cultivated groundnuts are resistant to the virus. Breeding programmes have been initiated to develop tolerant varieties and promising lines have been identified. Interestingly, the virus has largely been occurring in the groundnut growing belt of Anantapur district. Very little progress has been made so far in breeding for resistance to the insect pests.

Breeding for Quality Characteristics

Reasonably good progress has been made in India in breeding for large seeded types valued in the export market. As many as nine cultivars with seed mass ranging from 60 g/100 seed to 80 g /100 seed have been evolved during 1995 and 2010 and many more are in the pipe line. However, all these are very sensitive to the soil characteristics, moisture and micronutrient levels. Most of these varieties suffer easily from non-filling of pods when the conditions are not right.

Similarly, a good progress has been made in developing cultivars with more than 50% oil content. But these achievements are mostly after-the-fact results because even now there is no firm means of selecting for high oil content, which is a highly environment-sensitive trait, in segregating generations.

High oleic acid (O) and low linoleic acid (L) in oil is desired by the industry for longer storage. An O:L ratio up to 6.93 was found among ssp. *hypogaea*. Moore and Knauft (1989) reported a high (80%) oleic and low (20%) linoleic acid natural mutant which was controlled either by one or two recessive genes. The microsomal enzyme, delta-12-desaturase catalyses the addition of a second double bond to oleic acid at C-12 position, to convert it into linoleic acid. Two genes, ahFAD2A and ahFAD2B have also been identified, (which codes for the enzyme) (Jung *et al.*, 2000; Chu *et al.*, 2009). However, no high oleic acid cultivar has so far been released. No tangible success has been obtained anywhere in exploiting whatever resistance to aflatoxin has been identified so far especially the testa based resistance. Accumulation of the quantitative variation in resistance both the testa base and otherwise will require a selection methodology that works successfully in the segregating generations. More concerted efforts for identification of resistance of both types in the cultivated and wild species, understanding the infection and aflatoxin production process in terms of the anatomical fine structure, plant-pathogen interaction and signaling process and the molecular biology of aflatoxin production will be required for identifying one or more effective markers. Application of molecular biology may ultimately provide the way out.

Exploitation of Other Traits

Nothing seems to have happened and no progress has been made for traits like efficient adaptation to different cropping system, nutrient use efficiency and for developing groundnut as fodder. The situation of groundnut with respect to these traits is not much different from other crops though.

Breeding Methods

As already mentioned, there has been no substantial progress in advancement of science in breeding for groundnut improvement. The difficulty stems from the knowledge that molecular markers such as restriction fragment length polymorphisms (RFLPs), and random amplified polymorphic DNA (RAPDs) have not been successful in detecting variations in groundnut. Low genetic diversity among cultivated accessions is likely due to the origin of tetraploid peanut from a single, relatively recent allopolyploidy event that is thought to have involved genotypes of the wild diploid species *Arachis duranensis* and *A. ipaensis* (Burow *et al.*, 2009). Hence, exploring interspecific sources of variation must receive a high priority in groundnut.

Interspecific Hybridization in Groundnut Improvement

Majority of the closely related wild *Arachis* species are diploids. *Arachis* species in other seven sections (except in section Rhizomatosae) are distantly related and are diploids. Cultivated groundnut is tetraploid. An effective way of utilizing the wild species for improvement of cultivated groundnut is by developing tetraploids through either amphidiploidy or autotetraploidy. Development of tetraploid groundnuts will enhance the feasibility of introgression/transfer of useful traits and broadening of the genetic base.

For meeting this objective, synthetic amphidiploids have been developed that have easy crossability with cultivated tetraploid groundnut. Development of synthetic groundnut is a feasible way of utilizing important wild species gene pool of *Arachis*, closely related to cultivated groundnut, for improvement of the crop (Mallikarjuna *et al.*, 2012). About 25 new tetraploids synthesized employing diploid *Arachis* species are available at ICRISAT for utilization in the crossing programmes to broaden the genetic base of groundnut. The first synthetic groundnut originated from a cross between *A. cardenasii*, *A.*

diogoi and *A. batizocoi*, and it was utilized to develop backcross progeny (Simpson *et al.*, 1993). Two groundnut cultivars namely 'Coan' (Simpson and Starr, 2001) and 'NemaTAM' (Simpson *et al.*, 2003) were released utilizing this source. Recently, an amphidiploid was constructed utilizing *A. ipaensis* and *A. duranensis* (Fa´vero *et al.*, 2006). This amphidiploid is being used by ICRISAT, Brazil and Senegal to develop backcross population. Preliminary mapping data indicated low level of marker segregation distortion in F_2 population utilizing this amphidiploid (Dwivedi *et al.*, 2008).

Biotechnology in Groundnut Improvement

Transformation studies

Till now, no transgenic variety in groundnut has been released for cultivation. Major efforts in developing transgenics have so far been for developing virus resistant transgenics expressing coat protein genes of important viruses and Bt proteins among others.

Marker assisted selection (MAS)

Molecular markers such as restriction fragment length polymorphisms (RFLPs), and random amplified polymorphic DNA (RAPDs) have not been effective for detecting variations in groundnut. Due to the lack of polymorphism at the DNA level, the crop has not been subject to marker-assisted selection or resistance gene cloning. As a consequence, there is need to pursue genomic strategies in cultivated groundnut, with the specific goal of increasing the availability of useful molecular tools. Unfortunately, emphasis on genomic studies in groundnut in India has not received the thrust that Chinese and the USA institutions are giving.

However, the use of simple sequence repeat (SSR) markers, at more than 5,000 assayable loci, many of which are published may be useful for detecting and using variation in groundnut (Gautami *et al.*, 2009; Moretzsohn *et al.*, 2005). Genetic linkage maps composed of SSR markers have been constructed for wild diploid genomes AA (Moretzsohn *et al.*, 2005) and BB (Moretzsohn *et al.*, 2009), as well as for a tetraploid synthetic AABB genome derived from a cross between cultivated and wild amphidiploid species (Hong *et al.*, 2008). From these genetic maps, it was relatively easy to identify polymorphic markers because the corresponding populations incorporated high rates of polymorphism that are typical of wild *Arachis* species. Recently, low resolution linkage maps have been produced for purely cultivated tetraploid AABB genome crosses that harbor much lower polymorphism (Varshney *et al.*, 2009). In these linkage maps, polymorphic markers contain less than 200 SSR markers. Although individual SSR markers could be valuable for marker-assisted selection in cases where they are linked to traits of interest, in general, the number of polymorphic SSR markers is insufficient for routine genetic analysis. There is a need to expand the density and availability of genetic polymorphisms in groundnut, and to survey the status of polymorphic alleles across peanut germplasm. Concerted and broad research efforts on this line are needed.

QTLs for many traits have already been reported in groundnut and this may be useful in groundnut improvement in the coming years in combination with conventional breeding methods (Sujay *et al.*, 2012). The usefulness of QTLs will depend on their individual magnitude of effect, repeatability and ease of detection. These are still in the realms of speculation.

Marker assisted selection has been followed so far for traits like high oleic acid/linoleic acid ratio (Chu *et al.*, 2009) and disease resistance.

REFERENCES

Burow, M.D., Simpson, C.E., Faries, M.W., Starr, J.L. and Paterson, A.H. (2009). Molecular biogeographic study of recently described B- and A-genome *Arachis* species, also providing new insights into the origins of cultivated peanut. *Genome*, 52: 107–119.

Craufurd, P.Q., Wheeler, T.R., Ellis, R.H., Summerfield, R.J. and Vara Prasad, P.V. (2000). Escape and tolerance to high temperature at flowering in groundnut (*Arachis hypogaea* L.). *The Journal of Agricultural Science*, 135: 371–378.

DAC 2008 http://eands.dacnet.nic.in/At_Glance_2008.htm

Dwivedi, S.L. and Nigam, S.N. (1995). Breeding for improved seed quality in groundnut (*A. hypogea* L.) with special reference to export promotion: Issues and opportunities, Paper presented at workshop on" Status of Confectionery Groundnut Research in India and Strategies for Export Promotion, BARC, Trombay, Mumbai. p.10.

Dwivedi, S.L., Upadhyaya, H.D., Stalker, H.T., Blair, M.W., Bertioli, D., Nielen, S. and Ortiz, R. (2008). Enhancing crop gene pools of cereals and legumes with beneficial traits using wild relatives. *Plant Breed. Rev.* 30: 179–230.

Dwivedi, S.L., Nigam, S.N. and Nageswara Rao, R.C. (2000). Photoperiod effects on seed quality traits in peanut. *Crop Sci.,* 40: 1223–1227.

Fa´vero, A.P., Simpson, C.E., Valls, J.F.M. and Yuksel, Y. (2006). Study of the evolution of cultivated peanut through crossability studies among *Arachis ipaensis, A. duranensis*, and *A. hypogaea*. *Crop Sci.,* 46: 1546–1552.

Farquhar, G.D., Ehleringer, J.R. and Hubick, K.T. (1989). Carbon isotope discrimination and photosynthesis. Annual review of physiology. *Plant Molecular Biology*, 40: 503–537.

Gautami. B., Ravi, K., Lakshmi, N.M., Hoisington, D.A. and Varshney, R.K. (2009). Novel set of groundnut SSRs for genetic diversity and interspecific transferability. *Int. J. Integr Biol.,* 7: 100–106.

Hong, Y.B., Liang, X.Q., Chen, X.P., Liu, H.Y., Zhou, G.Y., Li, S.X. and Wen, S.J. (2008). Construction of genetic linkage map based on SSR markers in peanut (*Arachis hypogaea* L.). *Agric. Sci. in China*, 7(8): 915–921.

Jung, S., Swift, D., Sengoku, E., Patel, M., Teule, F., Powell, G., Moore, K. and Abbott, A. (2000). The high oleate trait in the cultivated peanut [*Arachis hypogaea* L.]: I. Isolation and characterization of two genes encoding microsomal oleoyl-PC desaturases. *Mol. Gen. Genet.,* 263: 796–805.

Ketring, D.L. (1984). Temperature effects on vegetative and reproductive development of peanut. *Crop Sci.,* 24: 877–882.

Ketring, D.L., Brown, R.H., Sullivan, G.A. and Johnson, B.B. (1982). Growth physiology. *In*: Peanut Science and Technology (*Eds*. Pattee, H.E. and Young, C.T.). American Peanut Research and Education Society, USA. pp. 411–450.

Mallikarjuna, N., Srikanth, S., Vellanki, R.K., Jadhav, D.R., Das, K. and Upadhyaya, H.D. (2012). Meiotic analysis of the hybrids between cultivated and synthetic tetraploid groundnuts. *Plant Breeding*, 131: 135–138.

Manivel, P., Bandyopadhyay, A., Mathur, R.K. and Rajgopal, K. (2000). Trend in yield components of groundnut cultivars released in India over years. *In*: National Seminar on Oilseeds and oils-research development needs in the millennium from 02-04th February, 2000 at the Directorate of Oilseeds Research, Hyderabad (A.P.) p. 106.

Manivel, P., Bandyopadhyay, A., Mathur, R.K., Samdur, M.Y. and Misra, J.B. (2004). Early generation selection for pod size in groundnut – A case study. *In*: Current trends of Research on groundnut in India: Short papers presented at the National Symposium on Enhancing productivity of groundnut for sustaining food and nutritional security. October 11-13, 2004. National Research Centre for Groundnut. pp. 2–5.

Miranda-Abilay, R., Herrera, M.F. and Wynne, J.C. (1988). Effects of shading in agronomic and morphophysiological characters of peanut varieties. 4th Sci. Meet., Fed. Crop Sci. Soc., Philippines, Davao City, Philippines, Apr. 27-30. (Abstr).

Moore, K.M. and Knauft, D.A. (1989). The inheritance of high oleic acid in peanut. *J. Hered.,* 80: 252–253.

Moretzsohn, M.C., Barbosa, A.V.G. Alves-freitas, D.M.T., Teizeira, C., Leal-Bertioli, S.C.M., Guimaraes, P.M., Pereira, R.W., Lopes, C.R., Cavallari, M.M., Valls, J.F.M., Bertioli, D.J. and Gimenes, M.A. (2009). A linkage map for the B-genome of *Arachis* (Fabaceae) and its synteny to the A-genome. *BMC Plant Biol.,* 9: 40.

Moretzsohn, M.C., Leoi, L., Proite, K., Guimaraes, P.M., Leal-Bertioli, S.C.M., Gimenes, M.A., Martins, W.S., Valls, J.F.M., Grattapaglia, D. and Bertioli. D.J. (2005). A microsatellite-based, gene-rich linkage map for the AA genome of *Arachis* (Fabaceae). *Theor. Appl. Genet.*, 111(6): 1060–1071.

Nageswara Rao, R.C. and Wright, G.C. (1994). Stability of the relationship between specific leaf area and carbon isotope discrimination across environments in peanut. *Crop Sci.,* 34: 98–103.

Nigam, S.N., Nageswara Rao, R.C., Wynne, J.C., WiUiams, J.H., Eitzner, M. and Nagabhushanam. G.V.S. (1994). Effect and interaction of temperature and photoperiod on growth and partitioning in three groundnut (*Arachis hypogaea* L.) genotypes. *Ann. App. Bioi.,* 125: 541–552.

Nigam, S.N., Dwivedi, S.L. and Gibbons, R.W. (1991). Groundnut breeding: Constraints, achievements, and future possibilities. *Plant Breeding Abstracts*, 61: 1127–1136.

Samonte, H.P. and Ocampo, A.M. (1989). Tolerance of mungbean, peanut, soybean and corn to acid soil conditions in the Philippines. *Phill. Agric.,* 72: 137–146.

Simpson C.E., Nelson, S.C., Starr, J.L., Woodard, K.E. and Smith, O.D. (1993). Registration of TxAG-6 and TxAG-7 peanut germplasm lines. *Crop Sci.,* 33: 1418.

Simpson, C.E. and Starr, J.L. (2001). Registration of COAN peanut. *Crop Sci.,* 41: 918.

Simpson, C.E., Starr, J.L., Church, G.T., Burrow, M.D. and Paterson, H.A. (2003). Registration of 'NemaTAM' peanut. *Crop Sci.,* 43: 561.

Singh, N.B. and Basu, M.S. (2004). Groundnut research in India: An overview. *In*: Basu, M.S. and Singh, N.B. (*Eds*.). Groundnut Research in India. National Research Centre for Groundnut, Junagadh, Gujarat, India. pp. 1–6.

Sujay, V., Gowda, M.V.C., Pandey, M.K., Bhat, R.S., Khedikar, Y.P., Nadaf, H.L., Gautami, B. Sarvamangala, C., Lingaraju, S., Radhakrishan, T., Knapp, S.J. and Varshney, R.K. (2012). Quantitative trait locus analysis and construction of consensus genetic map for foliar disease resistance based on two recombinant inbred line populations in cultivated groundnut (*Arachis hypogaea* L.). *Molecular Breeding*, 30: 773–788.

Upadhyaya, H.D. and Nigam, S.N. (1999). Detection of epistasis for protein and oil contents and oil quality parameters in peanut. *Crop Sci.,* 39: 115–118.

Upadhyaya, H.D., Mukri, G., Nadaf, H.L. and Singh, Sube. (2012). Variability and stability analysis for nutritional traits in the mini core collection of peanut. *Crop Science*, 52: 168–178.

Upadhyaya, H.D., Sharma, S., Singh, Sube and Singh, Murari. (2011). Inheritance of drought resistance related traits in two crosses of groundnut (*Arachis hypogaea* L.). *Euphytica*, 177: 55–66.

Vara Prasad, P.V., Craufud, P.Q., Summerfield R.J. and Wheeler, T.R. (1998). Sensitivity of fruit-set to high temperature episodes in groundnut (*Arachis hypogaea* L.). *Journal of Experimental Botany,* 49(Supplement): 30.

Varshney, R.K., Bertioli, D.J., Moretzsohn, M.C., Vadez, V., Krishramurthy, L., Aruna, R., Nigam, S.N., Moss, B.J., Seetha, K., Ravi, K., He, G.H., Knapp, S.J. and Hoisington, D.A. (2009). The first SSR-based genetic linkage map for cultivated groundnut (*Arachis hypogaea* L.). *Theor. Appl. Genet.,* 118(4): 729–39.

Wood, I.M.W. (1968). The effect of temperature at early flowering on growth and development of peanuts (*Arachis hypogaea* L.). *Aust. J. Agric. Res.,* 19: 241–251.

Wright, G.C., Hubick, K.T. and Farquhar, G.D. (1988). Discrimination in carbon isotopes of leaves correlates with water-use efficiency of field grown peanut cultivars. *Australian J. Plant Physiology*, 15: 815–825.

Wright, G.C., Nageswara Rao, R.C. and Farquhar, G.D. (1994). Water-use efficiency and carbon isotope discrimination in peanut under water deficit conditions. *Crop Science,* 34: 92–97.

Ye Chu, Corley Holbrook, C. and Ozias-Akins, Peggy (2009). Two alleles of ahFAD2B control the high oleic acid trait in cultivated peanut. *Crop Science*, 49: 2029–2036.

Zhu Xin-Guang., Long Stephen, P. and Ort Donald, R. (2010). Improving photosynthetic efficiency for greater yield. *Annual Review of Plant Biology*, 61: 235–261.

15

Cotton

B.M. Khadi, Y.R. Aladakatti and Rajesh Patil

University of Agricultural Sciences, Dharwad, India

Cotton is the most important fibre crop of the world with cultivation spread over 60 countries covering an area of 36 m ha. With about 33% of the world cotton area India occupies top position in cotton cultivation (Table 1). Of about 26.88 m tonnes of cotton produced globally, India contributes 22.5%. Among the five major cotton growing countries China's productivity is the highest (1300 kg ha^{-1}) while that of India stands at 495 kg lint ha^{-1}.

Table 1: Major cotton growing countries of the world

Country	***Area (m ha)***		***Production (m tonnes)***		***Productivity (kg ha^{-1})***	
	2010-11	***2011-12****	***2010-11***	***2011-12****	***2010-11***	***2011-12****
China	5.22	5.53	6.40	7.19	1226	1300
USA	4.33	3.91	3.94	3.43	903	877
India	11.14	12.19	5.53	6.05	496	495
Pakistan	3.00	3.24	1.91	2.08	637	642
Uzbekistan	1.48	1.43	1.38	1.30	932	909
Turkey	0.50	0.48	0.62	0.58	1240	1208
Australia	0.08	0.20	0.13	0.35	1625	1750
Brazil	0.89	0.91	1.40	1.51	1573	1659
Egypt	0.29	0.30	0.26	0.32	897	1066
Greece	0.37	0.34	0.31	0.33	838	970
Argentina	0.33	0.32	0.16	0.15	485	469
Others	5.93	7.16	2.76	3.59	–	–
Total	33.56	36.01	24.88	26.88	741	746

*Estimated (Provisional); *Source*: ICAC 2011, CAI, Cotton Statistics & News - Dec., 2011 & AICOSCA- Dec., 2011

V.L. Chopra (Ed.): Breeding Field Crops II: Advances, Studium Press (LLC), USA.

CULTIVATION OF COTTON IN INDIA

Area, production and productivity of cotton in India has progressively increased over the years (Table 2). A sharp increase in productivity was witnessed following introduction of *Bt*-transgenic cotton in 2002.

Table 2: Cotton cultivation scenario past and present in India

Year	*Area (m ha)*	*Production (m bales)*	*Productivity (lint kg ha^{-1})*
1950-51	5.88	3.04	88
1960-61	7.61	5.60	106
1970-71	7.61	5.60	125
1980-81	7.82	7.01	152
1990-91	7.40	11.50	224
2000-01	8.15	15.30	319
2005-06	8.82	24.40	478
2006-07	9.14	28.00	521
2007-08	9.56	31.50	560
2008-09	9.41	29.00	524
2009-10	10.33	29.50	486
2010-11	11.14	33.56	512
2011-12	12.19	36.01	502

(1 bale = 170 kg); *Source*: CAI, Cotton Statistics and News-Dec, 2011

Cotton is cultivated across the country under different agroclimatic and soil conditions. Accordingly, four zones of cotton cultivation are recognized in India (Table 3).

Table 3: Cotton growing zones in India

Zone	*States*	*Species grown*	*% of cotton area*	*% share of cotton production*
Northern zone	Punjab, Haryana, Rajastan	Intra- *hirsutum*, *Gossypium hirsutum*, *G. arboreum*, Intra- *arboreum* hybrids	13.90	14.49
Central zone	Madhya Pradesh, Maharashtra, Gujarat	Intra- *hirsutum*, Inter- *specific tetraploid, G. hirsutum, G. arboreum*, *G.herbaceum*, Intra- *arboreum, diploid* hybrids	64.18	57.95
Southern zone	Andhra Pradesh, Karnataka, Tamilnadu	All four species of cotton, Inter- *specific tetraploid,* Intra- *arboreum* hybrids	20.70	18.84
Eastern zone	West Bengal, Orissa, Bihar & North Eastern states	*G. arboreum*	1.21	1.15

The top three cotton growing states are Maharashtra (40.95 lakh ha) Gujarat (30.23 lakh ha) and Andhra Pradesh (18.54 lakh ha), which together contribute nearly 67% of lint to the Indian cotton output (Table 4). High productivity is seen in Gujarat (659 kg ha^{-1}) followed by Punjab (591 kg ha^{-1}), Haryana (561 kg ha^{-1}), Rajasthan (548 kg ha^{-1}) and Andhra Pradesh (499 kg ha^{-1}). A major contributor to higher productivity, witnessed in the last decade is the widespread adoption of *Bt*-transgenic hybrid cotton which now accounts for almost 90% of the area planted with cotton.

Table 4: State-wise area, production and productivity of cotton in India

State	***% irrigated area***	***Area (lakh ha)***			***Production (lakh bales)***			***Productivity (lint kg ha^{-1})***		
		09-10	***10-11***	***11-12****	***09-10***	***10-11***	***11-12****	***09-10***	***10-11***	***11-12****
Punjab	100	5.11	5.30	5.60	13.00	16.00	17.00	432	593	591
Haryana	98	5.07	4.92	6.05	15.25	14.00	17.00	511	587	561
Rajasthan	100	4.44	3.35	5.30	12.00	9.00	16.00	459	512	548
Gujarat	50	26.25	26.33	30.23	98.00	103.00	114.00	634	685	659
Maharashtra	20	35.03	39.32	40.95	65.75	82.00	69.00	319	379	310
Madhya Pradesh	30	6.11	6.50	7.06	15.25	17.00	17.00	424	462	426
Andhra Pradesh	25	14.75	17.84	18.54	54.50	53.00	48.00	628	566	499
Karnataka	33	4.55	5.45	5.49	12.25	10.00	12.00	457	346	405
Tamil Nadu	38	1.04	1.22	1.21	5.00	5.00	5.00	817	1003	1011
Orissa	–	0.54	0.74	1.02	1.00	2.00	2.00	314	470	341
Others	–	0.21	0.45	0.46	1.00	2.00	2.00	809	755	739
Loose supply	–	–	–	–	12.00	26.10	26.10	–	–	–
All India	–	103.1	111.42	121.91	305.00	339.10	345.10	502	517	481

* Estimated (Provisional). *Source*: Cotton Advisory Board (CAB)

Species composition of cultivated cotton in India has witnessed a major shift since independence (Table 5). In 1947, diploid *desi* cotton species *Gossypium arboreum* and *G. herbaceum* occupied nearly 97% area while tetraploid *G. hirsutum* contributed only 3%. The situation has reversed now with *G. hirsutum*, *intra hirsutum* and *hirsutum-barbadense* hybrids occupying 89% of area under cotton. Indeed, the successful development and largescale deployment of hand-cross based hybrids by Indian scientists is a landmark in the history of cotton improvement in the world. The current species composition involving about 11% area under *G. arboreum* and *G. herbaceum* species is regarded as healthy for maintaining the genetic base for tolerance to abiotic and biotic stresses such as moisture stress, resistance to leaf curl virus disease and jassids.

Table 5: Species composition of cultivated cotton in India

Species	***Percent area covered***					
	1947-48	***1955-56***	***1975-76***	***1996-97***	***2007-08***	***2010-11***
G. herbaceum	32.0	23.0	20.2	11.0	5.0	4.00
G. arboreum	65.0	36.0	26.1	17.0	10.0	7.00
G. hirsutum	3.0	41.0	52.5	35.9	19.0	2.00
G. barbadense	–	–	1.2	0.1	0.1	0.01
Hybrids	–	–	–	36.0	65.9	86.99

Shift in species composition had also a major consequence on the output in terms of total production and the composition of different staple types (Table 6). Prior to 1960, almost all cotton produced in India had <24 mm long fiber. Now, about 50% of cotton produced is of medium to superior-medium category (up to 24 mm length) and 40% belongs to long staple type (>24.5 mm staple length). Introduction of tetraploids from the new world has been responsible for the improvement in staple length. The range with regard to spinning potential has also widened (6s and 120s) as a result of introduction of hybrid cotton.

Table 6: Staple-wise production of cotton in India*

Year	***Staple length and class***			
	Short (< 20.0 mm)	***Medium/ Medium long (20.5 to 27 mm)***	***Long (27.5 to 32.0 mm)***	***Extra long (32.5 mm and above)***
1960-61	7.80	25.50	–	–
1970-71	8.10	36.40	1.6	–
1985-86	5.60	31.85	28.6	–
1990-91	10.70	48.45	29.1	–
1992-93	13.40	58.29	33.8	–
2000-01	9.50	74.00	52.0	4.5
2001-02	9.50	82.50	61.0	5.0
2002-03	9.00	72.00	51.0	4.0
2003-04	7.60	75.90	89.9	5.6
2004-05	7.11	136.80	94.1	5.0
2005-06	6.80	64.00	165.4	4.8
2006-07	6.00	54.00	216.2	3.9
2007-08	4.00	63.00	243.0	5.0
2008-09	5.00	65.00	206.0	2.9
2009-10	5.00	67.00	200.0	3.0
2010-11	4.00	68.00	201.5	3.5

Scale: Production in lakh bales. *Staple-group-wise production figures are estimated based on varietal composition. (1 Bale: 170 kg lint). *Source*: Ministry of Textile, GOI.

All India Coordinated Cotton Improvement Project (AICCIP) operating in 11 main centers and 16 sub centers located in different cotton growing states with head quarters at Coimbatore is the prime integrator of cotton research activities in the country. In addition, two ICAR Institutes, CICR and CIRCOT (see below) play a pivotal role in cotton research.

1. Central Institute for Cotton Research (CICR), Nagpur with regional stations at Coimbatore and Sirsa (Haryana) provides leadership in basic and strategic research apart from maintaining germplasm.
2. Central Institute for Research on Cotton Technology (CIRCOT), Mumbai has 13 regional stations spread over India and provides both service and lead with regard to fiber technology and value addition to cotton byproducts.

Cotton Breeding

The genus *Gossypium* contains about 50 species of which 45 are diploid and five are allopolyploid. Cultivated cotton comprises of two diploid species (2n=26) *G. herbaceum*, *G. arboreum* and two tetraploid species (2n=52), *G. hirsutum* and *G. barbadense*. Based on cytology and chromosome pairing behavior, eight genomes (A-F, G and K) are recognized in *Gossypium*. *G. herbaceum*, *G. arboreum* carry the A genome whereas *G. hirsutum* and *G. barbadense* have AD genome constitution. Cotton is said to be first domesticated in Southern Arabia from the *G. herbaceum* of South African descent. *G. arboreum* is derived from *G. herbaceum* following translocation, selection, and genome stabilization. The D genome species, which is a major contributor to fiber quality of tetraploid cottons, are distributed in Central and South America. The diploid species representing B, E and F genome are found in Africa and Arabia whereas those belonging to the C, K and G genome are restricted to Australia.

Cotton is mostly a self-pollinated plant with outcrossing frequency ranging from 2-10%. Cotton pollen is heavy, and wind and insect dispersal is low, thus favouring self-pollination. Breeding approaches in cotton have traditionally followed those recommended for self-pollinated crops. In the 1970s, hybrid cotton (both *intrapartum* and interspecific *G. hirsutum* × *G. barbadense*) were developed for the first time and popularized in India through large-scale production of hand-pollinated F_1 seed. This led to a shift in terms of both production method and productivity, and spawned a major cotton hybrid seed industry which is thriving today with the introduction of transgenic cotton.

Therefore, in recent years, methods suitable for heterosis breeding have increasingly been adopted. In the past, breeding objectives were mainly directed at disease pest tolerance and fibre quality, particularly, fiber length. In recent years, changes in spinning technology have called for stronger fiber and accordingly emphasis has been given to strength besides other fiber quality aspects such as micronnaire, length etc. A variety of breeding approaches have yielded cultivars, lines and hybrids for different zones and uses. These are summarized in Table 7.

Table 7: Major cotton varieties developed through different breeding approaches

Breeding method	***Line/variety***	***Remarks***
Plant Introduction	Sea Island Andrews	An Egyptian cotton accession introduced from West Indies directly released as a cultivar in Tamil Nadu
	Nectariless	An American introduction which lead to development of the first *intrapartum* hybrid cotton
	SB 289E	A Russian Introduction which formed the basis for the successful development of interspecific hybrid (*G. hirsutum* × *G. barbadense*) cotton in the world
Mass Selection	Bikaneri Narma	Developed from a local germplasm
	F 414 and H 777	Reselections from Bikaneri Narma
	M12	A selection from LRA 5166 and is the male parent of the hybrid Savitha
	G 22, HD 11, LD 133, Saraswati	*G. arboreum* cultivars
Pure Line Selection	MCU 5VT, CO_2 Kanchana	*G. hirsutum*
	Gourani varieties, Lohit	*G. arboreum*
	Sujatha	*G. barbadense*
Pedigree Method	Laxmi, LRA 5166, Khandwa-3	*G. hirsutum*
	AKA 4, Ekanath, Rohini, Sanjay	*G. arboretum*
	Digvijaya, V-797, G. cot 13 and Jaydhar	*G. herbaceum*
Heterosis Breeding	Hybrid 4, JKHY 1, H6, AHH 468, MECH 1, NHH44, Savitha, H8, CICR HH1, TM1312, Fateh, HHH 81, Raj HH 16, MDCH 201	Major *intrapartum* hybrids
	Varalaxmi, DCH 32, DHB 105, TCHB 213, HB 224	*G. hirsutum x G. barbadense* hybrids
	G. cot DH 7, G. cot DH 9, DDH2	*Desi* cotton hybrids

Attempts have also been made to utilize genetic male sterility (GMS) or cytoplasmic-genetic male sterility (CGMS) for hybrid seed production in cotton. As many as 10 GMS loci have been identified in *G. hirsutum*. Poor stability of male sterility, complex mating necessary for maintenance of male sterile lines or lack of suitable markers for distinguishing male sterile plants at the seedling stage, have prevented exploitation of GMS for hybrid seed production. Similarly, CGMS sources have been identified with cytoplasms of *arboreum, anomalum, harknessii, longicalyx, aridum, tribolum* etc. in the nuclear background of *G. hirsutum.* Many of the male sterility-inducing cytoplasm show yield penalty. Hence, CGMS-based hybrids have not been developed so far in cotton.

Major objectives of cotton breeding improvement are increasing lint yield, incorporating tolerance to pathogens, pests and abiotic stress tolerance, and improving fiber quality. Pests are the most serious problems in cotton cultivation which not only reduces yield but also the quality of the output. Transgenic approach has proved very effective in tackling major cotton pests (boll worm) and this strategy is likely to be employed for other pests too. Fiber quality comprises several components such as length, strength, fineness and maturity coefficient. These days, with the availability of high speed spinners, there has been an increasing demand for fiber strength. The strength to length ratio (S:L ratio) is of a greater importance and therefore, efforts to modify strength with respect to any length class of the cotton fiber are underway. A S:L ratio of 0.80 and above is targeted in *hirsutum* based cottons. Apart from this, specific requirements of the cotton fiber like absorbent cotton for medical purposes are also being bred for.

Transgenic Cotton

Bollworms *viz.*, American bollworm (*Helicoverpa armigera*), pink bollworm (*Pectinophora gossypiella*) and spotted bollworms (*Earias vitella*) attack cotton crop and cause significant yield losses. No source of resistance to the bollworms is available in global germplasm. Moreover, about 10% of insecticides on global basis and 45% in India are used for control of insects in cotton crop alone. Insecticides have adverse effects on (i) natural predators and parasites of bollworms, (ii) beneficial insects, (iii) human health, (iv) soil microflora (v) environment, and (vi) cost of production. Hence, there is need to develop bollworm resistant cotton. Genetic engineering provides a valuable avenue for developing transgenic resistance.

Bt Cotton in India

In India, the Government has approved commercial cultivation of *Bt* cotton hybrids since 2002 crop season. Four events have been approved.

These are: *cry* 1A(c) (Mon 531) developed by Mahyco Monsanto Biotech Ltd., *cry* 1A(c) and cry 2Ab (Mon 15985) stacked gene event developed by Mahyco Monsanto Biotech Ltd., *cry* 1Ac (Event 1) developed by J.K. Agri Genetics Seeds Ltd. (sourced from IIT, Kharagpur) and *cry* 1 Ab and *cry* 1 Ac (GFM event) developed by Nath seeds (sourced from China). Three more transgenic cotton events are under confined field trial *viz.*, cry1 Ac + cry1 F (3006-210-23) + (281-24-236) event of Dow Agro Science for bollworm resistance, cry1 Ac + cry2 Ab + CP4EPSPS (Mon 15985 + Mon 88913) of Mahyco for bollworm and herbicide resistance, and cry1Ac + cry1 EC (Event 1 + Event 24) of J.K. Agri Genetics for bollworm resistance.

The area under *Bt* cotton hybrids increased from 1.30 m ha in 2005-06 to 6.89 m ha in 2008-09. Thus within a span of seven years nearly 74% of the cotton area in India has come under *Bt* hybrids. Among the cotton growing states, Maharashtra leads with 2.57 m ha. under *Bt* cotton followed by Gujarat(1.45 m ha) and Andhra Pradesh(1.14 m ha). *Bt* cotton is being cultivated in all the three cotton growing zones of the country (Table 8). The released transgenics can be grouped into (i) Intra-*hirsutum* (ii) Inter-specific hybrids (*hirsutum* x *barbadense*)

Table 8: State-wise adoption of *Bt* cotton in India (Area in million ha)

State	***Year***									
	2002	***2003***	***2004***	***2005***	***2006***	***2007***	***2008***	***2009***	***2010***	***2011****
Punjab	–	–	–	0.070	0.280	0.580	0.48	0.61	0.72	0.75
Haryana	–	–	–	0.010	0.040	0.280	0.38	0.42	0.58	0.59
Rajasthan	–	–	–	0.002	0.005	0.038	0.12	0.20	0.40	0.41
Gujarat	0.0091	0.042	0.13	0.150	0.410	1.300	1.45	1.68	1.78	1.89
Maharashtra	0.0100	0.020	0.16	0.510	1.660	2.560	2.57	3.40	3.71	3.91
Madhya Pradesh	0.0020	0.020	0.09	0.140	0.300	0.470	0.52	0.62	0.61	0.65
Andhra Pradesh	0.0040	0.006	0.07	0.090	0.660	1.000	1.14	1.05	1.65	1.85
Karnataka	0.0020	0.003	0.04	0.030	0.080	0.150	0.17	0.27	0.37	0.42
Tamil Nadu	0.0003	0.008	0.01	0.020	0.030	0.060	0.07	0.11	0.12	0.18
Total cotton area	7.3900	7.840	8.97	9.160	9.160	9.560	9.37	8.36	9.94	10.65
% of total cotton area	0.3900	1.190	5.55	11.100	37.790	66.290	73.62	81.16	89.55	87.30

*Provisional figures. *Source*: Directorate of Economics and Statistics, 2011

Currently, more than 1000 *Bt* hybrids involving the four events and developed by more than 30 seed companies have been approved for commercial cultivation. Majority of the hybrids belong to intra *hirsutum* category and rest to the interspecific category (*G. hirsutum x G. barbadense).*

Development of *Bt* Kits

Central Institute for Cotton Research, Nagpur has developed diagnostic kits *viz.*, *Bt* Express, *Bt* Detect, *Bt*-Zygosity, *Bt* Quant for the detection of *Bt* toxin and these kits have been effectively deployed all over the country to verify the purity of *Bt* seed and ensure the supply of quality *Bt* hybrids seed to the farming community.

Molecular Breeding

Application of molecular tools for improving cotton is receiving increasing attention. The D genome of *G. raimondii* has been recently sequenced. The AD tetraploid genome draft is also available (Paterson, 2012). A BAC-based AD genome sequence may offer superior opportunities to elucidate the types and frequencies of changes that distinguish polyploid from diploid cottons and help in using genes from either ploidy class for the improvement of the other.

The use of RAPD, RFLP, AFLP, SSR, ISSR and SRAP markers for improving productivity and fiber properties of cotton is used routinely. Since SNPs are more accurate, throughput SNP genotyping technologies have helped improve lint percentage, fibre strength, fibre length and micronaire (Marc Lacape *et al.*, 2000; Lin *et al.*, 2005; Preetha and Raveendren, 2008).

The resurgence of erstwhile minor pests because of the imbalance in the pest control situation created by the *Bt* gene and emergence of new pests like the mirid bugs, tea mosquito bug and gall midge of cotton have necessitated molecular level research into host plant resistance. Engineering sucking pest tolerance is the next approach to reduce pesticide usage which is again on the rise in *Bt* cotton. Root knot nematode resistance is another area of current research. Of late, plant mediated RNAi technology has been useful in silencing the pest genes responsible in making them tolerant to gossypol. This could be the next big thing in pest management strategies for cotton.

Verticillium wilt is a soil borne disease caused by fungus *Verticillium dahliae* which cannot be controlled through fungicides. Resistance breeding is the only option. Since resistance genes are mostly confined to *barbadense* types, conventional hybridization to transfer these genes in to *hirsutum* background has not been easy.

Therefore, use of molecular means to incorporate specific genes is the method of choice. Genes involved in the synthesis of relevant compounds *viz.*, phytoalexins, PR proteins and defence protein genes related to plant structure have been identified, cloned and expressed in cotton to produce wilt tolerant plants (Cai *et al.*, 2009). Cotton leaf curl virus is also a global threat. Many ESTs are involved in signal transduction pathways which are selectively triggered upon infection. Some are also involved in the abiotic stress responses. Cytochrome reductase and ethylene-responsive factor genes, fasciclin-like arabinogalactan protein genes related to salt stress, NAC genes that respond to drought, cold and ABA stress are a few engineered genes against multiple stresses in cotton. The herbicide tolerant cotton containing the tfdA gene from *Alcaligenes* bacterium producing the monooxygenase enzyme against 2,4-D will have to be replaced if resistance starts building up. The second gene in *Bt* is a case in point (Boopathi *et al.,* 2011).

The various *omics* and their inclusive techniques are now being realized in cotton. All these omics *i.e.*, genetical genomics, transcriptomics, proteomics and metabolomics are being integrated with system biology to holistically understand the life processes in the cotton plant (Chen *et al.,* 2011). Next generation sequencing (NGS) platforms along with the use of SNPs, BACs and genome wide association (GWS) studies are underway to mine agriculturally important traits. The ultimate aim of biotechnology in cotton should be to enhance productivity and quality of the fiber, seed oil and cake through encouraging sustainable use of natural resources including genetic resources.

REFERENCES

AICOSCA: All India Cotton Seed Crushers Association, Mumbai. Cotton National. AICOSCA News Letter (December, 2011): 23. (www.aicosca.com)

Boopathi, N.M., Thiyagu, K., Urbi, B., Santhoshkumar, M., Gopikrishnan, A., Aravind, S., Swapnashri, G. and Ravikesavan, R. (2011). Marker-assisted breeding as next-generation strategy for genetic improvement of productivity and quality: Can it be realized in cotton? *International Journal Plant Genomics*, (doi:10.1155/2011/ 670104)

CAB: Cotton Advisory Board, Ministry of Textiles, Govt. of India Report 2011-12 (www.texmin.nic.in).

CAI: Cotton Association of India. Indian Cotton Crop Estimates for the season 2011-12. Cotton Statistics and News-Market Report, 2011 (www.caionline.in).

Cai, Y., Xiaohong, H., Mo, J., Sun, Q., Yang, J. and Liu, J. (2009). Molecular research and genetic engineering of resistance to *Verticillium* wilt in cotton: A review. *African Journal of Biotechnology*, 8: 7363–7372.

Chen, X., Guo, W. and Zhang, T. (2011). Cotton omics in China. *Plant Omics Journal*, 4: 278–287.

Directorate of Economics and Statistics, (2011). Department of Agriculture and cooperation, Ministry of Agriculture, Govt. of India, 2011. *Bt* cotton in India-Statistics. (www.eands.dacnet.in)

ICAC: International Cotton Advisory Committee, Washington DC, USA. Press Release, December 31, 2011 (www.icac.org).

Lin, Z.D., He, X., Zhang, Y., Nie, X., Guo, C., Feng, J. and McStewart, D. (2005). Linkage map construction and mapping QTL for cotton fibre quality using SRAP, SSR and RAPD. *Plant Breeding*, 124: 180–187.

Lacape, M.J., Nguyen, T.B. and Grivet, L. (2000). Use of DNA molecular markers in cotton breeding: Development of a marker assisted selection program at CIRAD/ France *in Plant and Animal Genome VIII Conference San Diego, CA, January* 9-12, 2000.

Ministry of Textiles, Government of India, Cotton Production Statistics. (www. texmin.nic.in).

Paterson, A. (2012). Repeated polyploidization of *Gossypium* genomes and the evolution of spinnable cotton fibres. *Nature*, 492: 423–428.

Preetha, S. and Raveendren, T.S. (2008). Molecular marker technology in cotton. *Biotechnology and Molecular Biology Review*, 3: 32–45.

Subject Index